ADVANCES IN CONTROL EDUCATION 2003
(ACE 2003)

A Proceedings volume from the 6th IFAC Symposium,
Oulu, Finland, 16 – 18 June 2003

Edited by

J. LINDFORS
Applied Information Technology Unit,
University of Lapland, Finland

Published for the

INTERNATIONAL FEDERATION OF AUTOMATIC CONTROL

by

ELSEVIER LTD

ELSEVIER Ltd
The Boulevard, Langford Lane
Kidlington, Oxford OX5 1GB, UK

Elsevier Internet Homepage
http://www.elsevier.com

Consult the Elsevier Homepage for full catalogue information on all books, journals and electronic products and services.

IFAC Publications Internet Homepage
http://www.elsevier.com/locate/ifac

Consult the IFAC Publications Homepage for full details on the preparation of IFAC meeting papers, published/forthcoming IFAC books, and information about the IFAC Journals and affiliated journals.

First edition 2003

Library of Congress Cataloging in Publication Data

A catalogue record for this book is available from the Library of Congress

British Library Cataloguing in Publication Data

A catalogue record for this book is available from the British Library

ISBN 978-0-080-43559-6

ISSN 1474-6670

Printed and bound in the United Kingdom

Transferred to Digital Print 2010

To Contact the Publisher

Elsevier welcomes enquiries concerning publishing proposals: books, journal special issues, conference proceedings, etc. All formats and media can be considered. Should you have a publishing proposal you wish to discuss, please contact, without obligation, the publisher responsible for Elsevier's industrial and control engineering publishing programme:

Christopher Greenwell
Publishing Editor
Elsevier Ltd
The Boulevard, Langford Lane   Phone: +44 1865 843230
Kidlington, Oxford   Fax: +44 1865 843920
OX5 1GB, UK   E.mail: c.greenwell@elsevier.com

General enquiries, including placing orders, should be directed to Elsevier's Regional Sales Offices – please access the Elsevier homepage for full contact details (homepage details at the top of this page).

6th IFAC SYMPOSIUM ON ADVANCES IN CONTROL EDUCATION 2003

Sponsored by
International Federation of Automatic Control (IFAC)
Technical Committee on Control Education

Co-sponsored by
Academy of Finland
Comsol Oy
National Instruments
University of Oulu
VTT Industrial Systems

Organized by
Finnish Society of Automation

International Programme Committee (IPC)
Leiviskä, K. (FI),   Chairman (Member of Advisory Committee)

Albertos, P. (ES) (Member of Advisory Committee)
Antsaklis, P. (USA)
Atherton, D. (UK) (Member of Advisory Committee)
Bars, R. (HU)
Cao, X. (HK)
Dinibütün, T. (TR) (Member of Advisory Committee)
Dourado Correia, A. (PT)
Fasol, K.H. (DE) (Chair of Advisory Committee)
Furuta, K. (JP) (Member of Advisory Committee)
Galluzzo, M. (IT)
Groumpos, P.P. (GR)
Hämäläinen, J. (FI)
Horacek, P. (CZ)
Jämsä-Jounela, S-L. (FI)
Jantzen, J. (DK)
Jörgl, H.P. (AT)
Koivo, H. (FI)

Kucera, V. (CZ) (Member of Advisory Committee)
Lautala, P. (FI)
Lee, T.H. (SG)
Lindfors, J. (FI), Editor
Malinowski, K. (PL)
Marlin, T. (CA)
Masten, M. (USA)
Middleton, R. (AUS)
Roberts, P. (UK)
Sa da Costa, J. (PTl)
Tso, S. K. (HK)
Verde, C. (MX)
Vlacic, L. (AUS) (Member of Advisory Committee)
Ylinen, R. (FI)
Yliniemi, L. (FI) (Member of Advisory Committee)
Zia, O. (USA)

National Organizing Committee (NOC)
Yliniemi, L.   (Chairman)
Hautala, H.   (Secretary)

Hämäläinen, J.
Kippo, A.
Leiviskä, K.
Marttila, P.
Peltola, E.

FOREWORD

The 6th IFAC Symposium on Advances in Control Education (ACE2003) took place in Oulu, Finland, 16 – 18 June 2003. The technical program took place at the Linnanmaa Campus of the University of Oulu and a short visit was arranged to the further training institute "POHTO" with social activities taking place in Oulu and its surrounding areas.

ACE symposiums are organized triennially under the auspices of the International Federation of Automatic Control (IFAC). Previous events were held as follows:

First symposium, Swansea, 1988, chaired by Professor Atherton (UK)
Second symposium, Boston, 1991, chaired by Professor Rabins (USA)
Third symposium, Tokyo, 1994, chaired by Professor Furuta (Japan)
Fourth symposium, Istanbul, 1997, chaired by Professor Joergl (Austria)
Fifth symposium, Gold Coast, 2000, chaired by Professors Brisk and Vlacic (Australia)

Universities are facing new challenges that also effect Control Engineering laboratories. The environment is changing and requirements for openness, flexibility and efficient operation are common. Networking of activities and actors increase, and technology is frequently used in eliminating the distance between teachers and students, researcher groups and partners. New technology for this is coming out almost every day and the pace of development is fast. Also the research in pedagogy concerns with the increasing utilization of new learning technology. Anyway, control engineers have always been pioneering also in the area of learning technology, and in this sense the new challenges look controllable for us.

ACE2003 - the 6th IFAC Symposium on Advances in Control Education was an international forum for scientists and practitioners involved in the field of control education to present their latest research, results and ideas. The symposium also aimed to disseminate knowledge and experience in alternative methods and approaches in education. In addition to three plenary lectures and the technical visit, the symposium included 12 regular sessions and panel discussion session on the topic "web- with or without". Technical sessions concentrated on new software tools in control education especially on the role of interaction in Control Engineering education, web-based systems and remote laboratories and on laboratory experiments.

We hope that papers given in ACE2003 and also these Proceedings were and will be interesting and also professionally rewarding.

Kauko Leiviskä
IPC Chairman

Leena Yliniemi
NOC Chairman

CONTENTS

PLENARY PAPERS

SIMULATION AND ANIMATION IN WEB

DEVELOPMENT IN CONTROL LABORATORIES

REMOTE LABORATORIES AND EXPERIMENTS

FUTURE AND CHALLENGES FOR CONTROL ENGINEERING CURRICULA

WEB COURSES

TEACHING CONTROL THEORY

ADVANCED CONTROL

LABORATORY EXERCISES AND LEARNING BY DOING

SIMULATION

CONTROL

POSSIBILITIES OF COMPUTER NETWORKS IN TRAINING

SOFTWARE TOOLS FOR CONTROL EDUCATION

THE ROLE OF INTERACTIVITY IN CONTROL LEARNING

S. Dormido Bencomo

Dept. Informática y Automática. Facultad de Ciencias. U.N.E.D.
Avenida Senda del Rey 9. 28040 Madrid. Spain
Email: sdormido@dia.uned.es

Abstract: The scenario for control education is changing and we must adapt to the new situation. Information technology opens a whole new world of real opportunities. Computers show a great potential to enhance student achievement, but only if they are used appropriately as part of a coherent education approach. Computers do not change in the way books or labs do, they allow us to go deeper and faster. This paper presents the personal experience of the author in the use of interactive tools in order to make students more active and involved in their own control engineering learning process. Some examples, with different degrees of complexity, have been selected in order to show how we can use the control visualization concept in a new family of interactive tools for control education. *Copyright © 2003 IFAC*

Keywords: Education. Control education. Interactive approaches. Computer-aided control system design. Teaching.

1. INTRODUCTION

The present scientific and technological environment offers unprecedented challenges and opportunities in order to apply the impressive advances produced in the last years in information technology to control education.

Because of the amazing progress made in computer technology, today it is possible to design "control education tools" with the following characteristics:

- Better man-machine interaction
- Natural and intuitive graphical user interfaces
- High degree of interactivity

Automatic control ideas, concepts and methods are so rich in visual contents that they can be represented intuitively and geometrically. These visual contents can not only be used for presenting tasks and handling concepts and methods, but also manipulated for solving problems. Our feeling is primarily visual and thus it is not surprising that visual support is so present in our teaching process. Control educators very often make use of symbolic processes, visual diagrams and other forms of imaginative processes in their explanations.

The basic ideas in automatic control often arise from very specific and visual situations. All teachers know how useful it is to go back to this specific origin when they want to explain skillfully to the students the corresponding abstract objects.

This way of acting with explicit attention to potential specific representations in order to explain the abstract relations that are of interest to the control expert is what we term *control visualization* The fact that visualization is an especially important aspect in the control expert's activity is something completely natural if we bear in mind the applied mathematics nature of control theory.

Visualization thus appears to be something profoundly natural in the origins of automatic control, in the discovery of new relations between mathematical objects, and in the transmission and communication of our control knowledge.

One of the important tasks for teachers in control engineering is to transmit to our students not only the formal and logic structure of the discipline but also, and certainly with much more emphasis, the strategic and intuitive aspects of the subject (Heck, 1999). These strategic and intuitive aspects are probably much more difficult to make explicit and assimilate for students precisely because they lie very often in the less conscious substrata of the expert's activity.

Interactive tools, which are accessible to students at any time on the Internet, are considered a great stimulus for developing the student's engineering intuition. These interactive tools attempt to "demystify" abstract mathematical concepts through visualization for specifically chosen examples. At the present time, a new generation of software packages have created an interesting alternative for the interactive learning of automatic control (Garcia and

Heck, 1999; Poulis and Pouliezos, 1997; Szafinicki and Michau, 2000).

This paper presents my personal experience in the use of interactive tools in order to make students more active and involved in their own control engineering learning process. The strategy that I will follow will be to show different examples of how the control visualization concept can be used in a new family of interactive tools for control education. The examples have a different degree of complexity but all share a common philosophy: the use of interactivity to improve the learning process in control engineering. The ultimate goal is to facilitate comprehension of the concepts that we are trying to transmit to our students.

The paper is organized as follows. First the importance of interactivity and visualization in control education is analyzed. Following this the paradigm of the interactive design is discussed. In the rest of the paper some examples showing the utility of interactivity in control are presented. Section 4 shows how an interactive tool can help the students in the understanding of the "aliasing phenomenon". Bifurcations are scientifically important because they provide models of transitions and instabilities when some control parameter is varied. In section 5 the simplest examples that is to say, bifurcations of fixed points for flows on the line are presented from an interactive point of view. Section 6 presents LSAD (Linear System Analysis and Design) where a great part of the iteration involved in the design of classical controllers can be graphically carried out. The pole-placement design method for the single-input-single output case is showed in section 7. When nonlinear systems are excited by a periodic forcing function, it is sometimes observed that the amplitude of the periodic output changes abruptly. An interactive example showing in an interactive way the jump-resonance phenomenon is presented in section 8. Finally section 9 offers some conclusions.

2. INTERACTIVITY IN CONTROL EDUCATION

Bearing in mind these general considerations, the computer can be regarded as a tool that allows us to visualize and manipulate, in an interactive way, control objects.

In order to design technical systems or simply to understand the physical laws that describe their behavior, scientists and engineers often use computers to calculate and graphically represent different magnitudes. In control engineering, these quantities include among others: time and frequency responses, poles and zeros on the complex plane, Bode, Nyquist and Nichols diagrams, phase plane, etc. Frequently these magnitudes are closely related and constitute different visions of a single reality. The understanding of these relationships is one of the keys to achieve a good learning of the basic concepts and it enables students to carry out control systems design accurately.

Many tools for control education have been developed over the years. Innovative and interesting ideas and concepts were implemented by Prof. Åström and coll. at Lund. In this context we should highlight the concepts of *dynamic pictures* and *virtual interactive systems* introduced by Wittenmark (Wittenmark *et al.*, 1998). The main objective of these tools is to make students more active and involved in control courses.

In essence, a dynamic picture is a collection of graphical windows that are manipulated by just using the mouse. Students do not have to learn or write any sentences. If students change any active element in the graphical windows an immediate recalculation and presentation automatically begins. In this way they perceive how their modifications affect the result obtained. Dynamic pictures cannot only be effective in presenting engineering concepts in the classroom but also beneficial in extending student experience in analysis and design assignments. This invitation to creativity can be most useful where specialized control-engineering student projects are concerned.

This strategy causes us to "think small and simple". This is justified by a frank assessment of our limited knowledge for designing educational software as well as by practical considerations about how to manage incremental innovation. As dynamic pictures are fairly easy to create and deploy, they provide a means for rapidly prototyping and testing control principle ideas. In particular, they can be used as sharp tools for investigating precisely what it takes to make a control concept known to students. In this way, the "virtue of simplicity" becomes an issue in learning research on the design and use of this kind of tools.

These tools are based on objects that allow direct graphic manipulation. During these manipulations, the objects are immediately updated, so that the relationship among the objects is continuously maintained. *Ictools and CCSdemo* (Johansson *et al.*, 1998; Wittenmark *et al.*, 1998), developed at the Department of Automatic Control at Lund Institute of Technology, and *SysQuake* at the Institut d'Automátique of the Federal Polytechnic School of Lausanne, (Piguet, 1999; Piguet, *et al.* 1999) are good examples of this new educational philosophy for teaching automatic control.

For those that begin learning in this field some concepts are initially difficult to grasp, due to the fact that their properties are expressed in two different domains: time and frequency. Transient behavior, such as settling time, overshoot, and the risk of saturation are analyzed typically in the time domain; while concepts like stability, noise rejection, and robustness are expressed more easily in the frequency domain. The basic mechanisms that connect them and other phenomena like, for example, the effects of sampling and nonlinear elements, to mention just a few, can be illustrated in a very effective way using these tools (Dormido *et al.* 2002).

3. THE PARADIGM OF INTERACTIVE DESIGN

Design involving a human being also benefits from a fast and intuitive approach because it lets the designer understand what is happening. Boring trial and error search from the best set of parameters is avoided. From a pedagogical point of view, effects deduced from simple systems and controllers can often be generalized and lead to an intuitive understanding of the underlying mathematical basis. From the point of view of the control design, the evolution in the use of interactive graphics as help can be divided into three phases.

1. *Manual calculation.* This period is previous to the availability of digital computers as a tool in the process design. The procedure consisted in the calculation of a few numerical values. Control engineers developed sets of rules in order to draw graphics by hand.

2. *Computers as an auxiliary tool.* With the advent of digital computers it was possible to make the creation of graphics much easier. Control engineers had the possibility of tune the design parameters using a trial and error procedure following an iterative process. Specifications of the problem are not normally used to calculate the value of the system parameters because there is not an explicit formula that connects them directly. This is the reason for dividing, each iteration, into two phases, as it is shown in Figure 1. The first one, often called *synthesis*, consists in calculating the unknown parameters of the system taking a group of design variables (that are related to the specifications) as a basis. During the second phase, called *analysis*, the performance of the system is evaluated and compared to the specifications. If they do not agree, the design variables are modified and a new iteration is performed.

3. *Computers as an interactive tool.* It is possible however to merge both phases into one and the resulting modification in the parameters produces an immediate effect (see Figure 2). In this way, the design procedure becomes really dynamic and the students perceive the gradient of change in the performance criteria given for the elements that they are manipulating. This interactive capacity allows us to identify much more easily the compromises that can be achieved.

Interactive design with instantaneous performance display goes one step further. In many cases, it is not only possible to calculate the position of a graphic element (be it a curve, a pole or a template) from the model, controller or specifications, but also calculate a new controller from the position of the element. For instance, a closed loop pole can be computed by calculating the roots of the characteristic polynomial, itself based on the model and controller; and the controller parameters can be synthesised from the

set of closed loop poles if some conditions on the degrees are fulfilled (Crutchfield and Rugh, 1998).

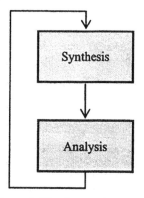

Figure 1. Non-interactive approach

This two-way interaction between the graphic representation and the controller allows to the manipulation of the graphical objects with a mouse in a very natural form. Since a good design usually involves multiple objectives using different representations (time domain or frequency domain), it is possible to display several graphic windows that can be updated simultaneously during the manipulation of the active elements.

The philosophy of interactive design with instantaneous performance display offers two main advantages when compared with the traditional procedure (non-interactive approach). First, it introduces from the beginning the control engineer to a tight feedback loop of iterative design. The designers can identify the bottlenecks of their designs in a very easy way and can attempt to fix them. Second and this is probably even more important, not only is the effect of the manipulation of a design parameter displayed, but its direction and amplitude also become apparent. The control engineer learns quickly which parameter to use and how to push the design in the direction of fulfilling better tradeoffs in the specifications. Fundamental limitations of the system and the type of controller are revealed (Åström, 1994, 2000), giving a way to find an acceptable compromise between all the performance criteria. Using this interactive approach the students can learn to recognize when a process is easy or difficult to control.

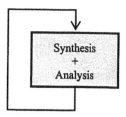

Figure 2. Interactive approach

Here are the main characteristic of the interactive design:

- The modification of the parameters produces immediately an update of the graphical scopes.

- The design process is completely dynamic.

- Students feel the gradient of the performance change in relation with the elements that they are handling.

- This interactivity allows to identify more easily the compromises can be reached.

4. ALIASING

Sampling is a fundamental property of computer-controlled systems because of the discrete-time nature of the digital computer. Sampling a continuous-time signal simply means to replace the signal with its values in a discrete set of points. The sampling instants are often equally spaced in time, that is $t_k = kh$. This case is called *periodic sampling* and h is called the sampling period. The corresponding frequency $\omega_s = 2\pi/h$ (rad/s) is called the sampling frequency. It is also convenient to introduce a notation for half the sampling frequency $\omega_N = \pi/h$ (rad/s), which is called Nyquist frequency.

The sampling process and the reconstruction of the signal can be described schematically as in Figure 3. A sinusoidal signal of frequency ω and unit amplitude, $u(t) = \sin(\omega t)$, is applied to the A/D converter. This continuous-time signal is thus converted to a sequence of number, which is again converted to a continuous-time signal $y(t)$ by the D/A converter (signal reconstruction).

Although no any information processing has been made by an algorithm inserted between the A-D and D-A converters, the output $y(t)$ presents in certain circumstances a very different form from that of the input $u(t)$.

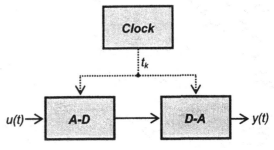

Figure 3. Sampling process

If we consider sinusoids only, the basic fact, is that to reconstruct a sinusoid from samples alone it is necessary and sufficient to have more than two samples per cycle (sampling theorem). If we have many more than two samples per cycle the task of fitting the sinusoid is easier; but if we have less than two samples per cycle, then a lower frequency sinusoid can be fitted to the same data and we cannot tell which is the true signal from the samples alone.

Most reconstructive processes will fit the lowest frequency that is consistent with the data. The process whereby a fast signal is sampled and then reconstructed with a lower frequency signal having the same samples is called *"aliasing"* (Åström and Wittenmark, 1997: Goodwin *et al.*, 2001).

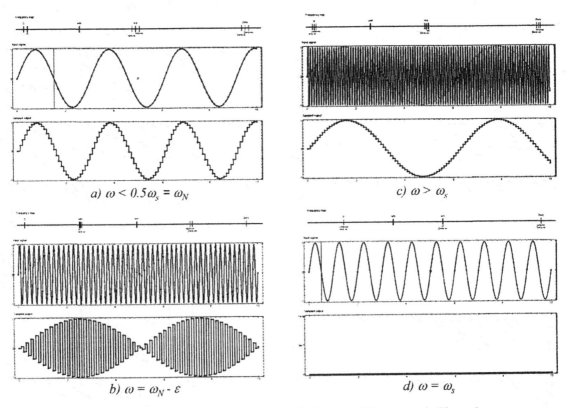

a) $\omega < 0.5\omega_s = \omega_N$

b) $\omega = \omega_N - \varepsilon$

c) $\omega > \omega_s$

d) $\omega = \omega_s$

Figure 4. Sinusoidal excitation and sampled output of the system in Figure 3

We have the surprising observation that the sampling procedure does not only introduce loss of information but may also generate new frequencies. The new frequencies are due to interference between the sampled continuous-time signal and the sampling frequency ω_s. This interference can also introduce beating in the sampled-data signal

To explain this phenomenon by using interactive concept we need to describe in a simple way the modification of the signal frequency ω and the sampling frequency ω_s. The metaphor has been a *"frequency map"* where the user can drag and modify both frequencies ω and ω_s and see the effects on the output $y(t)$ in an immediate way.

This is illustrated in Figure 4, which shows the output for four different frequencies. For frequencies smaller than the Nyquist frequency ($\omega < \omega_N$), the sampling theorem is fulfilled (Figure 4a). At frequencies close to the Nyquist frequency, ($\omega = \omega_N - \varepsilon$), there is a substantial interaction with the first alias, $\omega_N + \varepsilon$. Typical beats of ε (rad/s) are thus obtained (Figure 4b). For frequencies higher than the sampling frequency ($\omega > \omega_s$) creates a signal component with the alias frequency ($\omega_s - \omega$). The phenomenon of aliasing has a clear meaning in time. Two continuous sinusoids of different frequencies appear at the same frequency when sampled (Figure 4c). We cannot, therefore distinguish between those based on their samples alone. In the case that $\omega = \omega_s$ the output of the D-A converter $y(t)$ is identically equal to zero (Figure 4d).

5. BIFURCATIONS

One of the main goals of this example is to show how the students can be helped to develop a sound and practical understanding of the simplest bifurcations of fixed points for flows on the line (Strogatz, 1995).

5.1 Saddle-node bifurcation

The saddle-node bifurcation is the basic mechanism by which fixed points are created and destroyed. When a parameter is varied, two fixed points move toward each other, collide, and mutually annihilate. The prototypical example of a saddle-node bifurcation is given by the first order system

$$\dot{x} = r + x^2$$

where r is a parameter, which may be positive, negative or zero. When r is negative, there are two fixed points, one stable and one unstable. When $r = 0$, the fixed points coalesce into a half-stable fixed point at $x = 0$. This type of fixed point is extremely delicate because it vanishes as soon as $r > 0$, and now there are no fixed points at all. In this case, we say that a bifurcation occurred at $r = 0$, since the vector fields

for $r < 0$ and $r > 0$ are qualitatively different. The most common way to depict the bifurcation is to use the bifurcation diagram. Figure 5 shows the bifurcation diagram for the saddle-node bifurcation. The parameter r is regarded as the independent variable, and the fixed points are shown as dependent variables.

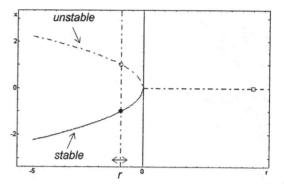

Figure 5. Saddle-node bifurcation diagram

5.2 Transcritical bifurcation

There are certain scientific situations where a fixed point must exist for all values of a parameter and can never be destroyed. For example, in the logistic equation and other simple models for the growth of a single species, there is a fixed point at zero population, regardless of the value of the growth rate. However, such a fixed point may change its stability when the parameter is varied. The transcritical bifurcation is the standard mechanism for such changes in stability. The normal form for a transcritical bifurcation is

$$\dot{x} = rx - x^2$$

where we allow x and r to be either positive or negative.

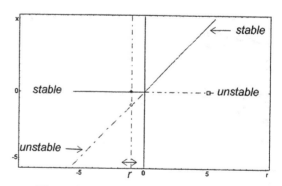

Figure 6. Transcritical bifurcation diagram

For $r < 0$, there is an unstable fixed point at $x = r$ and a stable fixed point at $x = 0$. As r increases, the unstable fixed point approaches the origin, and coalesces with it when $r = 0$. Finally, when $r > 0$, the origin becomes unstable, and $x = r$ is stable. We can say that an exchange of stabilities has taken place between the two fixed points.

There is an important difference between the saddle-node and transcritical bifurcations: in the transcritical case, neither of the two fixed points disappears after the bifurcation, instead they just switch their stability. Figure 6 shows the bifurcation diagram for the transcritical bifurcation.

5.3 Supercritical pitchfork bifurcation

This bifurcation is common in physical systems that have *symmetries*. For example, many problems have a spatial symmetry between left and right. In such cases, fixed points tend to appear and disappear in symmetrical pairs. There are two different types of pitchfork bifurcation. The simpler type is called supercritical. The normal form of the supercritical pitchfork bifurcation is

$$\dot{x} = rx - x^3$$

Note that this equation is invariant under the change of variables $x \to -x$. When $r < 0$, the origin is the only fixed point, and it is stable. When $r = 0$, the origin is still stable, but much more weakly so, since the linearization vanishes. Finally when $r > 0$, the origin becomes unstable. Two new stable fixed points appear on either side of the origin, symmetrically located at $x = \pm\sqrt{r}$. The reason for the term "pitchfork" becomes clear when we plot the bifurcation diagram (see Figure 7)

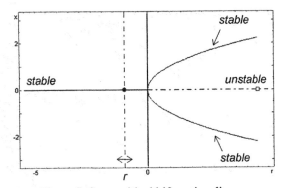

Figure 7. Supercritical bifurcation diagram

5.4 Subcritical pitchfork bifurcation

In the supercritical case, the cubic term is *stabilizing*: it acts as a restoring force that pulls $x(t)$ back toward $x = 0$. If instead the cubic term were *destabilizing*, as in

$$\dot{x} = rx + x^3$$

then we'd have a *subcritical* pitchfork bifurcation. Figure 8 shows the bifurcation diagram. Compared to Figure 7, the pitchfork is inverted. The nonzero fixed points $x = \pm\sqrt{-r}$ are unstable, and exist only below the bifurcation ($r < 0$), which motivates the term "subcritical". More importantly, the origin is stable for $r < 0$ and unstable for $r > 0$, as in the supercritical case, but now the instability for $r > 0$ is not opposed

by the cubic term. This effect leads to *blow-up*: one can show that x(t) → ±∞.

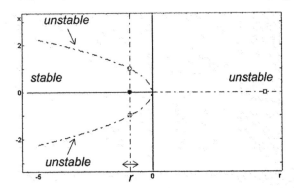

Figure 8. Subcritical bifurcation diagram

6. LINEAR SYSTEMS ANALYSIS AND DESIGN

Classical control design methods normally use simplifying assumptions in the design procedure; for example, root locus design is mainly based on the location of the dominant second order poles. In many occasions the controllers designed are good for giving students some insight into the problem and for designing an initial compensator. However the effects of the neglected dynamics can produce a significant deviation from the desired specifications. For this reason, students must simulate the system and iterate on their compensator until they get the fulfilment of the specifications. Unfortunately, most control textbooks ignore this iteration scheme, concentrating instead on the initial design methods.

From the student's point of view, the iteration can be very time-consuming and confusing. Students would benefit a lot from an interactive tool that automates much of the iteration and gives good insights about the progress of the design.

LSAD (short for Linear System Analysis and Design) is an interactive tool where a great part of the iteration involved in the design of classical controllers can be carried out in a simple way using an interactive and friendly graphical environment. Figure 9 shows the GUI windows of LSAD. The GUI windows can be modified in a very easy way in order to adapt the interface to the problem that must be solved. With this tool the students can solve in a graphical environment most of the exercises proposed in an introductory course on automatic control.

Compensator poles, zeros and integrators can be added or deleted through the Bode diagram, root locus, and Nyquist or Nichols plot. A pole or zero can be moved by first clicking on the "Move" button in the "Controller parameters" window, then clicking on the desired pole or zero (in whenever representation) and dragging it to another location. All the diagrams (Bode, Root locus, Nichols and Nyquist) update simultaneously during the move as well as the open loop and closed loop step response. Also some performance variables of the system are visualized:

gain margin, phase margin, critical frequency, cross-over frequency, peak value, time to peak, rise time, settling time, steady state error and velocity error.

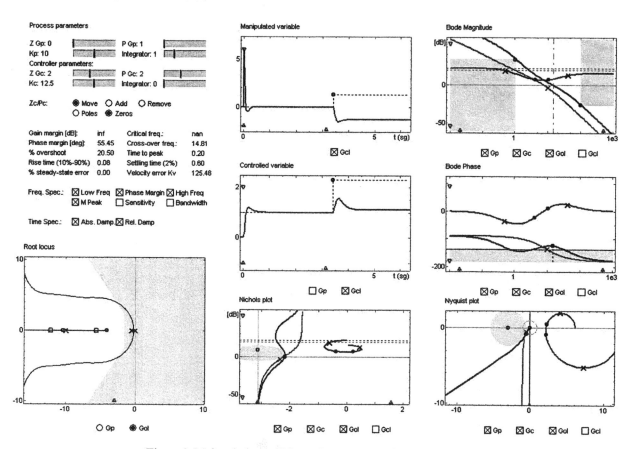

Figure 9. Main window of Linear System Analysis and Design

In order to determine the compensator different time and frequency specifications can be introduced using interactive elements. For example in Figure 9 the activation of the following specifications is shown: low frequency gain, phase margin, high frequency attenuation, M_p peak and absolute and relative damping. Thus students get immediate feeling on how moving a compensator pole or zero influences the design procedure. This enables them to gain a deeper knowledge into the tradeoffs in the design process rather than, using a manual time consuming procedure.

Next an illustrative example is presented that shows the interactive design technique and use of LSAD tool in enhancing the classical control design methods as they are now commonly taught.

6.1 Gain adjustment

The basic concept underlying both the gain-adjustment and lag compensation design is that the desired degree of stability (that is phase margin) may often be achieved by a reduction gain. In order to initiate the example, let us consider the use of a simple gain adjustment, although is seldom an adequate solution to a design problem. The system to be considered is shown in Figure 10a where the transfer function of the uncompensated plant is

$$G_p(s) = \frac{s}{s(s+10)}$$

The problem is to select the value of K_c such that the velocity-error constant K_v must be equal to 100. Since this is a type 1 system, K_v is found by taking the limit of $sG_p(s)$ as s goes to zero; the velocity-error constant of the uncompensated plant is equal to 10. In order to increase K_v to 100, K_c must be equal to 10. Using the GUI shown in Figure 9, the student can select the "Move" option (by clicking on the "Move" button). Once the button is clicked, the compensator gain can be dragged, in the Bode Magnitude, to obtain the desired specification (see Figure 10a)

6.2 Lag compensation

To establish the details of the lag-compensation design, let us consider the previous example in which the following specification is added: the phase margin must be equal to 45°. With the simple gain adjustment the phase margin is 17,9°; however, it is observed from the phase shift diagram in Figure 10a that for frequencies less than 10 rad/s the phase shift is less than 135°. Thus in addition to a gain adjustment, a lag compensation can be used to reduce the cross over frequency to about 8 or 9 rad/s.

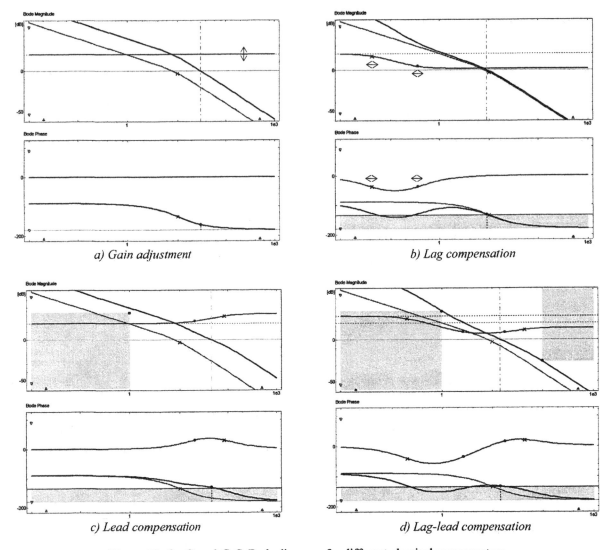

a) Gain adjustment

b) Lag compensation

c) Lead compensation

d) Lag-lead compensation

Figure 10. G_p, G_c and G_pG_c Bode diagrams for different classical compensators

The student can select the "Add" option (by clicking on the "Add" button) and then use the Bode diagram (magnitude or phase) to introduce the lag-compensator directly on the graph. Now the compensator pole and zero can be moved to obtain the phase margin specification by first clicking on the "Move" button in the window, then clicking on the desired pole or zero and finally dragging it to another location (see Figure 10b)

6.3 Lead compensation

To illustrate the lead-compensation technique, let us consider the previous example including the following specification: sinusoidal inputs of up to 1 rad/s should be reproduced with ≤ 2 percent error. Our first thought is the application of a lag compensator. However, the specification of a middle frequency gain requirement needs that $|G_c(j\omega)G_p(j\omega)| \geq 34$ for $\omega \leq 1$ rad/s. This require-ment is indicated by the yellow area in the Bode magnitude in Figure 10c. The use of lag compensation cannot be used, not because the phase

lag is too large at all frequencies but rather because the phase lag is too large in the frequency range that is necessary to meet the middle-frequency specification.

Although lag compensation may not be used as a solution to the current design problem, the situation is almost ideal for the application of lead compensation. Lead compensation may be successfully applied here because no high-frequency-attenuation specification is given and because the phase lag of the uncompensated plant increases slowly after crossover.

All the students need to do in order to interact with the tool is to interchange the relative position of the zero and the pole and move the location to the high frequency region (see Figure 10c)

6.4 Lag-lead compensation

In the two previous examples lag- and lead-compensation techniques have been showed as separate approaches. Here, by combining the two approaches to generate a composite lag-lead-compensation method, it is showed how it is possible to achieve better results than when either of the methods is used separately.

To illustrate the design procedure for lag-lead compensation, let us consider once again the plant discussed in terms of lag or lead compensation. To the previous set of specifications, let us add the high-frequency attenuation: sinusoidal inputs of greater than 100 rad/s should be attenuated at the output to 5 percent of their value at the input.

This specification requires that $\left| G_c(j\omega) G_p(j\omega) \right| \leq .05$

for $\omega \geq 100$ rad/s. Figure 10d shows this specification. In this case, the high-frequency attenuation specification rules out the use of lead compensation since no high-frequency gain may be used. We have seen in section 6-2 that lag can neither be used. However, this problem can be solved quite easily with the use of lag-lead compensation.

Following the same interactive design procedure the student can now select the "Add" option (by clicking on the "Add" button) and then use the Bode diagram (magnitude or phase) to introduce the lag compensator directly on the graph of the lead compensation. Now changing to the *move option* it is possible to interact with the lag-lead compensator poles and zeros in order to fulfil all the specifications.

Thus LSAD tool provides an excellent means of enhancing the standard "textbook" design approach for classical compensators. The images and the immediate feedback enable students to gain insight into the design very quickly.

7. POLE PLACEMENT DESIGN: A STATE-SPACE APPROACH

Pole placement refers to the fact that the design is formulated in terms of obtaining a closed-loop system with specified poles. In this section a simple regulation problem is presented. Further, only the single input-single output case will be discussed. Because computer control is considered, the control signals will be constant over sampling period. If the sampling period h is given, the process can be described by the following discrete-time system:

$$x(kh + h) = \Phi x(kh) + \Gamma u(kh)$$

The disturbances are assumed perturbations in the initial state of the system. The objective is to find a linear feedback law of the form: $u(kh) = -Lx(kh)$, so that the closed-loop system has a specified characteristic equation. With n parameters in the state-feedback vector, it is possible to place n poles arbitrarily, if the system is reachable (Åström and Wittenmark, 1997). This will guarantee that the disturbances decay in a specified way.

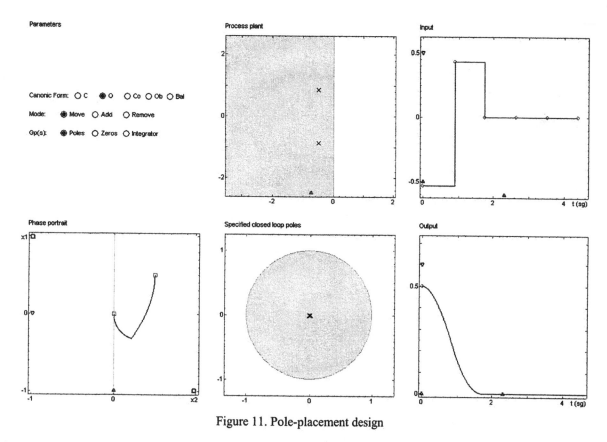

Figure 11. Pole-placement design

Figure 11 shows the basic screen of the developed interactive tool for the pole placement design. The

student can modify interactively the following parameters:

- The process plant
- The state variable representation
- The sampling period
- The specified closed loop poles
- The initial state

A deadbeat control of a second order system is depicted in Figure 11. To put the pole placement design method into practice, it is necessary to understand how the properties of the closed loop system are modified by the design parameters, that is the closed-loop poles and the sampling period. Using this tool the student can obtain a clear comprehension on how these parameters influence the response of the system.

8. JUMP RESONANCE

When nonlinear systems are excited by a periodic forcing function, it is sometimes observed that the amplitude of the periodic output changes very suddenly, or "jumps". A physical description of the observed phenomena is best given with the aid of Figure 12. If a constant-amplitude low-frequency sinusoidal forcing function is applied, the steady-state output response is periodic with the same fundamental frequency. When the input frequency is slowly increased the steady-state output amplitude normally increases slowly in a manner quite similar to that found in a linear system. However, at some critical frequency ω_1, it is found that only a small increase in frequency is required to cause a jump in amplitude from A to B (see Figure 12).

Further increases in frequency result in a gradual decrease in output amplitude as shown. Next, if the frequency is slowly reduced, the output amplitude repeats itself at the various frequencies until the critical frequency ω_1 is regained with amplitude at B. Now it is discovered that ω_1 is not a critical value for the decreasing frequency; i.e., the amplitude does not decrease in jump fashion at ω_1 but continues along an apparently normal resonance curve from B toward C. A second critical frequency ω_2 is discovered as the forcing function is decreased from B to C. At frequency ω_2 there is another discontinuous jump from C to D. D is found to lie on the original increasing-frequency section of the curve, and for further decreases in frequency this curve is retraced. The observed jumps in the amplitude of the frequency-response curve have given rise to the name "jump resonance" (Mira, 1969).

The explanation for the jump resonance phenomenon is the existence of a frequency range ω_2 to ω_1 for which the solution of the nonlinear differential equation is triple-valued. The upper and lower curve segments, from D to A and from B to C, represent stable solution values, and the segments from A to C represent unstable solutions which cannot exist physically. Thus, as the frequency is raised from D to A, the operation of the system is stable, and there is no tendency for the output amplitude to change, even

though two other solutions are possible in the frequency range from ω_2 to ω_1.

The best-known cases of jump resonances are associated with Duffing's equation, which is in the form

$$\ddot{y} + \dot{y} + f(y) = B \sin(\omega t)$$

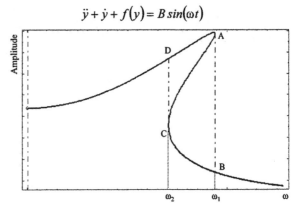

Figure 12. The jump-resonance phenomenon

Note that in Duffing's equation the nonlinearity is in the $f(y)$ term, which is commonly associated with the spring constant in a mechanical system. For a closed loop control system, this term is associated with the system gain constant (see Figure 13)

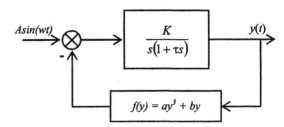

Figure 13. Non linear system

The non linear differential equation of this system may be written as:

$$\ddot{y} + 2\alpha\dot{y} + \omega_0^2 y\left(1 + \lambda y^2\right) = B \sin(\omega t)$$

where

$$\alpha = \frac{1}{2\tau}, \quad \omega_0^2 = \frac{Kb}{\tau}, \quad \lambda = \frac{a}{b}, \quad B = \frac{KA}{\tau}$$

Figure 14 shows how this new way of interactive control education provides practical insights into control systems fundamentals (Dormido, 2002). It is a dynamic picture in the sense mentioned earlier, and when the student manipulates some active element in the figure the new result is automatically produced. In order to understand the jump resonance phenomenon the student can modify interactively the following parameters:

- The parameters α, B
- The non-linearity parameters a and b.
- The initial condition
- The frequency ω

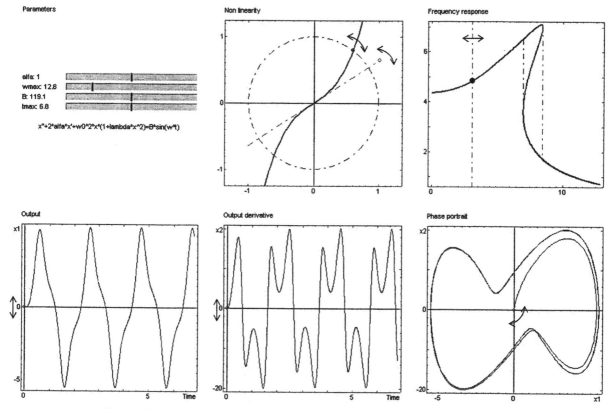

Figure 14. An interactive tool for the study of the jump-resonance phenomenon

9. CONCLUSIONS

One major issue has been addressed when developing these interactive examples: to maximize clarity and simplicity focusing each example on a single core concept. The objective is to design interactive tools that encourage our control engineering students to think and interact with the examples differently from the way they do in the case of textbooks or lecture presentations. Because the examples can be used in an unsupervised environment, simplicity and clarity are key features.

It is interesting to observe that new communication and information technology in education has always met with a certain resistance from the academy. This situation seems to be changing lately with a general enthusiasm about the Web, however difficult it may be to predict the future course of efforts such as ours to develop "interactive tools"(Antsaklis et al., 1999). Issues of licensing and commercialization are complex. Our plan is simply to continue exploration of concrete educational issues by focusing on interactive modules for concepts in control systems (Dormido et al. 2002; Dormido et al. 2003; Salt, et al. 2003; Sánchez et al. 2000; Sanchez et al. 2002, Tan et al. 2003).

The scenario for control education is changing and we must adapt to this changes (Kheir et al., 1996; Bissell, 1999, Bernstein, 1999, Dorato, 1999). Information technology opens a whole new world of real opportunities. Computers show a great potential to enhance student achievement, but only if they are used appropriately as part of a coherent education approach. Computers do not change in the way books or labs do, they allow us to go deeper and faster..

ACKNOWLEDGEMENTS

This work has been supported by the Spanish CICYT under grant DPI2001-1012.

10. REFERENCES

Antsaklis, P.; Basar, T.; DeCarlo, R.; Harris, N.; Spong, M.; Yurkovich, S.: (1999). "Report on the NSF/CSS WorkShop on new directions in control engineering education", *IEEE Control Systems Magazine*, **19**, n° 5, October, pp. 53-58.

Åström, K.J.: (1994). "The future of control", *Modeling, Identification and Control*, **15**, n° 3, pp. 127-134.

Åström, K.J.; Wittenmark, B.: (1997). *Computer controlled systems*, Prentice Hall 3rd edition.

Åström, K.J.: (2000). "Limitations on control system performance", *European Journal of Control*, **6**, pp. 2-20.

Bernstein, D. S.: (1999). "Enhancing undergraduate control education", *IEEE Control Systems Magazine*, **19**, n° 5, pp. 40-43.

Bissell, C.C.: (1999). "Control education: Time for radical change?", *IEEE Control Systems Magazine*, **19**, n° 5, pp. 44-49.

Crutchfield, S. G.; Rugh, W. J.: (1998). "Interactive learning for signal, systems, and control", *IEEE Control Systems Magazine*, **18**, n° 4, pp. 88-91.

Dorato, P.: (1999). "Undergraduate control education in the U.S", *IEEE Control Systems Magazine*, **19**, n°. 5, pp. 38-39.

Dormido, S. (2002) Control Learning: Present and Future. Plenary Lecture. *15th Triennial World Congress of IFAC, Barcelona, Spain*

Dormido, S.; Gordillo, F.; Dormido-Canto, S.; Aracil, J.: (2002). "An interactive tool for introductory nonlinear control systems education", *15th. IFAC World Congress* b'02, Barcelona, Spain.

Dormido, S.; Berenguel, M.; Dormido-Canto, S.; Rodríguez, F.: (2003). "Interactive learning of introductory constrained generalized predictive control", *IFAC Symposium on Advances in Control Education*, Oulu, Finland.

Garcia, R. C.; Heck, B. S.: (1999). "Enhancing classical controls education via interactive GUI design", *IEEE Control Systems Magazine*, **19**, n° 3, pp. 77-82.

Goodwin, G. C.; Graebe, S. F.; Salgado, M. E.: (2001). *Control system design*, Prentice Hall.

Heck, B. S. (editor): (1999). "Special report: Future directions in control education", *IEEE Control Systems Magazine*, **19**, n° 5, pp. 35-58.

Johansson, M.; Gäfvert, M.; Åström, K. J.: (1998). "Interactive tools for education in automatic control", *IEEE Control Systems Magazine*, **18**, n° 3, pp. 33-40.

Kheir, N. A.; Åström, K. J.; Auslander, D.; Cheok, K. C.; Franklin, G. F.; Masten, M.; Rabins, M.: (1996). "Control system engineering education", *Automatica*, **32**, n° 2, pp. 147-166.

Mira, C.: (1969) *Cours de systemes asservis non linéaires*, Dunod université

Piguet, Y.: (1999). "SysQuake: User Manual", Calerga.

Piguet, Y.; Holmberg, U.; Longchamp, R.: (1999). "Instantaneous performance visualization for graphical control design methods", *14th IFAC World Congress*, Beijing, China.

Poulis, D.; Pouliezos, A.: (1997). "Computer assisted learning for automatic control", *IFAC Symposium on Advances in Control Education*, Estambul, Turquía, pp. 181-184.

Salt, J.; Albertos, P.; Dormido, S.; Cuenca, A.: (2003). "An interactive simulation tool for the study of multirate sampled data systems",*IFAC Symposium on Advances in Control Education*, Oulu, Finland.

Sánchez, J.; Morilla, F.; Dormido, S.; Aranda, J.; Ruipérez, P.: (2000). "Conceptual learning of control by Java-based simulations", *IFAC/IEEE Symposium on Advances in Control Education*, Gold Coast, Australia.

Sánchez, J.; Morilla, F.; Dormido, S.; Aranda, J.; Ruipérez, P.: (2002). "Virtual control lab using Java and Matlab: A qualitative approach", *IEEE Control Systems Magazine*, **22**, n° 2. pp 8-20.

Strogatz, S. H.: (1995). *Nonlinear dynamics and chaos: With applications to physics, biology, chemistry and engineering*, Addison Wesley

Szafnicki, K.; Michau, F.: (2000). "New educational technologies applied to control education-example of resources sharing between engineering schools", *IFAC/IEEE Symposium on Advances in Control Education,* Gold Coast, Australia.

Tan, N.; Atherton, D. P.; Dormido, S.: (2003). "Systems with variable parameters; Classical control extensions for undergraduates", *IFAC Symposium on Advances in Control Education*, Oulu, Finland.

Wittenmark, B.; Häglund, H.; Johansson, M.: (1998). "Dynamic pictures and interactive learning", *IEEE Control Systems Magazine*, **18**, n° 3, pp. 26-32.

ELSEVIER
IFAC
PUBLICATIONS
www.elsevier.com/locate/ifac

INTERNET LEARNING IN CONTROL ENGINEERING: A FUZZY CONTROL COURSE

Jan Jantzen [1]

*Technical University of Denmark, Oersted-DTU,
Automation, building 326, DK-2800 Kongens Lyngby,
Denmark. Email: jj@oersted.dtu.dk*

Abstract: This educational study is based on a course taught over the Internet for seven years. The objective of the study is to evaluate the didactic method, e-mail tutoring. An average of about 45 students complete the course every year. The course evaluation is based on 42 - 51 student responses to a questionnaire, as well other statistics supplemented by personal observations. Student responses are generally positive. Such a course is vulnerable, however, to students dropping out. *Copyright © 2003 IFAC*

Keywords: teaching, fuzzy control, simulators, higher education, e-mail tutoring, pendulum

1. INTRODUCTION

A group of Italian students were taking the distance course, that this paper is about, and once wrote: *We are sorry if our work in the fuzzy field is getting slower, but Marco is in The Netherlands for four months and we have to interact by mail. So, this is a true internet course!* One of them had to go abroad as a part of his PhD studies, and they obviously took advantage of the flexibility that distance learning provides. Flexibility in time and space is the obvious benefit of Internet learning, and it benefits both sides: teacher and student.

The software tools available for distance teaching are excellent and the Internet is reliable enough. Technology is no longer a bottleneck. It is the didactic issues that make it difficult. For example, how does the teacher know how much work the student puts into the course, and how much and how well the student learns? Rather than discuss technological solutions, this paper reports on didactic components and learning outcomes

from the course *31361 Fuzzy Control (Internet Course)*. The course uses assignments set around a simulator of laboratory rig (Fig. 1). The course was first taught in 1996, and useful quantitative measurements, to be presented later, are available for the period 1999 - 2002. The aim is to show that its teaching method, *e-mail tutoring*, is one way to progress *from teaching to learning*, that is, a vision where students take responsibility for their own learning — in contrast to being pushed by the teacher.

How much have the students learnt in this course? There is no simple instrument to measure this, so we have to turn to the literature. First of all, it will be necessary to distinguish between 1) *assessment* and 2) *evaluation*. The former we

Fig. 1. Ball balancing laboratory rig. Length: 1.4 meters.

[1] The work is partially supported by the European Commission under contract IST-2000-29207 with the thematic network EUNITE, http://www.eunite.org

shall associate with the student, the latter with the course. The purposes of a student assessment are: to give a licence to proceed to the next stage or graduation, to classify the performance of students in rank order, and to improve their learning (Brown, 2001). Some methods of assessment are: essays, problem sheets, lab reports, presentations, projects, group projects, posters. Brown lists 23 methods of assessment and provides comments on advantages and disadvantages of each. A course evaluation scheme designed to find out whether the course and the teacher gave the student space and motivation for learning, indirectly indicates whether the student is seeking deep knowledge or not. The Australian government has developed questionnaires to this purpose. Each year, since 1993, graduates of Australian universities are invited to reply to 25 questions in a Course Experience Questionnaire, CEQ (see for example Ainley and Johnson, 2000). The 24 questions concern the quality of teaching, the clarity of goals and standards, the nature of the assessment, the level of the workload, and the enhancement of their generic skills. The 25th question asks graduates to indicate their overall level of satisfaction. All answers are on a five-point scale. The data are analysed in order to find tendencies. In the year 2000 there were 90,000 respondents. There happens to be a slight overlap between the CEQ and the questionnaire used in our fuzzy control course.

The fuzzy control course was completed by 174 students (97 dropped out) in the period 1999 - 2002. The statistical basis is regarded too small for drawing conclusions by means of statistical analyses. Instead the strategy of the present study has been to present as many quantitative measurements as possible, as well as personal observations and student testimonials. This type of evaluation is *portfolio* oriented, since it includes a variety of items, like an artist's portfolio. The conclusions drawn will thus be qualitative and more subjective than CEQ evaluations.

2. THE FUZZY CONTROL COURSE

The course under study concerns automatic control of an inverted pendulum problem, or more specifically, rule based control by means of fuzzy logic. A ball balancer (Fig. 2), implemented in a software simulator in Matlab, is used as a practical case study. After the course, students will be able to design their own controller and evaluate commercial design tools. The student work load is 60 hours, and the duration is at least 8 weeks. The course runs twice per year with a fixed end date according to the semester. It is rated at 2.5 credit units in the European Credit Transfer System, ECTS. There are similar courses (Yurkovich &

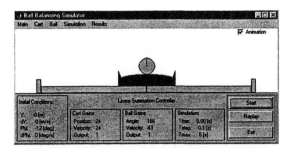

Fig. 2. Front panel of simulator Pendulum.

Passino,1999; Jurado, Castro & Carpio, 2002; Lo, Wong & Rad, 2002), but no distance learning courses in fuzzy control. A book by Hall (1997) describes several types of distance learning courses, and in that framework our course has these characteristics: correspondence by e-mail, text and graphics, interactive, and web-based administration and assessments. It does not make use of interactive multimedia, teleconferencing, desktop video conferencing, or chat rooms.

The inverted pendulum case study is an excellent benchmark problem and it is widely used as a didactic vehicle (e.g., Magana & Holzapfel, 1998).

2.1 Simulator

The pendulum in our course is a variant consisting of a steel ball on a curved bridge on a moving cart. It is fairly easy to control such that the ball balances on top of the bridge, but it is difficult to position the cart at the same time in a preset position. Students design a fuzzy controller that balances the ball on top of the bridge. The software simulator *Pendulum* (Fig. 2) is used throughout the course and can be downloaded from the World Wide Web [2]. To make it as independent as possible, the simulator only requires the student edition of Matlab and no toolboxes. If Matlab is unavailable, it is necessary to buy the student edition [3]. The students use a variety of platforms, but the same code runs on all. The course has a public home page, with information about how to get started and links to the course material. [4]

2.2 Didactic model

The course consists of 12 modules, and almost all modules are divided into three steps: 1) An objectives and preparation section, 2) an exercise, and 3) an assignment. In the first step the student downloads and reads the text material offline.

---

[2] http://fuzzy.iau.dtu.dk/download/pendul31.zip

[3] http://www.mathworks.com

[4] http://fuzzy.iau.dtu.dk/fuzcon.html

This can be done anywhere, on the bus for instance, independently of computers. In the second step, the student walks through an exercise. This must be done on a PC or workstation. The student only has to observe and explore. The third step is a matter of solving specific design problems related to the case study. The underlying didactic model is:

(1) Read
(2) Exercise
(3) Solve problem

The point is to make each step increasingly activating for the student. Step 3 is the main goal — to teach the design aspects — but it is difficult for a student to jump straight into designer mode; the model is meant to soften the approach to the designer mode.

The delivery of the course promotes a deeper learning compared to lectures. The e-mail assignment at the end of each module forces the student to be active and alert. Students must be precise, and they must think carefully, since they have to write their responses. The teaching style is *tutoring*, because the communication sessions are iterative. Compared to, say, a three-days crash course, the learning is deeper, because it is extended over a longer period and the student-teacher interaction is one-to-one.

2.3 Modules

The 12 modules have been changed and reorganised as a result of student feedback.

(1) Install simulator software
(2) Cart-ball model
(3) Linear controller design (optional)
(4) Controller test (optional)
(5) Fuzzy set theory
(6) Fuzzy logic
(7) Fuzzy controller design
(8) Fuzzy linear controller
(9) Fuzzy non linear controller
(10) Self-organizing controller
(11) Course evaluation
(12) Online exam

The first module is easy, and the objectives are to get started, to install the software, and to get the technology up and running. Then follows 4 proper modules each with a preparation phase, an exercise, and an assignment (see an example in the appendix). For each assignment, students download it as a text file, include it inside an e-mail, work with it, make notes about results and comments they want to pass to the instructor, write their answers directly into the file, and e-mail this to the instructor. The instructor will comment on their solution and return the e-mail note. The instructor may then urge the student to proceed with new questions related to the topics in the lesson.

Module 6 is an online self-assessment, similar to a quiz. It is anonymous, but students log on to a Web page with their user ID and answer a number of questions related to the textbook material. The result is returned automatically to the student's personal e-mail account. The didactic purpose is to establish what the student has learnt, and the practical purpose is to show the form of the online exam at the end of the course. Both modules 6 and 12 are multiple choice tests. The exam is marked, and so is the self-assessment, but the marks in the self-assessment are not recorded.

From a professional, fuzzy control engineering view, the organisation of the course reflects an unusual design approach:

(1) Build a conventional linear controller;
(2) build an exactly equivalent, linear, fuzzy controller; and
(3) make the linear fuzzy controller gradually nonlinear.

It is unusual, because it incorporates a linear fuzzy controller explicitly as a design aid. The idea is to stay in the linear domain as long as possible to exploit linear control theory and tuning rules. Leaving the firm basis of linear control theory when entering step 3 is like jumping into the unknown, but there is nevertheless a linear controller that can be used as a reference. The approach is described in details earlier (Jantzen, Verbruggen & Ostergaard, 1999).

2.4 Technology and communication

The course resides within the commercial product LearningSpace[5] by Lotus (IBM). From the course page, students access four different sections in the LearningSpace: Schedule, Media Center, Course Room and the Profiles section. In Schedule students can view the course modules and download them, as well as do the self-assessment. The Media Center is a library from where students can download software or articles. In the Course Room students and teachers can conduct open discussions and send messages of any kind. Finally, in the Profiles section participants can edit their profile and publish photos, their Web address, and other personal information. LearningSpace acts as a container for the files, and is more or less an administrative tool.

Even though there are messaging and discussion facilities inside of LearningSpace, the communica-

---

[5] http://www.lotus.com/products/learnspace.nsf/

tion mode in the course is e-mail. The teaching is *asynchronous*, which means that teacher and student can be online at different points in time. That is flexible, and it is an advantage when students are in various time zones. The files exchanged are plain text files, even the Matlab programs, in order to minimise the risk of virus.

2.5 Accreditation and student assessment

Students earn 2.5 credit points in the ECTS system, where one year's worth of student work is awarded 60 points. In the Technical University of Denmark, students are required to take 12 course units per year, therefore one course unit is nominally 5 points. A 2.5 points course is small, it corresponds to a student workload of 60 hours. For our course a student enrolls as a *guest student* — provided that he has submitted documents and proof that he is enrolled in another university — without showing up in person at the university. If he passes the course, the university issues a certificate, and mails it to the student. With this certificate in hand he may try and achieve a credit transfer, but that is a business between the student and his local university. The course is free of charge for students, while non-students have to pay a fee to the Technical University of Denmark.

The course uses two methods to assess the students: 1) the student must finish all modules, and 2) the student must pass an online multiple choice exam. The student can take the exam from any browser on the Internet. He is required to log in and thereby identify himself. It is conceivable that someone else could take his place and do the exam for him, since there is no means of visual identification by the teacher or anybody else. This loophole seems to be very expensive to fix, however, and so far it has been left unattended. Maybe it will be possible to fix with a digital signature in the future.

The exam takes 0.5 - 1 hour to complete — some students take several days, if they experience technical problems — and afterwards the instructor corrects the exam semi-automatically. The LearningSpace will grade the answers automatically, but the instructor is able to override the grading, before returning it to the student.

3. RESULTS AND DISCUSSION

The portfolio of evaluation items consists of responses to a questionnaire, free format comments from the students, grades collected from 86 students, statistics of students' time to finish the course, and teacher observations.

3.1 Course evaluation

Students post-evaluate every course at the Technical University of Denmark. They answer a questionnaire online on the Web. The response ratio is generally about 50 percent. The questionnaire is in three sheets: A) a general course evaluation, B) an evaluation of the teacher, and C) free format comments and criticism. The evaluation scheme was started some years ago, and data are available for our course from 1999 - 2002.

Q1: With respect to the course content, the teaching methods used were [not very suitable / good enough / well suited]

Q2: How well did the type of (final) examination or assessment suit the course content? [badly / not well / fairly well / well / very well]

Q3: How did you feel your dialogue/working relationship with the instructor was? [unsatisfactory / bad / satisfactory / good / very good]

Q4: How well did the instructor provide guidance? [very bad / bad / satisfactory / well / very well]

Q5: Did you get sufficient help when you asked for it? [no / pretty much / yes]

Q6: Did you receive relevant feedback on the assignments you handed in during the course? [no / a little / some / yes / a great deal]

There are more questions in the questionnaire, but they are irrelevant for this study.

Figure 3 gives an overview of the responses to the six questions. The overall impression is that most responses are on the positive side. It is interesting to note (Q1) that some students find the teaching method 'not very suitable', even though it is a distance learning course. This could be explained by the fact that some students enroll as a class as a part of their local curriculum — there are classes from Malmo (Sweden), Zwolle (The Netherlands), and Arhus (Denmark) that take the course year after year according to agreement with their local teacher. Those students will compare the teaching method with their other courses, and may find that the traditional teaching method is better. Especially if they are freshmen, that have seen only few different teaching methods.

Students that enroll voluntarily out of interest and need for the course find the teaching method good, since they realise it is the only way feasible.

Most students find the dialogue with the teacher good (Q3), but there are nevertheless students that find it bad. This is surprising, because they get maximal attention and the dialogue is one-to-one. It may be due to the fact that some students work in groups of two, at most three, and a member of a group may feel at a loss if the internal group communication is poor.

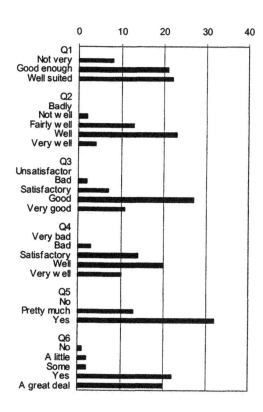

Fig. 3. Histogram of student responses to six questions Q1-6. There are between 42 and 51 respondents for each question.

Guidance, help and feedback is generally good (Q4-6), but again there are a few unhappy students, which is surprising. An explanation might be that the weaker students do avoid reading the textbook material, and therefore fall behind and into some frustration.

3.2 Assessment of students

The grading is based on points achieved from each question in the multiple choice exam. Fifteen questions are drawn randomly from a database of questions, so that it is unlikely that two students get the same set of questions. Each question has a number of points associated with it as a measure of its difficulty. It takes about an hour to do, and the student is required to get more than 50 percent right. For each student the number of achieved points is divided by the number of possible points for the 15 questions drawn, and the percentage of correct answers calculated.

Figure 4 shows the distribution of the grades of 86 students. Very few are below the passing limit. A few get top marks and there is a peak near 86 percent. The distribution does not seem unusual, except that there are small peaks here and there. This is perhaps a reflection of the three levels of students: undergraduate, graduate, and PhD. All

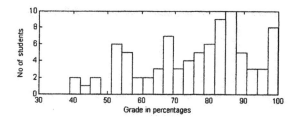

Fig. 4. Histogram of grades (86 students). The average grade is 76 percent.

three levels are admitted to the course. There is a distinct difference in their approach to the course. As a rough characterisation, the undergraduate students (Bachelor) are often less motivated, and try to answer the questions without reading the textbook material. When they fail to answer, they get frustrated. The PhD students, on the contrary, often investigate the material in great detail and suggest improvements to the course and the software. The graduate students (Master's) are a mixture of the two extremes. The level of difficulty of the course is aimed at the middle class, the graduate students.

Students come with diverse backgrounds, from different levels of education and cultural environments. To accommodate all backgrounds, it is necessary to make several modules voluntary, as in modules 3 and 4. Then the course can be customized to some extent.

3.3 Time consumption

The nominal length of the course is one semester, or about 15 weeks. The arrivals are difficult to control, though, and registration tends to become floating throughout the year. If a student falls behind, he will receive an e-mail with a reminder, and it is only when there is a revision of the course that the student will get kicked out. Figure 5 is a histogram of students' duration time. The average is about four months, which is acceptable. The figure clearly shows that some students, as expected, take a long time to finish. That can cause difficulties with the university administration, because certificates and marks are processed twice per year. Floating registration fits badly with a 'heart beat' of two pulsations per year. To overcome that problem, it is recommended to be more strict about starting the students at a certain deadline, so that they are likely to finish before a certain administrative deadline.

Not only do some of the students take a long time to finish, quite a few drop out. As mentioned briefly, 97 out of a population of 271 students dropped out, which is 36 percent. This may not seem alarming, but since the teacher's time consumption is proportional to the number of groups

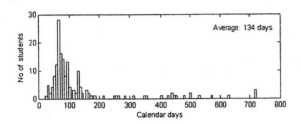

Fig. 5. Histogram of time to finish (170 students).

taught, each dropout is a direct loss of time. The dropout rate is the most difficult problem to manage.

A way to decrease the dropout rate has been to favour classes of students, rather than stray students from the Internet. For example, it is more efficient to agree with a colleague in another university to teach a class of students, that start together, follow the course at the same pace, and finish before a set deadline. Group pressure and the social environment greatly improves the likelihood of completion.

As a rule of thumb, the load on the teacher is about one tenth of the load on the student. That is, for a nominal workload of 60 hours for the student, the teacher spends 6 hours teaching. If two students agree to work in a group, the load on the teacher is still 6 hours. It is possible for one teacher to teach about 30 groups simultaneously. With two students per group, it is feasible to teach a class of 60 students. Because of the flexibility, it is feasible to teach two courses in the same time slot.

The direct load is rather high, and it is discouraging that it is proportional to the number of groups; there are no economies of scale. One time-saver, however, is to build a database of answers to the questions in the assignments. Each question in the assignment in the appendix is ended by a code, for instance '(fuzcon5.1)'. That code is automatically replaced by an answer, when the student returns the e-mail. The technical solution is to take the e-mail into Microsoft Word, and a short macro written for that purpose, finds the next code and replaces it by the answer (autotext). Only when there is a discrepancy between the student answer and the database answer, the instructor needs to supplement with further remarks. A student e-mail thus takes between 1 and 5 minutes to correct and return.

The large time-saver, however, comes the second time the course is taught: there is no or very little setup time. Distance learning requires such a strict organisation of the material, that next time around it will be more or less ready. In comparison to a lab course, for instance, the equipment has to be set up one or two weeks in advance. By relying on simulation instead —

and changing the teaching mode — that time is reduced significantly. If the textbook material is electronic and online, time can be saved on printing, copying, binding, and distribution.

The initial work load of setting up the course comprises: web pages, simulator, textbook material, assignments, and solutions. This can take from 2 weeks to many weeks, depending on how much is available beforehand. The cheapest possible solution uses a few (or no) web pages for organisation, an existing software package for simulator, a paper textbook which is distributed to the students, assignments in text files, and solutions. Building the assignments and solutions will be the major work in that case.

3.4 Student comments

The free format part of the students' course evaluation is a recommendable technique. While the student answers the set questions, he will sometimes find that there is some mismatch between the fixed choices and what he thinks. By the time the free format sheet comes up, it is an opportunity for the student to straighten that out; valuable comments come out that way. Some edited, but typical student comments are

(1) The course is an opportunity to try and tune a controller.
(2) Fuzzy on pendulum only gives marginal improvement; show another example that clearly shows the benefit of fuzzy.
(3) Would like more fun process, more theory.
(4) Make simulator more flexible.
(5) Sometimes we felt we did not quite understand the questions, sometimes we did not understand your answers, and sometimes you did not understand our answers
(6) The course is good for those who are interested.
(7) Do not save on explanations even it seems self-evident.
(8) Thank you for your always prompt answers.
(9) It is efficient if the modules are more or less decoupled.
(10) Good work / benefit ratio.

Comments 1 - 4 concern the contents of the course. The students seem to like practical work (tuning, simulator), but they would also like more theory. This is puzzling, because usually more theory makes a course harder and more boring. What is really behind, presumably, is the wish to see some powerful theory that can accomplish amazing results in practice. Comment 2 is a classical comment, and quite difficult to meet. On the one hand, the benefit of fuzzy is not necessarily a better performance, but a better (multivariable) structure and user interface. On the other hand, if one were to follow the wish of the student, it will be difficult to set an example small enough for educational purposes, that also shows a clear performance enhancement.

Comments 5 - 8 concern communication and motivation. The written e-mail communication in English is difficult for the weaker students and the same group of students have problems with motivation. They expect perhaps a more entertaining course, and find that Internet learning requires a high degree of self-discipline. Therefore it is recommended to avoid forcing a class of immature students into Internet learning; they must be motivated by their own need to learn. In the other end of the spectrum, PhD students may find the course too elementary.

Comments 9 - 10 concern efficiency. It is an advantage for the instructor, and apparently also for the student, that modules are independent; then the student can work on several modules at the same time to save some waiting time.

4. CONCLUSION

The objective of this study was to show that the teaching method described is one way to proceed from teaching to learning. There are negative student responses, but generally student responses are positive. E-mail tutoring, with 12 modules in a 60 hours course is activating for the student. The teacher constantly gets feedback from the student and knows exactly where the learning problems are. The student is forced to find the relevant information himself, and the learning is therefore deeper than traditional teaching provides with lectures, assignments, and a final exam.

The goal of having the student take responsibility for his/her own learning is still some distance away. For example, younger, immature students expect entertainment and a helping hand. When they find that the course is hard work rather than entertaining, they get frustrated and motivation decreases, and so does the learning. On the level of PhD students, assignments must be challenging enough, but then they find it natural to take responsibility for their learning.

E-mail tutoring as a teaching method works well, but a distance learning course, where participants join the course after finding it on the Internet, is vulnerable to dropouts. The teacher's load is proportional to the number of groups, and if a group drops out of the course, that time is wasted. It is necessary to make special arrangements to minimise the dropout rate, for example to teach a class of students. The course has been a lesser success than it was envisioned to be, partly because of the dropout rate, partly because the market was smaller than expected.

However, e-mail tutoring turns out to be efficient in ordinary classes, locally at the university. Teaching a class of up to 60 students is feasible.

For a future study, revising the questionnaire and its processing in the direction of the Australian Course Evaluation Questionnaire (CEQ) is one way of getting closer to a quantitative measure of how well students learn.

REFERENCES

Hall, Brandon (1997). *Web-Based Training*. Wiley.

Jantzen, Jan, Henk Verbruggen and Jens-Jørgen Østergaard (1999). Fuzzy control in the process industry: Common practice and challenging perspectives. In: *Practical Applications Of Fuzzy Technologies* (H.-J. Zimmermann, Ed.). Chap. 1, pp. 3–56. Dubois and Prade (Eds), The Handbooks of Fuzzy Sets Series. Kluwer. ISBN 0-7923-8628-0.

Jurado, Francisco, Manuel Castro and Jose Carpio (2002). Experiences with fuzzy logic and neural networks in a control course. *IEEE Transactions on Education* 45(2), 161–167.

Lo, C.H., Y.K. Wong and A.B. Rad (2002). Computer-aided learning of controller design: Focus on fuzzy logic control. *International Journal of Electrical Engineering Education* 39 issue 4, 358–370.

Magana, Mario E. and Frank Holzapfel (1998). Fuzzy-logic control of an inverted pendulum with vision feedback. *IEEE Transactions on Education* 41(2), 165–170.

Yurkovich, Stephen and Kevin M. Passino (1999). A laboratory course on fuzzy control. *IEEE Transactions on Education* 42(1), 15–21.

Fuzzy Set Theory   fuzzy05.txt

Please include this file INSIDE an e-mail (not as
an attachment nor Word file).

Assignments
----

1. Fuzzy sets

Give three examples of a fuzzy set. Example:
The set of tall people.

(fuzcon5.1)

2. More fuzzy sets

In the following list, which is NOT a fuzzy set:

a) The set of tall people ... ;
b) Possible speeds when the speed limit is 80
   kilometres per hour ... ;
c) Well done steaks ... ;
d) The set of trains expected to arrive between
   10:29 and 11:07 ...

(fuzcon5.2)

3. Membership functions

Model the following as fuzzy sets by means of
universes and membership functions:

a) what is your personal perception of the speed
   limit, when the road sign says 70 km per hour,

LIMIT = ... ?

b) hotel check out times, when the sign in the
   room says that guests must check out before
   11 am, i.e.,

CHECKOUT = ... ?

c) the set of large steaks, i.e.,

LARGE = ... ?

(fuzcon5.3)

4. Set operations

Given the following fuzzy sets:

a = [0 0 1 .2 .3 .4 .6 .8 0 1]

b = [0 .4 .6 .8 1 .8 .6 .4 0 0]

Determine the intersection and union.

(fuzcon5.4)

5. Probability vs possibility

In your opinion,

a. What is the probability that a basketball player
   is taller than 6'6" (1.98 meters)?

b. What is the possibility that a basketball player
   is taller than 6'6" (1.98 meters)?

(fuzcon5.5)

6. Probability vs possibility distributions

Define a probability distribution and a possibility distribution which could be associated with
the proposition "cars drive x mph on the
highways (of your own country)".

(fuzcon5.6)

Management questions

Your name(s):

Hours spent on this module (reading included):

Hours spent on hardware and software trouble:

What I liked about the module:

Suggestions for improving the module:

(suggestions will be collected in
http://fuzzy.iau.dtu.dk/tedfaq.nsf)

ELSEVIER

IFAC
PUBLICATIONS
www.elsevier.com/locate/ifac

THE FINNISH VIRTUAL UNIVERSITY

Pekka Kess

University of Oulu, Finland
Finnish Virtual University

Abstract: The building up of the Finnish Virtual University is based on the work to build an Information Society. All (21) Finnish universities decided to put their efforts together and to build a consortium the carry out the challenging task both technically, organizationally and financially.

The work for the virtual university is at the same time strategic discussion among the partner universities and within the project organization as well as operational project work to define, design and implement services needed.

The original Virtual University has been defined in terms of service portfolio designed using certain technical solutions. These are seen as very evolutionary solutions and when completed in 2004 the solutions will look slightly different from what they are today.

The project has taken the quality issues (both of the processes and of the service products) seriously. First steps have been taken but the will be a long way to go, before the best practice has been achieved. *Copyright © 2003 IFAC*

Keywords: virtuality, university collaboration, eLearning, higher education

1. INTRODUCTION

Many have become familiar with the Nokia phenomenon and their slogan, 'connecting people'. The Finnish Virtual University is happy with less. Connecting academic communities virtually would suffice.

The Finns populate the land strip of the same size as the Brits, with one tenth of the number of people only. The number of universities, however, totals 21 incremented by 36 polytechnics. This means that units are small and scattered around the country. It is difficult and costly to deliver quality education throughout the country under such circumstances.

Among many strengths of the Finnish higher education (HE) there are well-equipped and networked modern campuses with nomad students virtually grazing around with mobile gadgets. Another asset is the commitment of the government and the HE institutions to the shared vision of a knowledge-based society with a national virtual university in its pivot.

Moving from the current situation into the desired direction is not an easy and straightforward job. On the contrary, the list of challenges is dauntingly long and it tends to expand when we move on. Nevertheless Finland seems to be progressing. The federal funding for the project is ensured at least for the next year. The sustainability of the Finnish virtual university model will be put in the real mode test not until in 2004, when the pilot phase will be over.

In general, universities have a relatively good and well-functioning ICT infrastructure. Teachers have nearly enough PC's at their disposal. The greatest needs are in the area of student workstations. Another problem is insufficient technical and pedagogical support services

Information technology expertise possessed by individual students can, however, be a significant resource in the development of teaching applications. The problem often lies rather in the attitudes of teachers and even in their fear of losing authority.

According to a survey conducted, teachers mainly use information technology to prepare lectures and assignments, maintain contacts with other members of academia, acquire and process new information, and to conduct their research. Students mainly use information technology to complete their individual assignments, communicate with each other and their teachers, and acquire new information (Sinko, 2000).

The same survey showed that ICT has thus far not had a particularly profound effect on how teaching at the universities in general has been carried out. This does not mean, however, that new teaching practices have not been created through utilising these technologies. Many individual pilot projects show that when coupled with innovative pedagogical thinking, technology opens up interesting possibilities for revitalising higher education.

In this presentation the terms Finnish Virtual University, Virtual university and FVU mean the same entity. Term virtual university is used in connection to any other organisation formal or un-formal where so called virtual university activities take place.

2. RATIONALE FOR GOING VIRTUAL

The virtual university initiative was first outlined in the national information society strategy for education and research for the years 2000-2004 adopted by the Ministry of Education and the Finnish government. Based on the analysis of the global situation of Finnish HE, the vision for improving HE through going virtual was outlined and the plan for implementing it launched:

> *By the year 2004 a high-quality, ethically and economically sustainable network-based model of organising teaching and research will have been consolidated.*
> *There will be set up a virtual university by 2004 based on a consortium of several universities, business enterprises and research institutes. It will produce and offer internationally competitive, high-standard educational services.*

The aims have evolved from the original documents into actions and targets like

- to enable networking in teaching, studying and research
- to develop a new model of network based cross-university operation
- to diversify university studies
- to develop university curricula
- to improve the quality of teaching and studying in HE
- to make better use of the ICT networks and
- to improve the competitiveness of Finnish academia.

3. IMPLEMENTATION

The Virtual university will be established in stages. At the initial stage, the ministerial virtual university task force has coordinated the project. In connection with the negotiations on target outcomes between the ministry and the universities in spring 2000, the universities committed to establish the virtual university consortium.

The ministry of education chose then, based on applications from universities, about twenty specific inter-university projects to be funded until 2003. These projects are anticipated to play a key role in shaping the services of the virtual university and a substantial number of exemplary net-based courses and study programmes.

The development unit was set up in August 2000 to coordinate the start-up phase and emerging services. The consortium agreement between the universities was signed up, the steering committee elected and the action plan approved by the consortium in early 2001. The virtual university strategy was discussed thoroughly among the universities and was adopted by the consortium in early 2003.

All students (first degree students, post graduates or open university students) of any member university will be eligible for studies in the Virtual university or to put it more precise for studies in any member university of the Virtual university consortium. Students can take courses relating to their degree programs in the Virtual university, but it is the home university, which will award the degree.

The main distinctiveness of the FVU compared to other virtual university initiatives is the following:

- It is a national initiative that involves all of the country's universities;
- It is set in a context of a national information society strategy to improve the quality of teaching and learning at universities and offers learners greater access and flexibility through the integration of technology.
- It is not solely targeting the development and/or marketing of totally online courses to learners outside of its borders or responding to a competitive threat; and
- While not yet realized or fully planned, the FVU has a comprehensive vision of including teaching as well as research and support services.

An effective FVU was envisioned to offer the following advantages:

- greater flexibility of time and place
- flexible programs/courses and possibilities for individual additional studies
- international exchange of educational material
- co-operation and co-ordination of development work
- saving on space and facilities
- efficient use of time

The FVU was also seen as an opportunity to address more general weaknesses in Finland's higher education system such as the lack of a tradition in collaboration among universities and a slow progression of studies for students.

Due to the large expenditures required to develop virtual university programs, there was the need to pool limited resources to achieve economies of scale by connecting the work across Finnish universities. In addition to achieving economies of scale, the FVU was intended to address barriers to online learning applications including technical, pedagogical, social, administrative and regulatory issues. Key personnel involved with the FVU emphasize that the initiative does not require or promote full virtuality as the only model. Most of the persons met during the visit program were very resistant to the idea of offering fully online alternatives.

4. CHALLENGES

Increasing the share of students attending HE in Finland (like in all parts of Europe), increasing the number of students on certain popular fields while some areas suffer from shortage of students, calls for new pedagogical solutions. Increasingly heterogeneous student groups demand for improved didactics. Going virtual and meeting the more complex needs of students is one way of trying to address these problems.

HE institutions are facing serious resource problems. All these challenges have to be met with limited budgets. Dedicated extra funding covers only the piloting phase.

These facts place totally new demands on developing HE. Net-based education has been welcomed by many as, if not a panacea to all problems, but to bring substantial relief or at least alternative solutions to many ailments and challenges of HE. Online education, however, is by no means rendering easy solutions. On the other hand, there is no easy way of solving the pertinent problems HE is facing by continuing to develop campus-based education either. Face-to-face education is neither cheap, nor cost-effective. So educational policy-makers have to come to grips with the same fundamental problems of education whether seeking solutions from the net or inside the campus.

According to the action plan key issues to be addressed and hopefully to be turned into success factors will be:
- Support to collaborative design and delivery of net-based courses
- Solving IPR issues
- Integration of different modes of instruction
- Online tutoring
- Virtual mobility of students
- Finding a sustainable model of operation.

These issues are currently in active work-out by for instance through a number of inter-university projects. The aim of those projects is not only to design excellent courses to be run on the net successfully. It is hoped that they will serve also as examples of good practice to be scaled up. Moreover it is hoped that they will sow the seeds of new type of academic collaboration among professionals across institutions, faculties and research paradigms. The lack of adequate pedagogical and technical support also calls for sharing resources between institutions.

The IPR issues cannot be solved adequately within one national project, because of its global nature. The Virtual university has just issued a set of applicable contract models. In this endeavour FVU combined efforts with other educational institutions under the umbrella of the Ministry of Education, which provided best legal assistance in drafting model contracts and will hopefully provide certain level of online consultancy as well.

FVU is not aiming at full virtuality in the course offerings. Consequently attention will also be paid to improving current teaching and studying practices through incorporating online elements in any courses whenever appropriate. It will hopefully lead to a new practice optimising the use and the mix of different teaching and learning modalities in a flexible way. This gives the students an opportunity to choose from various methods of course delivery and different realization of the learning process. Flexibility should be stretched to its limits to allow maximum personalising and customising of learning environments and teaching arrangements.

A certain level of online tutoring is already available to open university students. The solutions developed there is a rich source for adopting and adapting the practice for degree students tutoring. It is believed that the tutoring services will largely be distributed and among member universities and the centralised services provided by the national portal will be quite thin in the beginning. In the long run when course design and course offerings will be provided on a large scale and when information systems and databases will be fully-fledged, there will be possibilities to provide more comprehensive services.

The Finnish universities are small. It means limited opportunities for students within their home campus. Regardless of high rates of student mobility, there is need for dramatically expanding virtual mobility of students. Providing students flexible opportunities to pick up courses from other universities without needing to engage in a lot of time and money consuming travelling, there is enormous potential for the FVU service provision. A task force has been collected to tackle the administrative challenges of the virtual mobility with credit transfers, financial transactions, student registering etc.

Even though the virtual university initiative is in the focus of the Finnish government information society strategy, it is not run top-down. Neither is it run bottom-up. It is clearly network based and managed. All the universities are stakeholders. The activities and services are thus defined, designed and will be run by innovative and enthusiastic academic networks assisted by the ministry owned company, CSC, which is responsible for the Finnish university network and scientific computing. The current organisational model will certainly need many modifications to run the virtual HE service successfully in the future. It remains also to be seen, how soon the consortium will manage to attract non-academic partners and extend its activities beyond national borders.

5. FINNISH VIRTUAL UNIVERISTY FUNDING AND GOVERNANCE

The Finnish Ministry of Education, has committed funding to the FVU until the end of 2004. The FVU has funding of approximately 10 million Euros per annum for the first years of its development. Approximately half of the sum has been awarded to the individual universities for their development and the other half to the 20 selected inter-university network projects. Additional funding of 1 million Euros has been raised for the development of the FVU portal. The Finnish Ministry of Education intends to make the FVU a permanent program and to consolidate the service provision developed through this initiative and the projects within the program.

All 21 universities in Finland have joined the consortium that is developing and managing the FVU. Projects, such as the scientific national electronic library FinELib and the Finnish open university, SUVI, will be closely integrated in the new initiative and collaboration is intended with the Virtual Polytechnic Initiative.

The consortium is an important mechanism to discuss and propose solutions to practical issues such as intellectual property rights, technical support required in the development work, and the development of teachers' knowledge and skills concerning online education.

To carry out the practical development and building work, in 2000, a Virtual University Development Unit was established as a joint service unit for the universities. The tasks of the Development Unit, which now has a staff of about 10 professionals, include:

- development of activities and the administrative structure of the virtual university consortium
- policy design and strategic planning regarding functions of FVU
- providing support to the projects initiated by universities and the consortium
- investigating and reporting on the activities of the FVU, monitoring and benchmarking relevant developments in other countries and reporting about them publicizing the activities of the virtual university and maintaining contacts with project partners in order to further develop joint activities
- ensuring efficient functioning of services on a practical level
- preparing model agreements for members (for example, agreements for partner networks, copyrights, and financial transactions)

The Development Unit works in close co-operation with the Ministry of Education's Virtual university task force and with a steering committee of a subset of consortium representatives. The Development Unit also creates and maintains contacts internationally, collecting and disseminating information on global trends in order to react quickly to changes in the environment while strengthening the operating capability of the network.

6. STATUS OF THE FINNISH VIRTUAL UNIVERSITY PROJECT

Parallel to the work of the FVU Development Unit, joint projects between universities were initiated that will produce the first online courses and form the basis for the activities of the FVU. These projects include 3 regional networks, 5 joint projects aiming at providing services ("meta-projects"), and 11 networks of specific disciplines ranging from social work (SOSNET) to a graduate training program for faculties of law. In addition, the funding provided directly to universities has been used to hire part-time or full-time support people to help in the training and support of teachers and staff who are developing online learning materials. The first programs started during the academic year 2001-2002.

A FVU portal is another major component here. The portal is intended as a functional and adaptable gateway to the FVU net-based services with the necessary functions and services for teaching and learning including:

- course selection
- course information including search engine
- registration tool
- student support services (guidance, advisory and information services, portfolio management)
- information and contact channels
- access to national electronic library services
- discussion forums and collaboration areas for teachers
- support services including versatile assessment tools.

The portal will be personalized to the various users. The portal will also have a tool for defining teachers' expertise profiles that can be used for defining individual, team, and organization specific training needs.

The major anticipated challenge in the portal's implementation is the development and funding of the mid-layer of functionality where the responsibilities of the universities and the development unit intersect.

The first set of functionalities of the portal was launched in the autumn of 2001 with additional functionality added by the end of 2002. The portal as currently envisaged is estimated to be fully functional in the autumn of 2004.

While the FVU is intended to facilitate all academic activity including research, these services have not yet been elaborated.

7. IMPLEMENTATION BENEFITS AND PITFALLS

While only officially launched in January, 2001, the FVU has been the subject of a much longer planning process that included a task force implementation plan that was ready at the end of 1999. The FVU has already had a major impact in Finland in the following ways:

Building of awareness of the use of computer networks for teaching

It is apparent that the announcement of the FVU and funding awards have raised awareness and interest among university faculty and staff in the practice of and research into the use of computer networks for learning. There is some confusion about the meaning of the terms "virtual course" or "virtual university" and a general belief that a face-to-face component will always remain necessary. However, the FVU has already had a beneficial impact in highlighting the opportunities provided through online learning and encouraging discussion and debate.

Putting the focus on teaching and learning

The prime goal of the FVU is to improve teaching and learning at Finnish universities. Through discussion of the use of online learning and teaching models, there is evidence of increased focus on teaching and learning.

Creating networks for collaboration, faster dissemination of best practices, and acceleration of staff capacity building

Each of the FVU projects that has been funded has linked faculty and staff across a number of Finnish universities. For example, one project called KASVI links the eight faculties of education in Finland. A number of projects have organized workshops and some of the workshops have attracted several hundreds of participants.

Some of the meta-projects in particular could have a major beneficial impact. The goals of four meta-projects are summarized below:

IT PEDA: To support universities developing strategies for virtual university activities including the creation and support of a network of pedagogical centres at participating universities. The project is also responsible for seminars for administrators on strategy development.

Tie Vie: Training of faculty in the pedagogical use of information and communication technologies.

IQ Form: Creating tools through which students can learn about themselves as learners and acquire skills to become more effective learners in virtual courses

OVI: Project for developing the virtual environment for assessing and study counselling of students including improving the skills and abilities required for successful studies, career counselling, and study management.

Creating a forum for discussion of key issues among administrators

Bringing together the top administrators of all of the universities in Finland as part of the consortium has had an immediate impact with much promise for the future. The issues raised at consortium meetings range from the required support services for virtual university activities to copyright issues. Much of the topics for discussion are not restricted to virtual university matters and will impact institutional policies for classroom teaching, such as credit transferability and copyright issues.

Creating economies of scale for content development, program delivery, and critical support services

Participation in one of the meetings of the FVU projects, the Eastern FVU project, clearly substantiated the benefits and necessity of pooling resources and efforts given the limited capacity and budgets of individual institutions. This project brings together three of Finland's smaller universities to offer jointly-designed courses, expand the course offerings, and ensure better quality and better use of local resources and expertise.

There is no doubt that the FVU will face significant challenges, especially given limited resources and a complex and difficult to change operating environment. Several major issues and challenges in implementing the vision of the FVU network are identified.

Managing Expectations

The announcement of the FVU and the funding award has raised huge expectations despite the relatively limited funding to carry out the initiative and a lack of history and widespread knowledge of how to implement online learning in Finland. The number of full-time staff devoted to the FVU is low and success will be dependent on continued efforts from other specialists as well and on goodwill. There are also different expectations and different definitions of

virtuality or effective virtual learning among the many participants. The impact of the FVU will probably be overestimated in the short term and, if successful, underestimated in the long-term.

To address the issue of managing expectations, the FVU needs to immediately:

- Identify the various stakeholder groups including funding groups and define their expectations as quantitatively as possible
- Develop and implement a communications plan and public relations strategy that includes short overviews addressing what the FVU is and what it is not. Illustrate the funding level through comparisons to investments made in other countries and
- Continue to utilize the FVU contact persons in each of the FVU participating organizations to inform academia and local interest groups and bring forward misconceptions and concerns.

Institutional commitment and strategic leadership

The level of knowledge and commitment of university administrators at the participating institutions varies. For many, eLearning is viewed as a nice demonstration of innovation or a source of some funding to shore up weak departments. In a few cases, virtual university courses or programs are seen as an integral component of the institution's primary mission or one part of a solution set to address a challenging issue. In part, the lack of attention and funding commitment may be due to the increasingly difficult budgetary environment for Finnish universities.

The FVU has begun to address the requirements to increase the knowledge of university administrators

Leadership on effective balance between pedagogical, administrative and ICT practices

The intention of the Ministry is to put pedagogy at the forefront of all funded and associated initiatives of the FVU. The FVU Development Unit is having a more balanced approach putting in addition to the pedagogical practices the admin and ICT practices into the big picture of the FVU. There is some resistance to "importing" models from other countries as there is a legitimate concern in adopting practices that may not be compatible with Finnish culture.

Most Finnish institutions have had some trial experiences in developing and offering online courses with a focus on videoconferencing classes or publishing course notes and useful course resources as opposed to implementing online learning activities. The departments of technical universities such as the Helsinki University of Technology and Tampere University of Technology have had prior experience with web-based offerings.

The FVU will in the future fund and encourage projects and activities that take advantage of situations where online learning can add real value. A number of projects are expanding video-conference classes or providing an online option to a face-to-face classroom activity without examining this value-added factor or best application areas.

The FVU will develop a plan to disseminate effective practices.

Learner Focus

There is a promising research that focuses on learners such as IQ FORM, and a survey has been carried out to determine and analyse the learner needs for the portal. A larger follow-up has been planned through a student and staff panel to evaluate the portal design.

International Strategic Alliances

With its reputation in the wireless industry, Finnish organizations receive a good reception from organizations around the world. The Finnish universities are fortunate in having opportunities to link with other EU countries in research and dissemination projects such.

Given the work involved in entering into and implementing strategic alliances, the FVU Development Unit will focus on entering into a few formal arrangements each year. These arrangements could include post-doc and faculty exchanges, cooperation in knowledge and tool exchange in areas such as learning object repositories, and faculty and expert visits. Further information will be posted to the FVU web site in other languages with projects encouraged to provide summaries and updates for posting.

Evaluation

The Ministry of Education has asked each FVU project to prepare a self-evaluation report that documents approaches used and results obtained at the end of the first year. The evaluation measures are those proposed by project participants and there are no "top-down" Ministry of Education measures.

Given the importance of evaluation in managing expectations and feeding into continual improvement of the FVU's investments. The FVU has developed an evaluation strategy for the initiative.

8. LATEST DEVELOPMENTS AND LESSONS TO OTHERS

One of the basic paradigms behind the Finnish Virtual University was the idea of giving the students an opportunity enhance their learning portfolio with courses from other universities in Finland. This has been achieved by
- setting up an agreement to include all Finnish universities, how the flexible studies are
  - planned
  - monitored
  - accepted
  - managed
  - financed

- building an information system to support these activities
- arranging extra funding from the Ministry of Education to cover the initial extra costs of these activities
- extensive training of students, teachers, administrative officers, etc about the new approach.

There are many developments in the virtual university setting that can be considered innovative and something to consider abroad as well. The key element is the learning service to facilitate the development of online courses and programs by encouraging:
- learners via one-stop shopping with a wide range of information, resources and services;
- participating institutions and their faculty members with an opportunity to take advantage of economies of scale by making available a wide range of services, knowledge and resources to support the development of online courses and programs; and
- participating institutions with an opportunity to take advantage of synergies and economies of scale in the marketing of their online courses and programs at home and abroad.

REFERNCES

Curry, J., The Finnish Virtual University: Lessons and Knowledge Exchange Opportunities to Inform Pan-Canadian Plans. Prepared for The Information Highway Advisory Branch, Industry Canada. 2001. 22 p.

Kess P. The Finnish Virtual University. 4th Business Meeting in Hagenberg, Austria 28.-29.6.2002. 4 p.

Kess P., The Creation of the Finnish Virtual University – First Three Years. ICT in Education, Rotterdam, 2.9.2002.

Sinko, M., ICT in Finnish higher education: impact on lifelong learning. Workshop on "Application of the new information and communication technologies in lifelong learning" Catania, 6 - 8 April 2000, www.virtualuniversity.fi

AN INTERACTIVE SIMULATION TOOL FOR THE STUDY OF MULTIRATE SAMPLED DATA SYSTEMS

J. Salt[†], P. Albertos[†], S. Dormido[‡], A. Cuenca[†]

[†]*Departamento de Ingeniería de Sistemas y Automática*
Camino de Vera 14, 46022 Valencia (Spain)
e-mail: {julian, acuenca, pedro}@isa.upv.es
fax : +34 963879579

[‡]*Departamento de Informática y Automática*
Avda. Senda del Rey s/n, 28080, Madrid (Spain)
e-mail: sdormido@dia.uned.es
fax : +34 913986697

Abstract: Dealing with practical applications of digital control, the assumption of a regular and uniform sampling pattern is questionable. On the other hand, the theoretical analysis of the controlled system performances is much more computationally involved. In this paper, an interactive simulation tool to deal with non-uniform sampling pattern situations is presented. It allows an easy understanding of the performance degrading or improvement when changing the sampling pattern parameters. *Copyright © 2003 IFAC*

Keywords: Non-uniform sampling, Multirate Control, Interactive Simulation, Control Education

1. INTRODUCTION

The assumption of linearity has been commonly used in the model-based study of the dynamic systems behaviour. Although linearity is not a perfect feature of real systems and it is accepted that the process behaviour is non linear, the simplicity in the mathematical treatment and the good approximation results of this approach in many applications have promoted the linear control systems theory. Consequently, all the basic courses in control engineering assume this linearity property in the models and, only in very simple cases –like relay control or the presence of saturation – a quick look to the nonlinear control is presented. Later on, students learn how to handle the plant non-linearities that cannot be deleted in the practical applications.

A similar trend was followed in the case of digital control. The neat synchronicity in the simple one-computer-based control systems recommended considering uniform and regular sampling in the study of discrete control systems. Most digital control concepts can be presented and discussed with this simpler approach, even it is also well known that there are fundamental issues that can not be covered with this setting. In any practical discrete control application, the sampling and updating pattern of the plant and controller variables is not simultaneous. There are always delays –computational, propagation or communication delays- and the rate of sampling or updating of signals is conditioned to the availability of resources or the convenience of capturing more detailed information. This results in a number of issues that can be determinant to achieve the expected controlled system performances. Other than the classical interest in analysing the intersampling behaviour of the continuous time plant variables, the influence of the parameters of the so-called non-uniform sampling pattern must be realized.

Again the problem is the computational complexity. There are many theoretical developments to deal with this topic but, from an educational point of view, it is quite difficult to grasp the main concepts and practical implications behind the unavoidable complexity of the proposed algorithms, (Salt and Albertos, 2000)

In particular, a multirate sampling system is defined as a hybrid system composed by continuous time elements, usually the plant, and some discrete time components, usually the controllers or the filters, where two or more variables are sampled or updated at different frequencies. It can be also considered that the discrete actions are not equally spaced on time and/or delayed.

A non-very restrictive assumption to simplify the treatment is to consider that the sampling pattern is periodic. That is, the process variables are sampled and/or updated at different and/or irregular intervals, but there is a global period T_0 with cyclic repetition. It may be also considered that there is a delay between the sampling and the updating of variables, but still a global periodicity is assumed. Much more complicated is the case of asynchronous sampling/updating, with a random occurrence of the discrete actions, and it will not be considered in this paper.

As previously mentioned, when teaching basic digital control courses, a perfect uniform sampling and updating pattern of the involved variables is assumed. But it should be pointed out that, in practical applications, the synchronicity of the set of discrete actions is not perfect or it can be modified in order to improve the performances. Thus, multirate digital control is an important topic not only for research purposes but to be included in control education curricula. This situation may be present in a wide range of simple applications and the students must known about their consequences in an easy way. For instance:

- Time sharing computer by means of several detection services (Jury and Mullin, 1959).
- Practical applications such as: aerospace (Halevi and Ray, 1988), robotic (Tsao and Hutchinson, 1994), chemical process control (Morant, *et al.*, 1986), computer hard disc control (Baek and Lee 1999).
- Missing and scarce data (Albertos, *et al.*, 1999).
- Distributed and multiprocessors control systems (Hovestädt, 1991).
- Real-time control systems (Salt, *et al.*, 2000).
- Multivariable control systems (Velez, 2000).

Moreover, in a great number of computer control applications the approximation of a regular pattern of sampled signals is assumed.

One of the goals of this work is to present a tool allowing the student to understand the influence of these phenomena in an intuitive environment without sophisticated formulas or requiring deep extra curricula contents.

From an educational viewpoint, the well-known book on computer-controlled systems, (Åström and Wittenmark, 1997), already pointed out the possibility of changing the position of the zeros of the discrete transfer function of a plant by allowing a faster updating of the control input.

But, in general, the complexity of the theoretical developments justifies the use of interactive simulation techniques that also allow for acting over a high number of parameters with hard crossed relations. The global and simultaneous dynamic visualization of different kinds of time and frequency diagrams allows grasping a clear understanding about the effects of the concerned topic.

2. THE INTERACTIVE SIMULATION IN LEARNING PROCESS.

In order to analyse and study the different characteristics of the dynamic behaviour of a system, it is common to use time and frequency techniques and tools. These will provide a complete and global picture of the system behaviour, showing up the interrelation among the different controlled plant performances.

The combined use of control system design tools and dynamic system simulation, leads to the computer-aided control system design environments, simplifying the design task. Additionally the possibility of simultaneous visualization in various windows of the effects in different performances of some design parameter changes helps to observe with more flexibility the change gradient over the system. This facility provides the understanding of the usual steps in a design procedure. The perception of synthesis and analysis phases is simultaneous with the consequent effort saving with relation to classical simulation environments.

In this sense some years ago, Åström and colleagues introduced some valuable concepts for control education tasks aid at Lund Institute. In this context the significance of concepts like dynamic pictures and virtual interactivity must be highlighted (Wittenmark *et al.*, 1998).

Basically a dynamic picture is a set of graphic windows that can be handled by the mouse. The windows are interrelated by objects in a way that some change in a parameter manipulated by the user implies the fast –practically immediate- consequence visualisation in graphics influenced by that object. One of the main advantages is that the student does not need the implementation of code sentences. The complete effort is leading to test and understand the systems control ideas and principles that the application involves.

The best examples of this new educational trend for teaching automatic control are the packages Ictools and CCSdemo (Johansson *et al*, 1998), (Wittenmark *et al*, 1998), developed at the Department of Automatic Control at Lund Institute of Technology, and Sysquake, at the Institut d'Automátique of the Federal Polytechnic School of Laussanne (Piguet, 1999; Piguet, et al 1999).

This is a new example of the efficient application of advanced information technology in fields as control education (Kheir, *et al.*, 1996).

3. DESCRIPTION OF THE APPLICATION

A SISO and MRIC (multirate input control) system has been assumed in the developed application. The system output is measured at low frequency and the control is updating at faster rate –this is the sense of MRIC-. The relation between both rates is N, which is called multiplicity. The output measurement is known as metaperiod T_0 because it fixes the interval where the cyclic repetition is produced. In this environment, when a controller is planned, various alternatives could be followed. Perhaps the basic one, introduced by Åström [Åström and Wittenmark, 1997], is the consideration of single rate loop but with the addition of a periodic gain $\tilde{K}_T$ that allows a specific location of zeroes of the process (figure 1). The discrete time controller operates at the global period but the hold device signal is multiplied by the T-periodic gain.

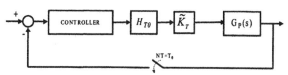

Figure 1:Basic Multirate loop with Periodic Gain.

In this application, an irregular distribution of the control updating instants inside the global period is also allowed. Thus, a general program feature is the possibility of change the sequence of control updating even assuming the irregular case. The restriction that was imposed is that every sampling period must be a multiple of a basic period, t_i, called "intersampling period" that is a kind of maximum common divisor of all the time intervals in the application. This is the same that to say that the relation among every sampling period –in a regular or irregular pattern- is rational.

There are many other multirate-controller structures that could be applied. A non-conventional controller structure with different rate parts could be assumed (Salt and Albertos, 2000). In every case the problem is to transform two (or more) different period discrete sequences. A simpler approach that we assumed here is the procedure shown at figure 2. The slow process output measurement is held with period NT and the controller is discretized with basic period t_i. Although the controller output is computed every t_i sec, due to hardware constraints the control action is

only updated with the selected pattern by means of the H_i hold.

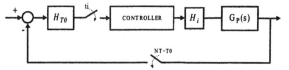

Figure 2: Multirate controller implementation with a special hold device.

An example of the implementation of the dual rate controller can be found in (Salt and Albertos, 2000). Here, in order to illustrate the simulation procedure, a simpler approach is followed, using a special hold device H_i whose operation is presented in figure 3. A special feature assumed in the behaviour of this device is that the control action energy is preserved. Specifically, the calculated control actions at the t_i instants where there is not a control updating imply an energy that will be added or substracted in the future control action. This extra energy is uniformly distributed in the time interval corresponding to the next future control action.

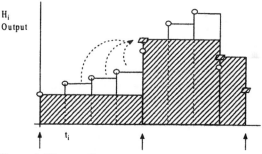

Figure 3: H_i operation.

In figure 3 the calculated control actions have been marked with circles and the implemented ones by rectangles, taking into account the missed energy. The arrows indicate when the control action is updated.

In the following, the main characteristics of the developed tool that has been programmed using SysQuake (Piguet, 2000) will be described. We must keep in mind that the main focus of the interaction is to make easier to the students the comprehension of control concepts and techniques which are difficult to understand (Dormido, 2002), (Dormido et al., 2002).

In this application, the main window is split into several parts showed in figure 4 (basic screen of the developed interactive tool). The parts are the following:

Parameters: In this window a number of parameters can be on-line tuned by means of sliders: the multiplicity factor of the multirate system N, the method of discretization of the transfer function Method, the number of poles and zeros of the transfer function Z TF and P TF, the PID parameters of the controller kp, ki, and kd including an optional derivative filter M, and the simulation time tmax.

Multirate map: This is a especially interesting window for this application. Three interactive variables are defined in the window by different colour lines. The metaperiod interval, (label T$_0$), is denoted by red. Green lines are used for the control updating periods (label T$_3$), and the most important parameter is the intersampling period –in blue- (label t$_i$). As previously explained this is the basis for scaling every sampling period in the application.

Transfer function: This window allows to drag graphically the poles (x) and zeros (o) of the linear transfer function in the *s*-plane.

Discrete transfer function: The corresponding discretized process, using the method selected in the *Parameters* window, is represented in this window.

Step response: The evolution of the closed loop system step response is shown in this window. The vertical lines give valuable information about the features of the response: the maximum overshoot and correspondent time and the settling time -95 and 98% of steady state-. To know these values, it is necessary to put the pointer of the mouse over the corresponding line and the magnitude appears at the state bar.

Root locus: This window shows the root locus diagram representing the evolution of the closed-loop poles, at the metaperiod z-plane, for the selected

magnitudes in the *Parameters* window.

Bode mag, Bode phase, and Nyquist curve: These windows show the Bode magnitude and Bode phase of the SISO MRIC non-conventional sampling system under study.

When one of the Bode diagrams is selected with the mouse pointer a vertical magenta line appears informing about the frequency (in rad/sec) of some point of the response. The value is shown in the state bar. This line has a correspondence with a circle of the same colour on the Nyquist plot. Two vertical green lines have been introduced in Bode's window. With the help of a horizontal auxiliary blue line, the gain and phase margin frequencies are pointed out. In this case, there is an analogy with the Nyquist curve. As it has been said before, these frequency values appear on state bar after selecting with the mouse the corresponding mark line at Bode diagrams.

All the figures reflecting the specific analysis techniques used for understanding these sampling cases have been based on Vectorial Switch Decomposition (VSD) by Kranc (Kranc, 1957). This methodology was reconsidered for computer applications purposes by Thompson taking into account matrix structures (Thompson, 1986). We have extended the results to irregular sampling patterns. Specific recurrent laws were obtained.

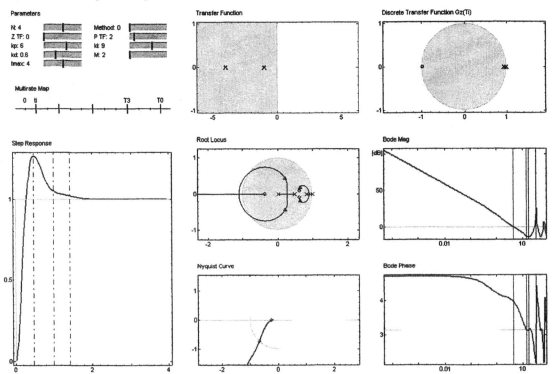

Figure 4: Sample window of the application.

4. ILLUSTRATIVE EXAMPLE

In this section, an example is shown in order to demonstrate some of the capabilities of the

application. It must be taken into account that its main characteristic, the interactivity, is difficult to be grasped in a written text. However, some of the potentials of the application will be illustrated.

Other examples can be found in (Salt et al., 2002).

Effect of the intersampling period t_i

This example shows how the variation of the intersampling period t_i changes the response of the multirate system. A second order system is considered and described by:

$$G(s) = \frac{4}{(s+1)(s+4)}$$

The multirate parameters are the following: $N = 4$, $T_1 = 3t_i$, $T_2 = 6t_i$, $T_3 = 9t_i$, (update control instants), $T_0 = 12t_i$ (metaperiod), $k_p = 6$, $k_i = 9$ and $k_d = 0.6$ (PID parameters). The simulation results are shown in Figure 5. As can be seen, the closed loop behaviour exhibits a more oscillatory response when t_i increases.

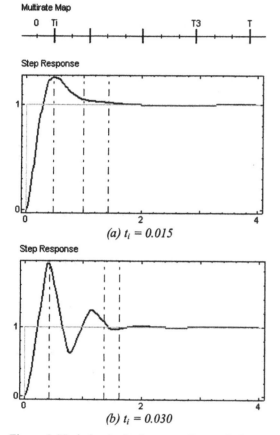

Figure 5: Variation in the intersampling period.

5. CONCLUSIONS AND FUTURE WORKS

An educational tool for teaching, in an interactive way, multirate sampled data systems has been presented in this paper. The tool developed in SysQuake helps the students to grasp the basic concepts of multirate sampled data systems and to gain an insight into the selection of the parameters and how the controlled system performance is influenced by this selection. Future works in this tool will include:

☐ Inclusion of the delay.
☐ Extension to others multirate schemes
☐ Extension to the multivariable case.
☐ Inclusion of measurable disturbances.
☐ Design of multirate control systems

Analysing the results from different parameter sets in the developed application, it is possible to conclude that:

- From the root locus it can be observed that a system performance variation is obtained by either a change of N magnitude or, for a fixed N, a change between regular or irregular sampling pattern. It is clear that a specific z-plane (metaperiod) area is covered by these changes. So, it is possible to assure that the control update period or sequence could be assumed as a new parameter design. The implemented program allows taking advantage about this procedure. As it can be shown in the step response plot window, the zeroes introduced by the control sampling sequence have an important influence and must be considered when a new design is planned.

- In the dual rate context the transformation from $N = 2$ to a greater multiplicity does not lead to a similar spectacular better system performance results that when the single rate to dual rate transformation is assumed. In the irregular pattern case the shortness between two successive control update sampling periods produces the same behaviour that assuming a lower N.

- When N grows up, a better magnitude and phase margins are reached. This is like a general rule in frequency domain that can be tested by the correspondent application windows (Bode magnitude and phase, Nyquist curve). A special attention must be observed again with the zeroes location introduced by non-conventional sampling.

- If the irregular sampling pattern is evaluated it must be shown that magnitude and phase margins grow up with the irregular pattern progress from the closest to single rate case to regular dual rate case. This result confirms the consistency between the frequency and time domain analysis environments. It must be remembered that, by means of step response, it could be observed that less maximum overshoot and settling times were obtained when the sampling pattern goes from irregular dual rate cases closer to single rate one to the regular dual rate case.

ACKNOWLEDGEMENTS

This work has been supported by the Plan Nacional de I+D, Comisión Interministerial de Ciencia y Tecnología (CICYT), under grants TAP98-0252-C02-02 and DPI2001-1012.

REFERENCES

Albertos P., Salt J., Tornero J. (1996). Dual Rate Adaptive Control. *Automatica*, **vol. AC-32, n. 7**, pp. 1027-1030.

Albertos, P., Sanchis, J., Sala, A. (1999). Output Prediction under scarce data operation: control applications. *Automatica*, **vol. 35, n. 10**, pp. 1671-1681.

Åström, K., Wittenmark, B. (1997). Computer Controlled Systems: Theory and Design. Third Edition. *Prentice Hall.*

Baek, S., Lee, S. (1999). Design of Multi-Rate Estimator and its Application to a Disk Drive Servo System. *Proceedings of the American Control Conference.* San Diego.

Dormido, S. (2002) Control Learning: Present and Future. Plenary Lecture. *15th Triennial World Congress of IFAC, Barcelona, Spain*

Dormido, S., Gordillo, F., Dormido-Canto, S., Aracil, J. (2002). An interactive tool for introductory nonlinear control systems education. *15th Triennial World Congress of IFAC, Barcelona, Spain.*

Halevi, Y., Ray A. (1988). Integrated Communication and Control Systems. Part I - Analysis. *Journal of Dynamic Systems, Measurement and Control*, **vol. 110**, pp. 367-373.

Hovestädt, E. (1991). Multirate Control Algorithms for time-variable sampling periods. *IFAC Design Methods of Control Systems*, pp. 105-110. Zurich (Switzerland).

Johansson, M., Gäfvert, M., Åstrom, K.J. (1998). Interactive tools for education in automatic control., *IEEE Control Systems Magazine*, **18**, n° 3, pp 33-40.

Jury, E.I., Mullin, F.J. (1957). The analysis of sampled-data control systems with a periodically time varying sampling rate. *IRE Trans. on Automatic Control*, pp. 15-21.

Kheir, N.A., Åstrom, K.J., Auslander, D, Cheok, K.C., Franklin, G.F., Masten, M.;Rabins, M. (1996). Control System Engineering Education, *Automatica*, **32**, n° 2, pp. 147-166.

Kranc G. (1957). Compensating Error-Sampled System by Multirate Controller. *Trans. AIEE*, **vol. 6, M. 6, part II**, pp.149-155.

Morant, F., Albertos, P., Crespo, A. (1986). Oxide composition control in a raw material mill. *IFAC Symp. on Microprocessor in Control.* Istambul.

Piguet, Y. (1999). Sysquake: User Manual, Calerga.

Piguet, Y., Holmberg, U., Longchamp, R. (1999). Instantaneous performance visualization for graphical control design methods. *14th IFAC World Congress*, Beijing, China.

Salt, J., Valera A., Cuenca A., Ibáñez A. (2000). Industrial Robot Multi-Rate Control with the VxWorks Real-Time Operating System. *IFAC AARTC'2000*, Palma Mallorca.

Salt J., Albertos, P. (2000). Multirate controllers design by rate decomposition. *30th Conference on Decision and Control*, Sidney.

Salt, J., Dormido, S., Cuenca, A. (2002). An interactive tool for teaching multirate sampled data systems. Internal Report Univ. Pol. Valencia/UNED.

Sklansky J., Ragazzini J.R. (1955). Analysis of errors in sampled-data feedback systems. *AIEE Trans.*, **vol. 74, part II**, pp. 65-71.

Thompson P.M. (1986): "Gain and phase margin of multirate sampled-data feedback systems", *Int. J. Control*, **vol. 44**, pp. 833-846, 1986.

Tsao, T., Hutchinson, S. (1994). Multi-rate Analysis and Design of Visual Feedback Digital Servo-Control. *Journal of Dynamic Systems Measurement and Control*, **vol. 116, n. 1**, pp. 45-55.

Vélez, C.M., Salt, J. (2000). Simulation of Irregular Multirate Systems. *Proceedings of 8th Symposium on Computer Aided Control System Design.* Salford. United Kingdom.

Wittenmark, B., Häglund, H., Johansson, M. (1998): Dynamic pictures and interactive learning, *IEEE Control Systems Magazine*, **18**, n° 3, pp 26-32.

ELSEVIER

IFAC
PUBLICATIONS
www.elsevier.com/locate/ifac

USING WEB SERVICES TO CONTROL
A LAB EXPERIMENT AND A POWER PLANT SIMULATOR

B. Šulc and J. Tamáš

Czech Technical University in Prague, Faculty of Mechanical Engineering
Institute of Instrumentation and Control Engineering
166 07 Praha 6, Technická 4, Czech Republic
sulc@fsid.cvut.cz
jan.tamas@fs.cvut.cz

Abstract: Internet control of a laboratory experiment was studied in the framework of a student project in 2001. The aim was to gather experience of this kind of remote control, which allowed effective use of the laboratory equipment and gave students the convenience of performing their laboratory experiments from anywhere. In the meantime, the experience gained and the need to establish common access to an operator training simulator of a power plant are the main reasons for considering a further development. New features and advantages provided by web services are described and discussed both for existing remote laboratory control and for planned introduction of the simulator into various types of courses. *Copyright © 2003 IFAC*

Keywords: Internet, man–experiment interface, simulator, temperature control, power plant, web services

1. INTRODUCTION

Web services are becoming a standard in application-to-application communication (Haas, 2002). While they bring long awaited unification of the programmatic interface between programs on different platforms, they also help programs to spread to new areas of use, thus bringing many new opportunities. They allow previously unthinkable solutions to emerge. However, they also bring many new challenges.

In a diploma project (Tamáš, 2001), Internet control of a laboratory experiment was designed on the following principles. The architecture consists of a client and a server application communicating with each other by means of a standard HTTP protocol running above the Internet protocol (TCP/IP). The experimental setup (Fig. 1) includes the controlled system and a PLC controller connected to a standard PC. The controlled system is a heat chamber with sufficiently slow dynamics to allow the delay of Internet traffic to be neglected. The controlled variable is the temperature inside a heat chamber heated by two bulbs, which are used as a manipulated heat source.

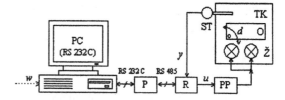

Fig. 1 Scheme of a remotely controlled laboratory experiment (heat chamber - TK, Honeywell UDC 1000 controller – R, heating bulbs – Ž, temperature sensor - ST)

The implemented solution proves to be fully functional and allows remote setting of all parameters of the PLC regulator, thus fully controlling the conditions in the heat chamber. However, the main disadvantage of this solution is the fact that a proprietary communication protocol between the client and server applications was designed and implemented. Further development took place to address and resolve this issue. Considering the possibility that the set of available functions of the experiment will increase in the future, one of the main requirements on the communication protocol was scalability, for the new functions that will be implemented. Another requirement on the newly developed solution was platform independency of the client application.

The server side was built as a web service to fulfil two requirements: a) to allow any client application on any platform to access the server's functions, thus fulfilling the platform independency requirement and, b) to allow future enhancement of the server and ease of deployment of new functions, thus fulfilling the scalability requirement. Web Services Description Language (WSDL) and Extensible Markup Language (XML) were used in the implementation.

This paper describes the use of web services in remote control of laboratory experiments. The possibilities that this architecture brings to the concept of remote use of a power plant operator training simulator will also be considered.

2. WEB SERVICES IN GENERAL

Web services (previously also known as application services) are services (usually including some combination of programming and data, and possibly including human resources as well) that are made available from a provider's web server for web users or other web-connected programs (TechTarget, 2002). Major advantages of web services are standardized protocols (TCP/IP and HTTP) and data formats (XML), which allow otherwise incompatible programs and platforms to communicate with one another.

The goal of web services is to provide services to other programs or individuals. These services can range from booking a hotel or renting a car to providing the current value of a requested stock quote or forecasting weather for a given area.

2.1. A closer look at web services architecture

In order for the remote party to be able to access the data and understand it correctly, some rules need to be defined describing the format and meaning of data and a way of accessing it. The format for web services data has been set to be Extensible Markup Language (XML). XML is similar to HTML as it uses tags to describe data. Unlike HTML, XML does not have a predefined set of usable tags. Therefore the developer of a web service can define his/her own tags that will be used to describe the data.

Knowing how to access the data provided by the web services is just as important as being able to understand this data. Accessing web service functions is described using Web Services Description Language (WSDL). WSDL is an XML-based language used to describe the services one offers and to provide a way for a remote party to access those services electronically. (TechTarget, 2002)

If one wants to access a certain web service, he/she must know who provides such a service. Information about available web services and the functions that these services provide are contained in the UDDI registries. A UDDI (Universal Description, Discovery and Integration) initiative can be thought of as the yellow pages of web services. They include lists of registered web services. The services are described using WSDL.

2.2. Remote control of lab experiments

Controlling laboratory experiments remotely is not a new concept. It has been tested and implemented many times and the associated challenges have mostly been solved. With the recent spread of the Internet, remote control has become even easier to implement. Using the standard Internet protocols (TCP/IP with HTTP) is in many cases sufficient for satisfactory remote control. Lab experiments can now be controlled by students from their homes or by remote universities cooperating on projects.

The advantages of such a concept are quite obvious. Laboratories accessed remotely (sometimes called virtual labs) provide better accessibility to students and lower the costs associated with the preparation and maintenance of experiments. (Wu Sheng, et al., 2000). In the case of remote access to computer models, expensive hardware and software is not required on the client machines.

One of the problems with remotely controlled lab experiments is the fact that proprietary and often incompatible application layer protocols of the OSI/ISO reference model are used (Open Systems Interconnection reference model first proposed by the International Standards Organization). It is only possible to access such experiments from a given platform using given software used during development of such a remotely controlled system.

This limitation greatly reduces the potential benefits of such remote experiments, as they can only be accessed from a certain client program or a platform. Several proposed solutions of remotely controlled experiments presented at the Symposium on Advances in Control Education in 2000 (e.g. Wu, et al., 2000; Röhrig and Jochheim, 2000; Lindfors, et al., 2000; Corradini, et al., 2000; Sánchez, et al., 2000; Ling, et al., 2000 and Roth, et al., 2000) have been limited in this way. If, however, they had been designed using a web services approach (standard-based open interface), the benefit from such an approach would have been much greater, as experience would be gained not only by the deploying university but possibly also by other universities or research facilities interested in similar projects. New inter-university projects could be launched allowing cooperation in distance education to enter a new stage.

2.3. Remote control of lab experiments using web services

With web services, incompatibility problems can be overcome as the interface between the lab experiment and the remote party is standardized. Because WSDL allows future changes in the experiment without the previous implementations of the remote access software becoming outdated, this approach can prove to be very useful. Not only does it allow different client types to access the experiment, but it also allows future changes in the experiment without disabling the operation of the current clients. New functionality can be added to the experiment, allowing new clients to take advantage of it while at the same time allowing old clients to function the way they used to before the change took place.

2.4. Example of a lab experiment implemented as a web service

In a recent project, client and server software was developed that enabled remote control of a laboratory experiment at the Czech Technical University (Fig. 2).

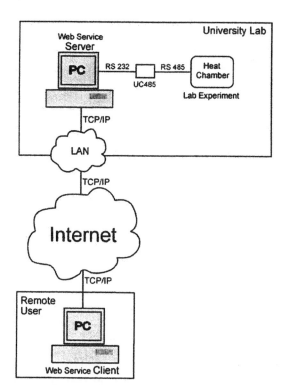

Fig. 2 Scheme of the connection of the remote user to the laboratory experiment (hardware point of view)

This experiment is now used for students to learn to tune the PLC and also to get a feel of what it is like to control a given variable (in this case temperature) in a system of the second degree. However, the use of a given remotely controlled experiment can go be-

yond the intended aims. Having a standardized application layer interface using web services architecture (XML and WSDL), this system can be used to "simulate" the tested behavior on a real system under specified conditions. Rather than running a simulation in Matlab, for example, one could run a test on a real system instead. (Although support for web services is not directly implemented in the current release of Matlab, it can be expected to be included in upcoming releases). Implementation of the safety precautions is included directly on the server side, therefore no harm can be done to the experiment even if unpredictable behavior of the remotely accessing party takes place or the connection is lost.

The following is a scheme (Fig. 3) of the functioning of the server.

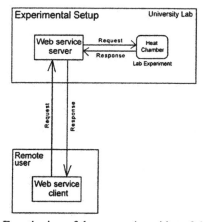

Fig. 3 Functioning of the server (provider of the web service) (communication point of view)

The server sends requests to the controller for the current temperature in a predefined time interval (or a request for a change in a parameter setup) and as a response obtains the current temperature in the heat chamber (or confirmation of a performed change in a parameter setup). The client sends a web service request to the web service provider (server) and in response obtains the requested data.

The communication can therefore be divided into two parts:
- Communication between the server (web service provider) and the experiment (controller)
- Communication between the web service consumer (client) and the web service provider (server).

The first part of the communication goes beyond the scope of this paper and it should suffice to say that the server stores the value of a time interval, in which it periodically requests from the controller the current value of the temperature in the heat chamber. The value of this time interval can be changed both locally and remotely by a web service request. Whenever a command arrives at the server (either from a remotely connected or from a locally connected user), this command is then sent directly to the controller. The response from the controller is returned to the sender (remote or local). This response can be - depending on the sender - either a web service response or a message on the local screen.

Originally, the second part of the communication was implemented using a proprietary communication protocol. Requests were sent in plain text and were described only in the diploma work. This approach was not very practical for future development of the project, and so transformation to a web service was undertaken.

Currently, the communication between the client and the server is built entirely as a web service. This means that the server contains a WSDL file describing the service, it accepts web service requests in the

web services:

- Unified cross-platform interface – based on TCP/IP, HTTP, XML and WSDL standards, this solution can be implemented on a variety of platforms
- Scalability - changes to the experiment (server side) can be made without having to change the client software
- Passes through most firewalls – HTTP protocol is used, therefore no changes need to be implemented in the setup of most firewalls, as HTTP traffic is allowed on most networks
- Allows cooperation between on-line users – connection of several users to the experiment at one time is allowed and better management of access rights is provided.

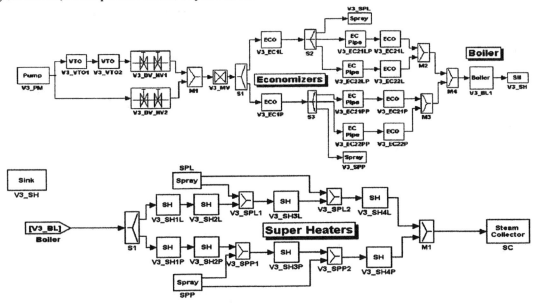

Fig. 4 Simulink models of the most important parts for a coal-fired power plant simulator

form of XML files, and it produces XML files as a response to these requests. The response depends on the type of request. It can be either a temperature in the heat chamber, confirmation of a successful change in the setup of the controller parameters, the current value of the set point variable, etc. An advantage of this approach is that new functions can be added in the future without affecting the function of the current clients.

Implementation of this solution not only solved a problem of scalability in one laboratory experiment, but the experience gained also led to progress with another project, an simulator currently under development at Czech Technical University.

2.5. Advantages of web services over proprietary implementation of remote control

The following are the main advantages of the use of

3. POWER PLANT SIMULATORS AND THE WEB

In recent years, coal-fired power plant simulators of various kinds and for various uses have been developed in co-operation with Energy Training Centre in Tušimice, Czech Republic. (Neuman P. et al, 2002, Jan, J. A., Šulc B., 2001, 2002a, b). The last development (Fig. 4) was on an operator training simulator using Matlab-Simulink as a tool for simulation and Wonderware InTouch as a tool for visualization and manipulation. These purely software realizations are supplemented in the industrial version, by an operator panel. Although the "school" version lacks the feeling of working in an operator room, (and this is the specific advantage of the industrial simulator, where the operator uses the real control panels in his operator room), it offers a broader range of activities to be performed. Besides providing acquaintance with the work of the operator, this solution allows students to test advanced control algorithms, perform identification experiments and other control engineering tasks otherwise hardly performable in a real environment. In normal operator simulators these functions are not available, as they go beyond the responsibility of the operator. The scope of these tasks cannot be fully estimated in advance and so a well designed interface between the student and the simulator is necessary. Web services will play a key role in the design of such solutions as they provide optimal support for this task. The design of this interface will form part of a dissertation work currently being undertaken at Czech Technical University.

Remote control of this simulator is possible by means of native support of InTouch software. However, this solution is somewhat complex, especially from the point of view of deployment of the client side, as it requires InTouch to be installed on the client machine.

From our experience with web services in the heat chamber experiment came the idea that we might also design this simulator as a web service. Such a solution would allow any client program to access the simulator. InTouch would no longer need to be installed on the client machines.

4. CONCLUSION

Good experience with the use of web services presented in this paper opens a new way for a significant cooperation among universities and other outside users in exploiting existing laboratory experiments, especially those providing rare features.

Those interested in accessing the experiment can fully define the use of the experimental equipment according to their needs. By reading the WSDL file describing the service, remote users gain access to all the necessary information relating to the user-available functions and data (all measured and manipulated quantities and parameters), and are thus able to design the experiment based fully on their intentions, taking into account their hardware and software capabilities. This approach also provides an easy way for future upgrades and implementation of new functions during the development phase.

Experience gained in the application of web services in the control of the heat chamber experiment will be fully developed and used in establishing remote control of an operator training simulator of a coal-fired power plant, which has lately been under development at the Czech Technical University.

5. ACKNOWLEDGEMENT

This research has been supported by Research Project of MSMT of the Czech Republic J04/98: 21200009.

6. REFERENCES

Corradini, M. L., T. Leo, S. Longhi and G. Orlando (2000): Control Education over the Internet: Performing Experiments in a Remote Laboratory. ACE 2000, Gold Coast, Australia.

Haas, H. (2002): Web Services Activity. www.w3.org/2002/ws.

Hrdlička, P., Neuman, P., Šulc, B. (2002): Process Instrumentation Modular Models of Thermal Power Plant for Operator Training Simulators, 15th IFAC World Congress on Automatic Control, Barcelona, Spain.

Jan, J. A. (2002): Soft Computing in Object Oriented Nonlinear Modelling and Control of Thermo-Fluid Dynamic Systems. PhD Thesis. Czech Technical University in Prague.

Lindfors, J., L. Yliniemi and K. Leiviskä (2000): Using WWW to Support Control Engineering Training. ACE 2000, Australia.

Ling K.V., Y.K. Lai and K.B. Chew (2000): An Online Internet Laboratory for Control Experiments. ACE 2000, Australia.

Neuman, P., B. Šulc and T. Dlouhý (2000): Nonlinear Model of a Coal Fired Boiler Applied to an Engineering Simulator. Preprints IFAC Symposium on Power Plants and Power Systems Control 2000, Brussels.

Neuman, P. and B. Šulc (2000): Engineering Models of a Coal Fired Steam Boiler in Teaching, Training and Research Applications. ACE 2000, Gold Coast, Australia.

Roth H., K. Schilling and O. Rösch (2000): Performing Control Experiments in Virtual Laboratories via Internet. ACE 2000, Gold Coast, Australia.

Röhrig, C. and A. Jochheim (2000): Java-based Framework for Remote Access to Laboratory Experiments. ACE 2000, Gold Coast, Australia.

Sánchez, J., F. Morilla, S. Dormido, J. Aranda and P. Ruipérez (2000): Conceptual Learning of Control by Java-based Simulations. ACE 2000, Australia.

Tamáš, J. (2001): Performing Laboratory Experiments in Control Engineering via Internet. Diploma Thesis (in Czech). Czech Technical University in Prague.

TechTarget (2002): IT-specific Encyclopedia. www.whatis.com.

Šulc B., J. A. Jan (2001), User Friendly Simulink Thermal Power Plant Modelling using Object Oriented Non-linear Dynamic Model Library, 5th IASTED International Conference Power and Energy Systems (PES 2001) Tampa, Florida, USA, pp. 253-256, ISBN: 0-88986-317-2.

Wu Sheng, Lim Choo-Min, Lim Khiang-Wee: An Integrated Internet based Control Laboratory, University of Singapore. ACE 2000 Final Program & Preprints, Gold Coast, Australia.

COMPUTER SIMULATION SUPPORTING CONTROL ENGINEERING TEACHING

Leena Yliniemi and Kauko Leiviskä

University of Oulu, Control Engineering Laboratory, BOX 4300, FIN-90014
UNIVERSITY OF OULU
firstname.surname@oulu.fi

Abstract: Teaching is a very complex and demanding human activity. The computers together with network technology have made it possible to utilise computer based learning material for supporting the teaching of control engineering. This has been found also at the Control Engineering in the University of Oulu, where computer based learning material called CALEXX has been developed during many years. Especially interactive examples, animations and simulations make the understanding of control engineering illustrative and easier, because the students can "experiment" with situations, which would be too expensive or too dangerous in real life.

Different computer simulators have been developed for examining the operation and control of different processes as rotary drying in mining and mineral industry, plate rolling mill in steel industry, freeness control in grinding mill in pulp industry and PVC manufacturing in chemical industry. The simulators are running in a Windows PC in the MATLAB or MATLAB Web Server environments. The simulators are an important part of the training material CALEXX. *Copyright © 2003 IFAC*

Keywords: education, control engineering, simulators

1. INTRODUCTION

For effective and efficient teaching a computer has proved an excellent tool for visualizing control engineering. Therefore at the Control Engineering Laboratory in the University of Oulu computer based learning material has been developed during many years for supporting the teaching of control engineering. The training material called CALEXX has been originally developed in the beginning of 90's in an EU project and it was introduced for the first time in the paper by Yliniemi *et.al* (1994). The material has been later completed and updated. It consists of several separate modules, the contents of

which include both basic control engineering and advanced control theory as measurements, process dynamics and modelling, adaptive and intelligent control methods and optimisation. Four process simulators play important role in the training material. The simulators represent different industrial sectors as mining and mineral industry, steel industry, chemical industry and pulp and paper industry.

In this paper, a general look to the structure and contents of CALEXX is presented. However, the main interest focuses on the simulators in CALEXX.

2. OVERALL STRUCTURE OF THE TRAINING MATERIAL CALEXX

The training material CALEXX forms an electronic library, which contains eight different modules " bookshelves". These bookshelves further contain "books" as ToolBook calls it's files. Asymetrix ToolBook, later Multimedia ToolBook has been used as the main development tool. When the user comes to the system, the library book is opened. Figure 1 presents the layout of the library book.

The eight bookshelves refer to eight topics the training material deals with:

1. Process measurements
2. Process dynamics and modelling
3. Controller design and tuning
4. Advanced control methods and optimisation
5. Pulp mill and freeness control simulation
6. PVC process control and simulation
7. Plate rolling mill control and simulation
8. Rotary dryer control and simulation

The number of books in eight bookshelves is more than one hundred, and the number of pages (computer screens) in the books is more than two thousands. The pages include different elements as hypertext, interactive and solved exercises, simulation models, animations and video sequences. In addition to the books in the bookshelves, the material includes two assistant books: "Introduction to the material" and "How to use" book which are common to the whole material.

A book is constructed so that the first page is a cover page containing the name of a book. This page is automatically wiped to left in order to get the second page, which is an index or navigation page. It contains the topics of a book.

The design of navigation is very challenging, when a hypermedia application is so large as CALEXX is. It can cause easily navigation problems including "lost in hyperspace"-feeling, having difficulty in gaining overview, not being able to find information that is known to exist and determining how much is left. For preventing "lost in hyperspace"-feeling several navigation tools as navigation buttons, a navigation menu, a map book, an index book and hot words have been used in the material. Also the background colour in the first page of a bookshelf is different in different bookshelves. Figures 2 and 3 show examples about information pages.

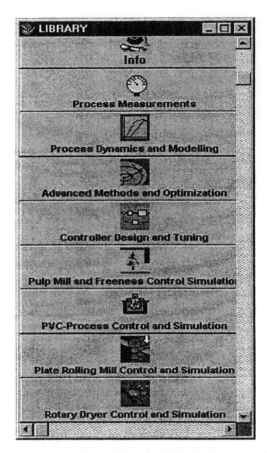

Fig. 1. Library of the training material CALEXX.

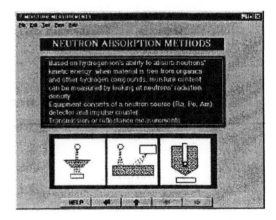

Fig. 2. Example about an information page in CALEXX.

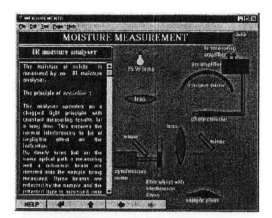

Fig. 3. Example about an information page in CALEXX.

The typical layout is that the hypertext lies on the left side or in the upper part of the page and animations, figures, diagrams etc on the right side or in the bottom part of the page.

3. PROCESS SIMULATORS IN CALEXX

The important part in the training material CALEXX forms four process simulators developed to different industrial sectors. The aim of the simulators is to examine and to visualize the operation of a process in open loop or in closed loop situations. The physical or experimental models are based on real processes used in companies or research institutes.

At first, each simulator includes quite detailed description about the process. The operation of the process can be simulated also without control. It makes e.g. possible to estimate the effect of the model parameters i.e. the sensitivity of the parameters on the process output. To the students at the Control Engineering Laboratory it is important to understand the operation of the process, because the laboratory belongs to the Department of Process and Environmental Engineering. This means that the students get basic knowledge about different processes with good knowledge about control engineering.

The controllers in the simulators are usually PI or PID controllers. The student has to tune the controllers for achieving the optimal result. The bookshelf "Controller design and tuning " gives more information from tuning methods. The student can link to this bookshelf and further to the book concerning tuning methods. The simulators run in a Windows PC under MATLAB or MATLAB Web Server environments.

3.1 Contents of freeness control simulator

Freeness is determined as the measurement for the filtering ability of groundwood pulp in a grinding mill. The essential objective is to find out the best parameters so that the freeness stays constant at the desired level in spite of varying operating conditions. The aim of the control strategy is also to maximize the production in the grinding mill. If the target freeness can not be reached within the energy limits of a grinder, then the grindstone must be sharpened.

The temperature of grinding has a very great effect on the type of pulp produced. Grinding temperature, when steadily maintained at any predetermined point, gives better control of consistency and freeness. This results in increased operating efficiency and maximum production output with less power required per ton to produce it. Temperature control also prevents any sudden deluge of cold water on a hot grinding stone, thus minimizing temperature strains and increasing the length of service. The need for frequent burring of the stone is also reduced.

The operation of the simulator is based on the experimental model, where the production rate of the grinder has been presented as the function of grinder load and grindstone sharpness. Figure 4 presents the layout of the user interface with the operating curves

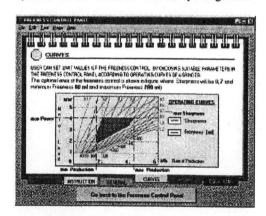

Fig. 4. The user interface of the freeness control simulator.

3.2 Contents of PVC manufacturing simulator

PVC (polyvinylchloride) is one of the few synthetic polymers that have wide application in commerce. Suspension polymerisation is the most important method to manufacture PVC. The manufacturing process can be divided into polymerisation, remaining gas removing, drying of resulting polymer, product storing.

The simulator is based on the experimental model of the temperature control in PVC polymerisation reactor. Deviations from the correct temperature of polymerisation have significant effects on the quality of PVC. The temperature control system of the polymerisation reactor is based on cascade control, which contains a main controller and slave controllers. The simulation includes the following four steps:

1   Tune up the PID controllers. Set up the optimal parameter values for the controllers. In the cascade control the slave controllers are tuned first.

2   Set up the values for process inputs.

3   Simulate.
Simulation is based on the calculations of heat to be needed and reaction conversion.

4   Check the values for the process outputs.
The final results are presented in the relation to the polymerisation time.

The operation of the simulator is divided into three phases:

1   Simulation from the beginning of the polymerisation

2   Simulation from the steady state

3   Examination of the results.

The simulator was originally developed in the MATLAB environment, but it was later translated to a web environment using a MATLAB Web Server. The process and its operation in the web environment has been described in detail in the thesis by Vandezande (2000) and in the paper by Lindfors *et.al.*(2000).

Figure 5 describes the layout of the PVC manufacturing simulator with controllers.

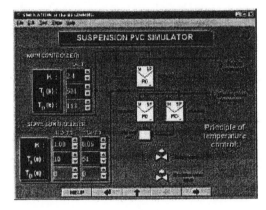

Fig. 5 The layout of the PVC manufacturing simulator with controllers.

3.3 *Contents of plate rolling mill simulator*

The plate rolling mill simulator is quite versatile including in addition to the real simulator also the detailed description about the steel plant and the operation of the process itself with its instrumentation and control systems. Figure 6

presents the contents of the bookshelf "Steel simulator"

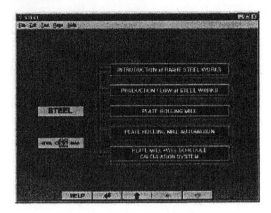

Fig. 6. The index page of the plate rolling simulator.

The simulator is based on the plate mill pass schedule calculation system. This model has been developed in the Finnish steel plant. The simulation is started by selecting a plate which it is wanted i.e. by giving the plate type and dimensions. The model selects the adequate slab from which the plate is rolled.

Figure 7 presents the user interface of the simulator.

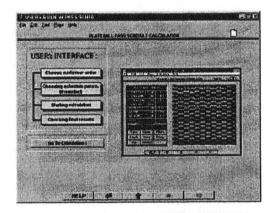

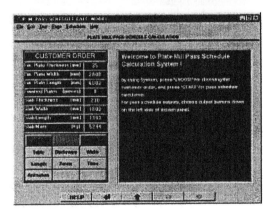

Fig. 7. The user interface of the plate rolling mill simulator

The rotary dryer simulator is used very actively in different courses given at the Control Engineering Laboratory. It depends on that the simulator describes the operation of the plant rotary dryer located in the laboratory. The model is based on heat and mass transfer equations.

The simulator includes both process simulation and control simulation. The operation of the process is examined by changing operation conditions or model parameters. Step disturbances to various inputs can be made. Different control systems have been constructed for the dryer. The students examine primarily the operation of PID control, but also fuzzy control and fuzzy control together with PID control can be examined.

The simulator is described in detail in the thesis by Maes (1999). Figure 8 presents the contents of the rotary dryer simulator and Figure 9 the control interface of the simulator.

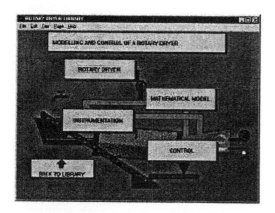

Fig. 8. Index page of the rotary dryer simulator.

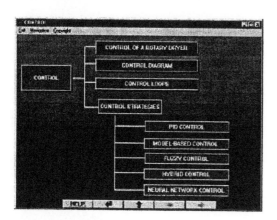

Fig. 9. Control interface of the rotary dryer simulator.

4. CONCLUSIONS

It is well known that teaching is very demanding and difficult task. For visualizing the teaching of control engineering computer based training material, especially computer animations and simulations help to understand control engineering, because they make possible to experiment situations, which are expensive or dangerous in real life.

The focus in this paper is aimed on the contents and use of the simulators. The structure and contents of the computer based training material CALEXX is described. The material includes eight modules called bookshelves, four modules of these include basic control engineering and advanced control methods and four process simulators, which represent different industrial sectors as mining and mineral, chemical, steel and pulp and paper industry. The material is running in a Windows PC under Multimedia ToolBook. The simulators run under Matlab or MatlabWeb. Each simulator runs standalone. The use of simulators is free for the students in the Control Engineering Laboratory, but for others it is chargeable.

REFERENCES

Lindfors, J., L.Yliniemi and K. Leiviskä (2000). Using WWW to support control engineering training. In: Preprints of 5th Symposium on Advances in Control Education - ACE2000. Goald Coast, 17-19 December, 2000

Maes, M. (1999). Modelling and Simulation of a Rotary Dryer . M.Sc.Thesis. Group T Leuven Institute of Technology.

Vandezande, L. (2000) Simulation of the temperature control system of a PVC suspension polymerisation reactor. M.Sc.Thesis. Group T Leuven Institute of Technology.

Yliniemi, L. and K. Leiviskä (1994). Training of Process Automation with Hypermedia. In: Preprints of 3rd Symposium on Advances in Control Education – ACE1994. Tokyo, 1-2 August, 2000.

ELSEVIER
IFAC
PUBLICATIONS
www.elsevier.com/locate/ifac

COMPUTER-AIDED EDUCATION: EXPERIENCES WITH MATLAB WEB SERVER AND JAVA

Vesa Hölttä, Heikki Hyötyniemi

Control Engineering Laboratory,
Helsinki University of Technology,
P.O.Box 5400, FIN-02015 HUT, Finland

Abstract: Teaching material has recently become increasingly available on the Web. However, reading text and looking at static images is not very motivating when compared to hands-on experimenting. The latter may also provide the student with deeper understanding of the subject. Two tools for creating interactive mathematical teaching material on the Web are discussed: With the MATLAB Web Server one can create easily Web pages that call MATLAB functions and scripts. Java applets are much more time consuming to implement, but they are advantageous in e.g. simulations and other "real-time" applications. The two technologies can be used together to combine their advantages. *Copyright © 2003 IFAC*

Keywords: education, educational aids, software tools

1. INTRODUCTION

Among other things, the World Wide Web has provided means for publishing and distributing teaching material. This works well with text and pictures and in subjects such as history and geography where learning does not involve personal experimenting. However, in a mathematical context the student should be encouraged to do calculations himself of herself as well, in order to better digest the material that has been presented.

Obviously, calculations can be done by pencil and paper. However, if the theory and examples are in the Web, why not do the exercises with the computer as well? Hard work is always needed for learning new things, but why not try to provide an easy to use and motivating environment for this? Struggling with theory may become, if not interesting, at least meaningful. (Hyötyniemi, 1994) What would be a good technology for creating such a learning environment so that it does not require tremendous amounts of work?

A simple solution to the question above is to provide on a Web page exact instructions for using some software and let the student do the work with the non-web-based program. This approach has been used to teach MATLAB and Simulink to control engineering students e.g. by Tilbury, *et al.* (1998), who gave code to copy and paste and simulation models to download. For learning how to use the specific software this works well, but what if one wants to teach the theory rather than the tool? And what if the students do not have access to the software in the first place?

Being dependent of a certain application can be avoided by using Java applets like Crutchfield and Rugh (1998) have done. On their Web site the student can find on the same Web page the mathematical background, e.g. for sampling continuous signals, and an applet that illustrates the theory. Applets can be easily used with most Web browsers and can be tried out without the student needing to write code or to do other time consuming preparations.

Because of drawbacks in the Java approach, to be discussed later, other methods would also be of interest. It seems that a promising alternative exists: A product called MATLAB Web Server (MWS) has become available some time ago and it can be used to create a user interface to MATLAB via the Web. This paper introduces the MWS, compares MWS and Java applets and gives practical experience of using the two technologies in control engineering education. In this paper, a way to use both approaches together is also presented.

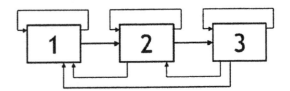

Fig. 1. Typical organization of a Web-based lesson of the course: 1. Create data, 2. Apply the method, 3. Visualize results. The arrows represent possible steps in the session.

2. APPLICATION: COURSE ON MULTIVARIATE STATISTICS

The Control Engineering Laboratory of the HUT offers a course on multivariate statistical methods where the textbook by Hyötyniemi (2001) is used. The course concentrates mainly on linear regression methods — starting from simple multivariate regression, and ending in *independent component regression* and *subspace identification*. It turns out that the conceptual tools and methods are rather simple after one has grasped the basic ideas. However, understanding the essentials is not easy because the characteristic phenomena of the multivariate methods take place only in high-dimensional spaces — and managing dimensions higher than three is extremely difficult to a novice in the field. A computer equipped with the capability of mathematical manipulations and interactive graphics can help the student in gaining intuition. For example, there are the following phenomena that can efficiently be visualized using the computer:

- Seeing the effects of different data preprocessing methods on the data distributions.
- Understanding the correspondences between data distributions and covariance eigenstructures.
- Recognizing the effects of latent structures on the resulting models.

Another important lesson is that — after all — theory can be far from practice. The theoretical results, derived with assumptions on noise normality, etc., can be misleading, and the results can be sensitive to non-idealities in practical situations. This kind of deficiencies in theories are extremely difficult to capture theoretically in a convincing way; it is through computer simulations that such phenomena can best be demonstrated. When considering nonlinear models and methods, such surprising behaviors can easily be found, but there exist various pitfalls also what comes to linear theory:

- Neural networks can give non-optimal models and independent component analysis can fail.

- There exist a difference between mathematical optimality and physical plausibility (for example when implementing *balanced reduction*).
- System properties can have drastic effects on the robustness of recursive identification.

When a computer and the Web are used in education, two main problems are to be considered:

(1) **Control of the session.** When implementing hypermedia environments, it has been noticed that it is easy for the user to get lost in the "hyperspace", in which case reaching the final conclusions of the lesson cannot be guaranteed. There are perhaps two main guidelines to follow here: First, consistency in the session should be reached, so that proceeding through the topic is logical. The session outlook should also remain similar from session to session to avoid confusion. Second, there should not be too much material given to the student at the same time, and the session should proceed only little by little. Luckily enough, in this kind of a learning environment these objectives can be reached. On this course the structure of a typical session is always the same: generate data, construct model, visualize results. In order to keep the student on track of what is being done, a good approach is to organize the lesson as a linear series of HTML pages with distinct steps and the possibility to restart when needed, relevant routes being visible to the user at all times (Figure 1).

(2) **Transparency of the tool.** The second main question is how to reach interactivity between the user and the application, without the tool disturbing this communication too much. Dynamic, fast creation of content according to the student's intentions and instructions is needed. This can be implemented with the technologies presented below.

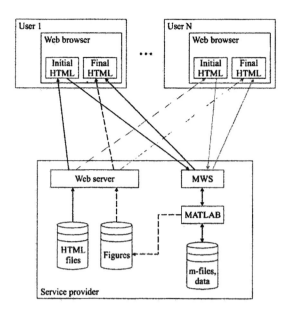

Fig. 2. Network structure used by the MATLAB Web Server (MWS).

3. MATLAB WEB SERVER

3.1 Concept

The MATLAB Web Server (MWS) is a commercial product that belongs to the MATLAB product family from MathWorks. The MWS makes possible to use MATLAB functionality via the Web. In spite of not being free of charge, MWS was selected as a technology to try because because MATLAB itself is widely used by the mathematical and the engineering communities.

To say it short, the MWS is a cgi-bin application that is used to launch MATLAB script files (m-files) remotely. The user can give parameters that are used during the m-file execution. With the MWS, almost all built-in and self-written MATLAB functions can be called from the m-file, including functions that produce plots to be shown to the user.

3.2 Technology

The way the MWS works is illustrated in Figure 2. Web pages that are used to launch the MWS contain a HTML form. If the application needs parameters from the user he or she fills them to the form before submitting it. When submitting, the browser contacts the MWS and passes parameters, if present. The MWS executes the m-file linked to the application and as a response it creates a new HTML page containing the result in a format specified by the service provider. The new page is transmitted to the user's Web browser.

Figures are saved separately to the hard drive of the server in the jpeg format. In order to show the figures to the user, the HTML page created by the MWS contains references to the figures and the user's browser downloads them directly from the server. A more detailed description of the technology can be found in Díez, *et al.* (2002).

3.3 Example

In order to gather experience in using the MWS, suitable exercises in (Hyötyniemi, 2001) were implemented (see *www.control.hut.fi/Kurssit/As-74.191/*). Figure 3 is an example of a user interface that was created. After the problem statement the user has access to further information on the routines used via hyperlinks. In the middle there are text fields for the parameters to be modified by the user and at the bottom is the MATLAB code to be executed. When the user submits the form a new window (Figure 4) opens and the resulting plot is drawn after a couple of seconds of data transfer and processing delay. In this exercise the distribution of the sizes of eigenvalues corresponding to the eigenvectors of a dataset covariance matrix is plotted and the user can change the degrees of freedom and the noise level in the data. Random data with the predefined statistical properties is generated each time the routine is called.

3.4 Discussion

The MWS can be used to create easy to use demonstrations and exercises that are based on the use of MATLAB. Making an application involves writing HTML code for the input and output pages and the m-file to be executed. With some experience, this can be done in a couple of hours or, if the necessary algorithms are already implemented or if only standard MATLAB functions are used, in even less than an hour.

Any Web browser can be used as the user interface and the user does not need to install any plug-ins or additional programs. The service provider needs knowledge of MATLAB to perform the calculations and/or create plots. Basic HTML skills or a program that generates HTML is required in order to create the user interface.

All the MATLAB functions and toolboxes that have been installed to the server can be used with the MWS. This might be convenient in the case that there is a single license for an expensive toolbox, since by using the MWS it can be shared by several users. It is also possible to run a previously created Simulink model and to study the results.

In the example above the result was a plot. However, it could be several plots, scalars or

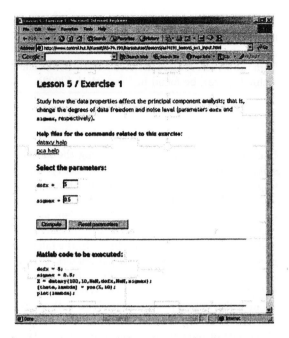

Fig. 3. Example input for the MATLAB Web Server: selection of data properties.

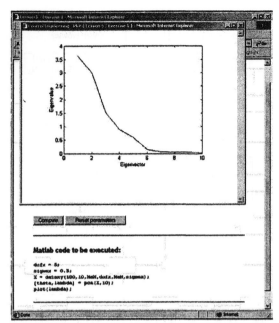

Fig. 4. Example output from the MATLAB Web Server: covariance matrix eigenvalues.

matrices or any combination of these. Other forms of output can be imagined as well: The MWS could create a downloadable data file to be used in a system identification homework assignment. The system parameters could be based on the student number so that each student would get unique results making collaboration less easy.

The most serious weakness of the MWS is that it lacks the interactive features of MATLAB. This means that plots cannot be zoomed or rotated as in MATLAB and functions such as `ginput` that require input from the user during their execution cannot be used. Simulink and other graphical user interfaces cannot be used either.

In the MWS concept, all calculations are performed centrally in the server running the MWS. The user's computer is used only to download data from the server. If there are many users running simultaneously computationally demanding applications, substantial load is created to the server and users will experience slower responses.

4. JAVA APPLETS

4.1 Concept

Java is an object oriented programming language that was created by Sun Microsystems to be a programming language for the Web. It is distributed free of charge and there is plenty of code samples, tutorials and related material available.

Applets — programs written in Java for the Web — can be embedded into an HTML page. Some browsers may need an additional program

component (the Java Virtual Machine) to be able to show applets, but modest computer skills should be sufficient for installing one.

4.2 Technology

Compiling a Java program results in bytecode that is not dependent of the operating system or processor. When a user accesses a Web page containing an applet, the Java Virtual Machine converts the bytecode to instructions that can be run on the user's computer.

Applets are loaded from the server to the user's computer (Figure 5). Compared to the MWS network structure, this is the only time data transfer is needed, resulting in a simpler structure.

4.3 Example

An example of a Java applet is shown in Figure 6. The applet is on a Web page that contains information on the properties of second order systems. The applet has on the left the step response of a second order system and the user can change system parameters with the sliders on the right. The response is updated immediately as the user changes the parameters.

4.4 Discussion

Java applets can interact with the user in a way not possible to the MWS. Since all calculations are performed on the user's computer the workload

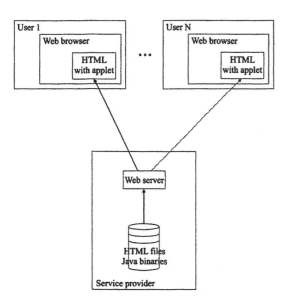

Fig. 5. Network structure used by Java applets.

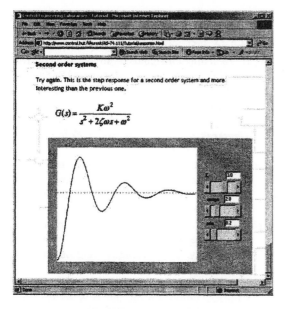

Fig. 6. An applet to demonstrate the step response of a second order system. The step response can be seen on the left and system parameters can be changed using the sliders on the right.

of the server is reduced. Java can be used free of charge.

Because Java is a general purpose programming language, essentially anything can be implemented. However, it is by no means meant for being used for scientific computing, so that it does not have e.g. a built-in support for complex numbers. Similarly, all other algorithms needed, like the matrix inverse or the fast Fourier transform, must be written from scratch while MATLAB would provide functions that are ready to use. Java source code is available on the Web to some extent, but code quality and ease of use may not be good. If it takes a few hours and basic

HTML and good MATLAB knowledge to create a MWS application, it takes few days and good Java knowledge to write a simple applet.

5. COMBINING THE TWO

5.1 Concept

An interesting question is how to combine the good sides of both technologies presented in this paper, namely the mathematical diversity and easiness of MATLAB and the interactivity and visualization capabilities of Java. Figure 7 provides one answer to this question. The exercise at hand illustrates the differences between the results of K-means (KM) and expectation maximization (EM) algorithms used for clustering data. The user can change parameters from a page similar to the one in Figure 3. MATLAB then does the clustering and creates own files for the results of both algorithms.

The output presented to the user (Figure 7) contains two instances of the same applet. The one on the left downloads the KM clustering result and the one on the right the EM result. The clusters that were found are shown in different colors in a three-dimensional Cartesian plot and the user can rotate the plot with respect to all three coordinate axes using the sliders that are next to the plots. When comparing the clustering results, the user should be able to notice that the EM algorithm gives results that are intuitively more plausible (see encircled clusters in Figure 7).

5.2 Technology

MATLAB is able to create and use native and user-defined Java objects and to write them in binary files. In the example, the clustering results are saved to the hard drive of the web server. To give structure to the data an object hierarchy was created: The base of the hierarchy is an abstract class with methods for rotating objects in three-dimensional Cartesian coordinates and for projecting the three-dimensional data to the two-dimensional computer screen. All objects (points, labels, coordinate axes etc.) to be drawn to the applet window extend this class.

5.3 Discussion

The key idea of this approach is to use Java only where the MWS cannot be used in order to minimize the amount of time consuming Java programming. To rationalize even further, the applets should be written as generally as possible so that the same applet can be used with several

51

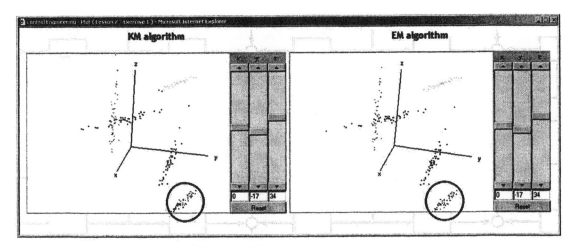

Fig. 7. Java applets are used to visualize the output of the MATLAB Web Server. Points that have been assigned to the same cluster have the same color. The algorithms give different clustering for the encircled points (the circles have been added afterwards — on a color display the difference is easier to spot).

similar MWS applications. The applet above was designed to be able to visualize any collection of points and lines of different color in a three-dimensional space. From the point of view of the applet the process that created the data is irrelevant; the result could be any three-dimensional object. The strength of Java in visualization and interactivity compared to the MWS is thus taken advantage of while the data creation can be implemented in MATLAB quickly with existing mathematical routines.

Most MATLAB users are not familiar with creating Java objects in MATLAB. This can be avoided e.g. by writing the data to a text file that is downloaded by the applet. In this case the applet must be provided with tools for handling text files and interpreting their contents.

6. CONCLUSION

The MATLAB Web Server and the Java programming language can be used separately or together to create an interactive learning environment. Java is a more versatile choice than MWS, but requires also more expertise and more work. While a simple Java application might take a couple of days to be finished, a MWS application can be created in a couple of hours.

The MWS does not offer means to use the interactive features of MATLAB via the Web. It also involves an amount of data transfer delay that may be frustrating to the user. Java applets used alone or in combination with the MWS solve some of these problems, but are more laborious to implement.

REFERENCES

Crutchfield, S.G. and W.J. Rugh (1998). Interactive learning for signals, systems, and control. *IEEE Control Systems Magazine* 18, 88–91.

Díez, J.L., M. Vallés and A. Valera (2002). A global approach for the remote process simulation and control. *15th Triennial World Congress of the International Federation of Automatic Control.*

Hyötyniemi, H. (1994). Hypertechniques in control engineering education. *Preprints of the IFAC Symposium on Advances in Control Education* pp. 185–188.

Hyötyniemi, H. (2001). *Multivariate regression — Techniques and tools.* Helsinki University of Technology, Control Engineering Laboratory. Report 125. Espoo, Finland.

Tilbury, D., J. Luntz and W. Messner (1998). Controls education on the WWW: tutorials for MATLAB and SIMULINK. *Proceedings of the American Control Conference* 2, 1304–1308.

AN EXPERIMENTAL LABORATORY FOR UNDERGRADUATE AUTOMATIC CONTROL COURSES

Alberto Leva

Dipartimento di Elettronica e Informazione
Politecnico di Milano
Via Ponzio, 34/5 - 20133 Milano (Italy)
Voice (39) 02 2399 3410 - Fax (39) 02 2399 3412
Email leva@elet.polimi.it

Abstract: This paper presents a laboratory used at the Politecnico di Milano for the experimental activity of undergraduate Automatic Control courses. A peculiarity of the laboratory is that many different control problems can be experimented with by means of a single apparatus, reasonably priced and easy to maintain. Peculiar is also the large number of students involved, up to 150 of them being managed simultaneously with 72 experimental workstations. Great attention is also paid to the integration between experimental activity and lectures. *Copyright © 2003 IFAC*

Keywords: Experimental laboratory; automatic control; process control.

1. INTRODUCTION

This paper describes a laboratory designed and implemented at the Politecnico di Milano and a set of assignments for undergraduate courses in Automatic Control. The aim is to give students the opportunity to integrate theoretical knowledge with practical experience, where the role and relevance of each concept becomes evident. The laboratory also tries to recreate the 'look and feel' of real-world process control situations. The expected result of laboratory activity, closely coupled with lectures, is that students receive, *at the same time*, the necessary control-theoretical foundation and at least the basic practical skills that must be based on it, as is nowadays considered essential for control professionals (Bialkowski, 2000; Bristol, 1986; EnTech, 1994). The range of experiments is wide, because mastering single-loop control concepts is fundamental, as loops are the basis of any industrial control system, but also proficiency in the use of control structures is important, because most processes

are multivariable in nature, and by no means is knowledge of single-loop control sufficient to cope with them. Needless to say, if it is also required that the laboratory be 'hands-on', designing it is a nontrivial challenge. This is particularly true at the Politecnico di Milano, where it is necessary to serve about 1000 students per year (in 7-8 classes,) and the courses expose students to practical issues very early—an attitude that is consistent with current education trends (Amadi-Echendu and Higham, 1997).

Most experimental setups used in undergraduate Automatic Control laboratories deal with single-loop problems. Others allow to treat more articulated control structures, but tend to be correspondingly complex and expensive. At the author's knowledge, no simple and reasonably priced setup is available that allows to treat a sufficiently wide variety of control problems and structures. The work presented is an attempt to fulfill this need, as far as process control is considered. Different control structures can be experimented with

Fig. 1. One of the laboratory rooms (a) and the workstation (b).

the same, simple and inexpensive setup. In particular, it is possible to treat problems involving single-loop control, decentralized control, feedforward compensation, cascade control, and decoupling control.

At present, the laboratory activity is done in the course titled 'Fundamentals of Automatic Control' (FAC, 1st semester of the 2nd year,) and from the 2002/2003 academic year it will extend to 'Engineering and Technology of Control Systems' (ETCS, 1st semester of the 3rd year.) However, for space limitations, only the activity for FAC is described here; that for ETCS is just sketched, essentially to illustrate how many different experiences can be made with such a simple setup, and will be presented in detail in future works.

2. THE LABORATORY

The laboratory has 72 workstations (18 in each of four rooms) so that a maximum of three students share one. Each workstation is composed of a PC with A/D and D/A cards, the apparatus presented here and described in (Leva, 2002) with enough detail to allow replicating it, and the necessary control software. The laboratory room and workstation layout (without the apparatus) are shown in Fig. 1. The task consists in posing and solving some control problems, described in the following sections, and applying the solutions obtained to the apparatus; each problem is conceived so that students achieve proficiency in solving it and, at the same time, learn some general lesson.

The apparatus is the simple temperature control shown in Fig. 2. The two transistors heat a small metal plate while the fan provides cooling. The outputs are the measurements of the temperatures of the transistors and of the plate, while the inputs are the commands to the transistors and to the fan. The apparatus was initially designed by the author. After a first prototype had proven satisfactory, a specialized firm designed, engineered, and constructed the final product (at a cost of US\$ 65 per item.) The software, written by the author in the LabVIEW programming language, allows open-loop experiments with various inputs and closed-loop control with different structures.

Legend:
1 Enclosure (without the transparent top)
2 Base (printed circuit)
3 Power supply circuitry
4 Metal plate
5 Transistors
6 Temperature probes
7 Fan
8 BNC connectors

Fig. 2. Photos of the apparatus.

Experimental data are recorded in ASCII format, for subsequent processing in the Matlab environment (that students already know from the mathematics course of the 1st year.)

Due to the number of students involved, the laboratory activity must be guided very closely preserving, at the same time, student autonomy. Thus, a PowerPoint presentation was prepared for each assignment, some slides indicating the task activity steps (making actions and/or observing results,) others giving explanations; different slide title colors were used to facilitate comprehension. The presentations are conceived in such a way that the reading and brief commentary of the 'explanatory' slides allows the action to occur over a period compatible with the apparatus behavior. In so doing, each laboratory room can be effectively managed by two instructors: one leads the experience with the presentation, the other stays among the workstations to help the students; Fig. 3 shows a sample of the first FAC assignment slides. Every occasion is exploited to provide physical evidence for system- and control-theoretical concepts (see, e.g., the outlined words 'movement' and 'equilibrium' in Fig. 3.) At present, plans are also underway to offer access to a (limited) number of apparatuses via the web.

3. THE MODEL OF THE APPARATUS

A dynamic model of the apparatus can be derived in a straightforward way assuming that (a) the three temperatures of the two transistors and the plate are individually uniform; (b) heat transfer depends linearly on temperature difference; (c) the thermal powers generated by the two transistors (P_{g1} and P_{g2}) are nonlinear algebraic functions of the two transistor commands, denoted by Q_1 and Q_2 and being 0-100, i.e., $P_{g1,2} = P_{max}f(Q_{1,2}/100)$, where $f(0) = 0$, $f(1) = 1$, and P_{max} is a suitable constant; (d) the transistors-to-plate heat transfer coefficients (γ_{tp}) are equal and constant; (e) the transistors-to-air heat transfer

From our point of view the transistors can be considered simply as electric heaters. They generate thermal power varying from zero to a maximum, proportional to the command received.

Given this, the phenomena involved in the experimental apparatus are essentially thermal:

First, thermal power is generated in the (two) transistors

Then, all the bodies store the thermal energy

There are internal thermal exchanges: transistors-probes, transistors-plate, plate-probe

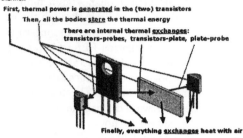

Finally, everything exchanges heat with air

What did we do in the previous step?

We have set the manipulated inputs to constant values (CS, i.e. Q_1, to 20, Q_2 to 0 e V_f to 0), recording the movement of the apparatus to establish whether or not the temperatures reach constant values

Since constant temperatures are reached, we say that the apparatus has reached an equilibrium

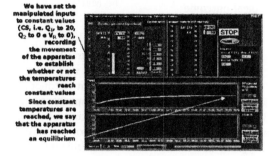

Fig. 3. Explanatory (top) and action (bottom) slides fom the first FAC assignment.

coefficients (γ_{ta}) are equal; (f) γ_{ta} and the plate-to-air heat transfer coefficient (γ_{pa}) depend on the fan command Q_f, varying linearly from a minimum ($\Gamma_{ta0}, \Gamma_{pa0}$) to a maximum ($\Gamma_{ta100}, \Gamma_{pa100}$) as the fan command goes from 0 to 100; (g) the air temperature is constant. The model equations are

$$
\begin{aligned}
C_t \dot{T}_1 &= P_{g1} - \gamma_{tp}(T_1 - T_p) - \gamma_{ta}(T_1 - T_a) \\
C_t \dot{T}_2 &= P_{g2} - \gamma_{tp}(T_2 - T_p) - \gamma_{ta}(T_2 - T_a) \\
C_p \dot{T}_b &= \gamma_{tp}(T_1 + T_2 - 2T_p) - \gamma_{pa}(T_p - T_a) \\
P_{g1} &= P_{max} f(Q_1/100) \\
P_{g2} &= P_{max} f(Q_2/100) \\
\gamma_{ta} &= \Gamma_{ta0} + (\Gamma_{ta100} - \Gamma_{ta0}) Q_f/100 \\
\gamma_{pa} &= \Gamma_{pa0} + (\Gamma_{pa100} - \Gamma_{pa0}) Q_f/100
\end{aligned}
\tag{1}
$$

where C_t and C_p are the thermal capacities of the transistors and of the plate; T_1, T_2, T_p and T_a are respectively the temperatures of the two transistors, the plate and air; the other symbols have the meaning stated above.

4. THE REGULATORS' BLOCKS

The control structures dealt with in the laboratory are composed of PID regulators and (low-order) transfer function blocks. Focusing attention on the PIDs, it is important that students learn about some implementation-related facts that play an important role when they are applied to the process, especially as part of a con-

trol structure. The PIDs employed are in the 2-d.o.f., output-derivation ISA form (Aström and Hägglund, 1995), i.e.,

$$
u = w + K \left[br - y + \frac{r - y}{sT_i} - \frac{sT_d y}{1 + sT_d/N} \right], \tag{2}
$$

where r, y, u, and w are, respectively, the Laplace transforms of the set point, the controlled variable, the control signal, and a bias signal, K is the PID gain, T_i and T_d are the integral and the derivative time, N is the ratio between T_d and the time constant of a second pole required for properness, and b is the set point weight in the proportional action. These regulators include antiwindup, bumpless auto/manual switching, direct/reverse action, normal/velocity form, local/remote set point selection, two logical inputs that force the manual mode and the local set point mode (C_{man} and S_{loc},) two logical inputs that prevent the control signal from increasing or decreasing (F_+ and F_-,) and four logical outputs that signal the high and low saturation (C_{hi} and C_{lo},) the local state of the set point (S_{isloc},) and the manual state of the regulator (C_{isman}.) For every control structure treated, a LabVIEW program was created; students just use these programs, thus no knowledge of the LabVIEW language is required. In the FAC course only experiment-based modeling and single-loop control are dealt with, and the window of the program used appears in Fig. 3 (bottom).

5. ACTIVITY FOR THE FAC COURSE

The activity for FAC lasts 15 hours out of a course total of 105, and is organized as follows.

5.1 Experiment-based modeling

After the model (1) has been written and explained, the students perform a set of guided open-loop tests, recording several step and sine responses as shown in Fig. 4. Then, they are asked to translate (1) into a Simulink scheme selecting $f(Q) = k_1(Q/100) + k_2(Q/100)^2$, where k_1 and k_2 are suitable constants, and to set the scheme parameters so as to reproduce the recorded data. Instilled in them is the concept that experimenting must be on a systematic basis (i.e., for example, changing only one parameter at a time.) Though their success is typically limited, students learn that parameterizing a model of the entire process under control is very difficult, even in a simple case like this. Hence, they are helped to firmly understand why the practice often suggests to derive a 'partial' model for every control problem, basically from I/O data and sticking to the linear, time-invariant domain.

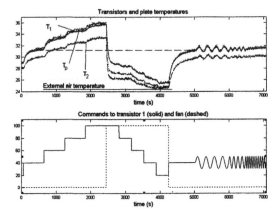

Fig. 4. Open-loop experiment.

5.2 Single-loop control

The first control problem tackled in the FAC course, for which the first partial model is derived, is to design a PID regulator for controlling T_p acting on Q_1, the other inputs being disturbance. From the linearization of (1) it turns out (computations are omitted for brevity and the symbol 'δ' denotes variations with respect to steady-state values) that the transfer function $G_{1p}(s)$ from δQ_1 to δT_p is second-order with no zeros. Since the order of (1) is 3, this means that the symmetry assumptions in that model make it non completely observable. Students are then directed to look at the recorded step responses (Fig. 4) to see if this fact is important, and the expected answer is yes because those responses exhibit a quite apparent slope change, calling (in the linearized domain) for the presence of the zero that in (1) is lost by cancellation. Hence, it is advisable to select the structure

$$G_{1p}(s) = \frac{\mu_{1p}(1 + s\tau_{1p1})}{(1 + sT_{1p1})(1 + sT_{1p2})(1 + sT_{1p3})}. \quad (3)$$

Students identify $G_{1p}(s)$ 'by hand': they set its parameters so that its step response fits the measured one, getting in this way awareness of which characteristics of the transfer function influence which aspects of the response. They also parameterize a first-order model, $\tilde{G}_{1p}(s)$, that approximates the response but has no objective structural relationship with physics. In fact, the instructors point out that the measured response 'appears to be first-order' except for the slope change; therefore, also a very simple model like $\tilde{G}_{1p}(s)$ can perhaps be useful. A possible outcome is shown in Fig. 5 (left), where

$$G_{1p}(s) = \frac{0.1325(1 + 80s)}{(1 + 155s)(1 + 35s)(1 + 15s)}, \quad (4)$$
$$\tilde{G}_{1p}(s) = \frac{0.1325}{1 + 120s}.$$

Note that in the frequency domain (Fig. 5, right) the models look far less similar than they do in the time domain: $G_{1p}(s)$ reproduces the magnitude

and phase values found with the sine tests better, its structure being connected with physics (albeit linearized.) Each student then designs three regulators: (a) a PI tuned on $\tilde{G}_{1p}(s)$ by cancellation, with a 20-30 dB high-frequency gain limit that becomes a response speed limit through the regulator gain (T_i is obliged to equal the model time constant;) (b) a cancellation PI for $\tilde{G}_{1p}(s)$ with a looser high-frequency gain constraint, about 40-50 dB; (c) a PID not tuned by cancellation, so that all the parameters (including N) can be employed for the tradeoff between disturbance rejection and control upset caused by noise, keeping the high-frequency gain around 30 dB. All these regulators are designed as 1-d.o.f. (i.e. $b = 1$,) respectively denoted by $R_{1l}(s)$, $R_{1h}(s)$ and $R_c(s)$, and with a typical outcome

$$R_{1l}(s) = 10\frac{1 + 120s}{120s}, \quad R_{1h}(s) = 100\frac{1 + 120s}{120s},$$
$$R_c(s) = 20\left(1 + \frac{1}{25s} + \frac{4s}{1 + 4s}\right),$$
$$(5)$$

whose high-frequency gains are about 20, 40, and 32 dB. Finally, the synthesized controllers are applied to the apparatus and results are discussed, a typical outcome with the three regulators (5) being shown in Fig. 6. It is shown that experimental results can be forecast with the identified models, the estimate being better if the more accurate model $G_{1p}(s)$ is used. The students also test their regulator on another apparatus, to see the importance of a robust design.

5.3 PID tuning recipes

The students tune a PI from a first-order model identified with the method of areas from the recorded step response, and with the IMC (Internal Model Control) tuning rules (Morari and Zafiriou, 1989). They also tune a PI and two PIDs by applying the open-loop Ziegler-Nichols rules (Ziegler and Nichols, 1942). Space limitations do not allow to show the outcome of all these regulators, see (Leva, 2002) for a complete description, but the students are led to some important conclusions. First, they become aware that the capability of making a good model employing physical laws and experimental data, and of tuning a regulator by considering all the aspects of interest (including robustness,) always yields the best results. Secondly, they realize that recipes can be most helpful, but one must understand whether or not the recipe fits the problem. If it does not work, the control engineer must be capable of diagnosing why. The third conclusion is that some final on-site adjustments are almost always necessary, especially if tight control is an issue, but conscious tuning remains fundamental.

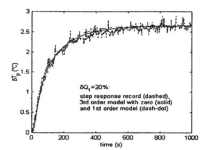 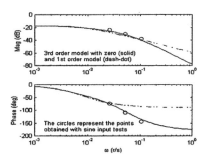

Fig. 5. Comparison of models $G_{1p}(s)$ and $\tilde{G}_{1p}(s)$ with data in the time and frequency domains.

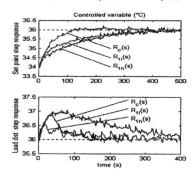 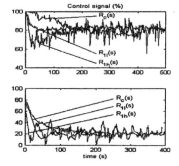

Fig. 6. Closed-loop experimental results with $R_{1l}(s)$, $R_{1h}(s)$ and $R_c(s)$: response to a 2°C set point step and to a 35% load disturbance step.

6. ACTIVITY FOR ETCS (SKETCH)

The activity for the ETCS course refers essentially to the use of control structures (Shinskey, 1981). It takes profit from the fact that the students already know the apparatus from the FAC course, and is organized as follows.

Decentralized control. The objective is to control T_1 and T_2 with two independent PIDs acting on Q_1 and Q_2. In tuning the regulators, the students experience that the two loops interact, so that (a) they disturb one another and (b) when one is opened, for example by setting the PID to manual, the dynamics seen by the other PID changes. They learn to quantify interaction by means of the Relative Gain Array (RGA,) defined in the present case as

$$RGA := \left| \begin{array}{cc} \dfrac{\partial T_1/\partial Q_1|_{Q_2}}{\partial T_1/\partial Q_1|_{T_2}} & \dfrac{\partial T_1/\partial Q_2|_{Q_1}}{\partial T_1/\partial Q_2|_{T_2}} \\ \dfrac{\partial T_2/\partial Q_1|_{Q_2}}{\partial T_2/\partial Q_1|_{T_1}} & \dfrac{\partial T_2/\partial Q_2|_{Q_1}}{\partial T_2/\partial Q_2|_{T_1}} \end{array} \right|, \quad (6)$$

by computing its elements symbolically in Matlab - from (1) - and by 'measuring' them through step tests. It is noted that a precise measurement of (6) is almost impossible because it requires ideal regulations. Hence, the students see that, for setting up a multivariable control system, at least some model knowledge is necessary.

Feedforward compensation. The goal is to control T_p acting on Q_1, while Q_2 is a disturbance, and $Q_f = 0$. Since the apparatus is symmetrical,

it can be assumed that Q_1 and Q_2 act on T_p in the same way. Therefore, if Q_2 is known, to compensate for it one can connect the opposite of its value to the 'Bias' input of the PID that controls T_p. If Q_2 is unknown, (1) suggests that the compensation be made based on T_2. To determine the compensator, the students apply a step to Q_2, recording T_2 and T_p, and see that Q_2 can be *approximately* reconstructed from T_2 by filtering the latter with the inverse of what appears to be the transfer function from Q_2 to T_2, augmented with the necessary number of high-frequency poles to achieve properness and noise attenuation. The approximation is crude, but the idea of taking T_2 as a representative of Q_2 does come from (1). The lesson, then, is that joining model knowledge and clever empirism leads to solutions that are simple and, in most practical cases, could hardly be improved. It is also noted that some logic is necessary: when the PID is in manual mode, the compensator must be disabled; when switching to automatic, it must be (re)initialized to produce zero output, or the control will have a bump.

Cascade control. The objective is to control T_2 acting on Q_1 and having as internal controlled variable the difference $T_1 - T_2$; the choice is motivated by explaining that, in so doing, the transfer function seen by the outer loop's regulator is 'quite similar' to an integrator. Students learn to tune the two PIDs in the correct order, to separate the inner and outer loops' critical frequencies, and to understand the required logic. In fact, in this case the outer PID outputs a set

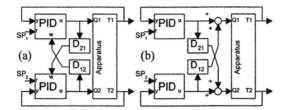

Fig. 7. Feedback (a) and feedforward (b) decoupling.

point for $T_1 - T_2$; for this quantity it is difficult to define the saturation values *a priori*, and it may happen, for example, that the outer PID raises the set point while the inner one is already in high saturation, causing a windup phenomenon. It is explained, however, that for this problem the individual PIDs' anti-windup mechanisms are useless, and that the solution is to connect the C_{hi} and C_{lo} outputs of the inner PID to the F_+ and F_- inputs of the outer one, so that when the inner PID saturates the outer one is allowed to move its output (i.e., the inner's set point) only in the direction that makes the inner PID move away from the saturation. In addition, when the inner PID is set to manual or its set point source is set to local, the outer one must be forced to manual; this is done by connecting the OR of the inner's S_{isloc} and C_{isman} outputs to the outer's C_{man} input.

Decoupling control. The goal is to control T_1 and T_2 with two PIDs and a decoupling network. The problem is not easy because, as can be guessed, here interaction is relevant. By means of simulations in Matlab, it is shown that it is preferable to arrange the decoupler's blocks in a feedback configuration, see Fig. 7(a), because in a feedforward one, see Fig. 7(b), there are four main problems: (a) the open-loop transfer functions of the decoupled loops are not the same as if there were no interaction, (b) the PIDs cannot be easily initialized prior to switching to automatic, (c) since the manipulated variables are not controller outputs, it is not easy to set up the anti-windup correctly, and (d) when one of the control variables hits a saturation limit, its controller still modifies the other control variable through the decoupler: hence, as both controllers are competing for the only unconstrained control variable, both loops are lost. The students identify the four process transfer functions involved through step tests, and experiment both static and dynamic decoupling. The lesson learned (apart from that of setting up a decoupler) is that, adopting the 'correct' (feedback) decoupling structure, there is no need for the very complex logic that would be necessary to tackle the mentioned problems in the feedforward case.

7. CONCLUDING REMARKS

A laboratory for undergraduate Automatic Control courses at the Politecnico di Milano has been presented. Guided experimental activity was integrated closely with lectures and made available to a very large number of students, all at the same time, under the guidance of a comparatively small number of instructors.

The laboratory serves the basic course (FAC) with the aim of providing experimental counterparts for the abstract concepts introduced, and of making students experience a real control problem as soon as possible. From 2002/2003 it will serve also the ETCS course, where more advanced concepts (like control structures) are taught, and for mastering them practical experience is necessary from the beginning. According to the students' impression and to the first outcomes, it appears that these goals are attained satisfactorily.

All the experiences, though covering a wide conceptual range, are made with the same, simple apparatus. Plans are currently underway to extend the laboratory scope beyond that of process control, including e.g. mechanical experiments.

REFERENCES

Amadi-Echendu and E.H. Higham (1997). Curriculum development and training in process measurements and control engineering. *Engineering Science and Education Journal* **1997**(June), 104–108.

Aström, K.J. and T. Hägglund (1995). *PID controllers: theory, design and tuning—2nd edition*. Instrument Society of America. Research Triangle Park, NY.

Bialkowski, W.L. (2000). Control of the pulp and paper making process. In: *Control system applications* (S. Levine, Ed.). pp. 43–66. CRC Press. Boca Raton, FL.

Bristol, E. (1986). An industrial point of view on control teaching and theory. *IEEE Control Systems Magazine* **6**(1), 24–27.

EnTech (1994). Competency in process control—industry guidelines, version 1.0.

Leva, A. (2002). A hands-on experimental laboratory for undergraduate courses in automatic control. *IEEE Transactions on Education (in press)*.

Morari, M. and E. Zafiriou (1989). *Robust process control*. Prentice-Hall. Upper Saddle River, NJ.

Shinskey, F.G. (1981). *Controlling multivariable processes*. Instrument Society of America. Research Triangle Park, NY.

Ziegler, J.G. and N.B. Nichols (1942). Optimum settings for automatic controllers. *Transactions ASME* **64**, 759–768.

DEVELOPMENT OF AUTOMATION EDUCATION IN THE INSTITUTE OF TECHNOLOGY

Hietanen, T., Kurki H., Jauhianen, O., Heikkinen, T., Kimari R., Poutiainen, R.

Oulu Polytechnic, Institute of Technology, Degree
Programme in Automation Technology
Kotkantie 1, 90520 Oulu, Finland
firstname.lastname@oamk.fi

Abstract: In the paper the degree programme in Automation technology is studied. Both the mission and the content of the education are described. The development of the education is based on connecting technical and non-technical aspects hierarchically to the teaching. The qualitative research method in the case study is applied. *Copyright © 2003 IFAC*

Keywords: automation technology, education, control

1. INTRODUCTION

The development in the automation technology has been fast due the development in the information technology. In the field of automation technology scientific research studies more complicated control algorithms as well as device vendors produce more intelligent high-tech devices. This has lead to the situation where the gap between new technology and implementations in the industry is rising. This development offers challenging task for engineering education organizations.

Despite the fact that technological aspect is essential in the engineering education also other skills are important. In addition to traditional technological activities, engineering is about reasonable risk taking, teamwork, change management, nontechnical decision making and marketing (Rainey, 2002). *Yeh* is proposing that the software engineering education can be divided into two aspects: human and professional. The human aspect has to do with how we educate each student to be a good and happy global citizen (Yeh, 2002).

Graduated automation engineers are mainly employed by the industry or industry related business. Therefore, the demands of the industry must be carefully considered. According the *Ollila* (Paaso, 1998), the capabilities of the graduated software engineers should be:
- they are top professionals,
- they have a general overall in their area,
- they have practical language skills, and international experience,
- they have wide all around education,
- they are profit seeking,
- they are innovative and have initiative,
- they are good communicators and
- they are co-operative (Paaso, 1998).

The above list of the general capabilities of graduates is very challenging – and difficult to be integrated into one person – and not all of the list items can be influenced. However, the providers of education should actively seek ways to create learning scenarios, which support the strengthening of these capabilities of the students. In the high-tech engineering sector, learning has changed from 'once-in-a- life-time-education' to 'life-long-learning' (Paaso, 1998).

The core of technological education is summarized according *Alamäki* as follows: technological development, control, and utilization have been and still are directly dependent on human abilities to innovate and solve problems; technology is human activity (Alamäki, 1999).

The essence is learning by doing, emphasizing learner activity, where a learner is creating a bridge between his own thinking and the external technological reality. In addition, there is a evidence that even young students are able to reach a higher level of a abstract thinking in special fields, especially when they are interested in something and when they begin to reflect to these phenomena. Furthermore, it is beneficial for learning that students have the possibility to apply their knowledge and skills in technological processes. This means processes where they design, develop, determine, control, utilize and assess technological systems by using different devices and tools (Alamäki, 1999).

In the institute of technology the automation technology has been on its own degree programme since 2000. In succession to the development of the education in the automation technology has taken huge steps. Due this development more new technology such a fieldbus technology, fuzzy control, higher-level control networks and information networks is added to the execution of the education. In addition, pedagogical methods supporting the higher demands of industry has been utilized. Teaching is more based on project learning, students are lead to be more self-direction and the methods for continuous feedback in education are developed.

In the paper the degree programme in Automation technology is studied. Both the mission and the content of the education are described. The qualitative research method in the case study is used. The result of the research is tool for transparent continuously transforming education.

2. DEGREE PROGRAMME IN AUTOMATION TECHNOLOGY

The Institute of Technology is a part of Oulu Polytechnic. In Finland, polytechnics form a sector of higher vocational education. This sector of higher education is young, and so is Oulu Polytechnic, which is set up officially in 1996. However, the Institute of Technology was founded in 1897.

The Institute of Technology provides education in the sectors of technology and transportation. Six different programmes leading to vocational qualifications (B. sc. Eng) are offered. The completion of one program requires 160 credit units (equal to 240 ECTS). The scheduled duration of the studies is four years.

The Automation technology was former located inside the information technology degree program as specialization, until it was separated as its own degree program in 2000. Each year, 30-40 new students start studies in Automation Technology.

The studies in Automation Technology can be divided according the table 1. At the beginning of the studies skills in natural science as well as in foreign languages are developed. The most part of the vocational studies take place in the third and fourth class. At the end of studies the students have to complete the bachelor thesis. During the first three study years most students work as a trainees in the industry.

Table 1. Content of the Degree program in the Automation Technology.

Content	Credits (ECTS)
basic studies	115,5
vocational studies	64,5
training	30
selective studies	15
bachelor's thesis	15

3. MISSION

The mission of the degree programme in Automation Technology is to produce skillful engineers for the challenging tasks in the Northern Finland industry, so that the engineers could develop experts in their own field. This demands that the engineer has both the adequate theoretical knowledge and the practical skills. Not the mention skills to be a global citizen, where foreign exchange programs are important.

A common problem in the engineering education is the following; how to get talented and highly motivated students? There is no definite answer, however it is important to generate marketing strategy to the potential students and to the industry, which probably will hire graduated engineer. The main goal of the marketing is to create and maintain a positive image

The education in the automation technology is described as a system block (fig. 1), where the input is the student and the output is a graduated engineer.

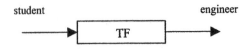

Fig 1. The transfer function of education.

The transfer function of education contains relationship network between student, society, industry and science, as seen in fig. 2. The control groups determine continuosly the content and the objective of education. The control groups are formed from the partners of the relationship network.

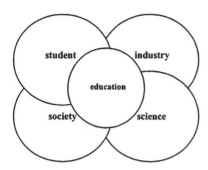

Fig 2. The partners of the relationship network.

The students are able to control the education, in addition to the direct and indirect feedback, in the teamwork with the teachers. The main goals of the control group containing experts from the industry and from the university are to follow the technological development and to determine the expectations for the graduated engineer.

The most important indicators for the education are the employment and the study days. The indicators are related to the funding from the society.

4. EXECUTION

The vocational studies are chronologically divided into the three classes. The content, the target and the method of implementation of the studies are also taken into account while determining the classes.

In the basic vocational studies students will adopt the concepts of automation technology. They will understand both the tasks and productional aspect of automation technology. Important prospect is also to build a bridge between natural science and automation technology. In the education teamwork, problem based learning and computer programming (Consonni, 2001) for example with the MATLAB is important (Nise, 1995).

In the advanced vocational studies students will learn more detailed the field of automation technology. They will acquire the implementation methods of the automation technology. In the advanced vocational studies main topics are, information networks, fieldbus technology advanced control methods (Jurado, 2002) and automation systems. After the advanced vocational studies students are able to work in the automation project as project engineers.

In the advanced special studies students will learn duties of the Automation engineer. In the studies of instrumentation student will learn ALMA (AutomationLifecycle-MAnagement) -programming. ALMA is a widely used program in Finland in the documentation of the instrumentation in industry. The configuration of the MetsoDNA-automation system is learned in the advanced studies of the automation systems. The project based programming with the logics (OMRON, Siemens) is learned also in the advanced special studies.

Approximately third of the vocational studies is carried out in laboratory. This consists both the computer laboratories and the actual laboratory exercises with the automation related devices.

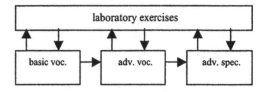

Fig. 3. The organization of the vocational studies in the Automation technology.

5. SUMMARY

In the beginning of the research the external demands of the automation engineering education are defined. Despite the fact that teaching of technology is essential also other demands are to be met. The mission of the degree programme in Automation Technology is to produce skillful engineers for the challenging tasks in the Northern Finland industry.

In the basic vocational studies students will adopt the concepts of automation technology. In the education teamwork, problem based learning and computer programming are important.

In the advanced vocational studies students will learn more detailed the field of automation technology. After the advanced vocational studies students are able to work in the automation project as project engineers.

In the advanced special studies students will learn duties of the automation engineer. Each step is strongly related to the laboratory exercises where both practical and co-operative skills are developed.

In the research hierarchical steps for the vocational studies are created. The aim of the research is to make it easier to control and to develop the studies. It also increases the information how the courses are connected and organized.

The division into the classes enables new aspects for the control groups comparing to the situation where the studies are presented as a separate courses. It should be also easier for the student to recognize how the vocational studies are organized.

6. CONCLUSION

The result of the research is hierarchical tool for transparent continuously transforming education. The tool offers detailed determination for the vocational studies. As well as increases non-technical information of the vocational studies.

Due to the fact that the degree programme in the automation technology is new, there is practically no statistical information available for the indicators; it is difficult to prove the advantages of the tool. However, the evaluation from the control groups is promising.

REFERENCES

Alamäki, A. (1999). How to educate students for a technological future: technology education in early childhood and primary education. Ph. D. Thesis, University of Turku, Finland. B 233.

Consonni, D. (2001). A Modern Approach to Teaching Basic Experimental Electricity and Electronics. *IEEE Transactions on education.* **Vol 44**. NO 1, February 2001. p6-15.

Jurado, F. (2002). Experiences With Fuzzy Logic and Neural Networks in a Control Course. *Transactions on education.* **Vol 45**. NO 2, May 2002. p6-15.

Nise, N. (1995). Control Systems Engineering. Second edition. Addison-Wesley. Menlo Park, California.

Paaso, J. (1998). Computer based teaching technology for software engineering education. Ph. D. Thesis, University of Oulu, Finland. C 123.

Rainey, V. (2002). Beyond technology – Renaissance engineers. *IEEE Transactions on education.* **Vol 45**. NO 1, February 2002. p4-5.

Yeh, R. (2002). Educating future software engineers. *IEEE Transactions on education.* **Vol 45**.NO 1, February 2002. p2-3.

www.elsevier.com/locate/ifac

REMOTE AND MOBILE CONTROL OF MULTIDISCIPLINARY EXPERIMENTAL SYSTEMS

Florin SANDU [1], **Seppo LAHTI** [2], **Seppo RANTAPUSKA** [2],
Ingmar TOLLET [2], **Juha LÖYTÖLÄINEN** [2], **Riikka KIVELÄ** [2]

[1] *"Transilvania" University of Brasov, Romania*
[2] *"EVTEK" Espoo-Vantaa Institute of Technology, Finland*

Abstract: The paper presents wireless access to remote experiments. Client-side development was performed by wap-/web- browser to send stimuli and control parameters. Server-side was accomplished at two levels: wap-/web- server, accessible by Internet and work-bench server, implemented by National Instruments LabView, for direct programming of test equipment, and for accessing the wap-/web- server by Intranet, in local network neighbourhood, for the results' page publishing. *Copyright © 2003 IFAC*

Keywords: measuring elements, tests, tele-control, communication systems, networks, protocols.

1. INTRODUCTION

The present paper was accomplished in the frame of the „Leonardo da Vinci" Pilot Program RO/01/B/F/PP141024 "Virtual Electro-Lab".

Recent research performed by the authors converged towards Internet publication not only of information but also of experimental resources [1]. Students, teachers and researchers from any location (even from home) and even on the move (see § 2) can access (by web/wap-browsers) laboratory equipment (driven by LabView ™ virtual instrumentation software) for real (not simulated) test/measurement and control/monitoring. After this remote-accessed measurement process and data tele-transmission, mathematical post-processing of results and experimental data-base creation can be accomplished, making this sort of laboratory a part of tomorrow's e-university.

The electronic core of this implementation, the work-bench (described in § 5), can be multidisciplinary extended/adapted by appropriate sensors and transducers at one edge and by drive elements that interface with external equipment, at the other edge.

2. TECHNOLOGICAL ASPECTS OF WIRELESS ACCESS

Sandu *et al.* (2002a) presents preliminary research on SMS-based remote transmission of experimental results, implemented with a GSM modem controlled by a twin-microcontroller development board (for data conditioning, acquisition & compression and dial-up respectively).

Recent accomplishments based on 3rd generation mobile phones are presented in Sandu *et al.* (2002b) with newest models implemented with CDMA (Code Division Multiple Access) and built-in Microsoft Mobile Explorer 3.0, supporting TCP/IP (Transmission Control Protocol / Internet Protocol), HTTP (Hyper Text Transfer Protocol), WAP (Wireless Application Protocol), PPP (Point to Point Protocol), POP3 (Post Office Protocol 3) & STMP (Simple Transfer Mail Protocol) and Java applets. These most versatile implementations were completed also to enhanced *mobile* configurations, using a notebook PC with the mobile phone operating as a high-speed wireless modem through a serial cable (could be also USB) or by infrared, doing complex experiments like the ones presented in Kayafas *et al.* and Sandu *et al.* (2000).

3. THE EXPERIMENTAL CONFIGURATION

The automated bench for moisture control at "EVTEK" is presented in figure 1. It includes also supplementary facilities that were not used in the concrete configuration (extra stand-alone dual indicator for the temperature and moisture sensor, extra air pumps that can blow air into the experimental enclosure through an electric heater). The parts that were used in the present implementation are interconnected as in figure 2.

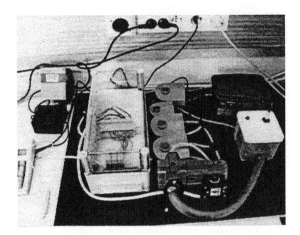

Fig. 1. The automated bench for moisture control. They can be observed, from left to right, the indicator for temperature and moisture (with its power supply), the box of for protected mains voltage connectors and opto-relay(s), the air pumps and the extra air heater, the main enclosure with moisture (and temperature) double transducer.

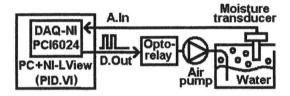

Fig. 2. Schematic of the test configuration. Its simple representation is very similar to that published on the wap page

The work-bench server is a Pentium III PC (Windows 2000) hosting a National Instruments PCI 6024 data acquisition and distribution board driven by LabView 6.1 (with extra PID Toolkit) in a virtual instrumentation assembly with Internet access.

A digital output of the board, controlled by the PID.VI, virtual instrument for proportional-integrative-derivative processing of error between the moisture set-point (SP) sent by the remote operator and process variable (AP –moisture actual point) in order to deliver PWM pulses to the control circuit of a special relay (with optical command for perfect galvanic separation from the controlled circuit) that feed directly, from mains 220 V AC, the air pump that blows air through the water basin, producing vapours that increase moisture. As mentioned, moisture decrease could be forced as well, but for the first experiments it was left natural.

4. MODE OF OPERATION

By operating his/her mobile phone the user can access the wap-page of the experiment (see figure 3) http://www.mobile.evtek.fi/pros_demo/index8.wml .

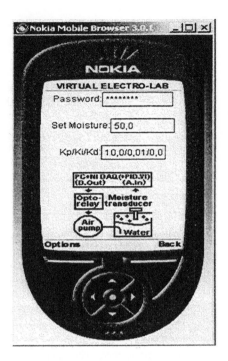

Fig. 3. The wap-page for the remote experiment, as tested on Nokia Mobile Browser 3.0.1. One can easily submit moisture set point and Kp/Ki/Kd PID constants (with values by default – as displayed in this particular screen – and suggested specific syntax).

As it can be seen from figure 3, the wap-page includes:
- a schematic (visual, *intuitive*, representation) of the system under test; for wap the format is .wbmp and aspect factors could be 96x48 points – to fit the screen of a normal mobile phone – or 96x96 points – to fit the screen of a personal digital assistant (PDA), e.g. Palm Pilot (see figure 4) that runs a wap browser (e.g. AU System Wap Browser 1.1, Edgematrix "WAPman", 4thpass KBrowser 1.1 etc.), connected by infrared, as "mobile terminal" to a mobile phone, used as "wireless modem";
- wap-/web- forms for remote submission of (stimuli)/inputs (post of the SP value), information for reconfiguration (by switching) or control parameters (e.g., in our implementation, set-point and constants of the PDI – proportional integrative derivative – control of the air pump); this can be done also by any other Internet terminal – e.g. by OPERA browser (web- and also wap-enabled) – see figure 5.

5. NETWORK ARCHITECTURE

The upper layer (the web-server – WS), is accessible in the Internet. The lower layer includes particular „work-bench servers" (WBS) – for each experimental configuration dedicated to remote control, like the one described in the present paper.
The wap-page can be developed with any wap-kit (e.g. the one in Nokia Mobile Internet Toolkit 3.1).

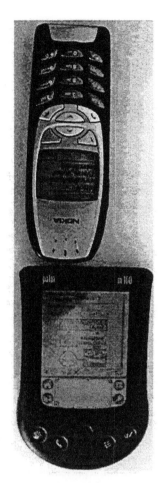

Fig. 4. Palm Pilot M100 (running AU System Wap Browser 1.1) connected via infrared to the Nokia 6310 GPRS mobile phone that is acting like a modem for dial-up connection to the server. Such PDA-s could be convenient to use due to larger and easy to use touch-screen.

Fig. 5. The wap-page of the experiment, as seen with OPERA universal browser.

5.1. Client Implementation

Development was done in the simplest way, for compatibility with most terminals. The source code of the above-mentioned ... /index8.wml page is:

```
<?xml version="1.0"?> <!DOCTYPE wml
PUBLIC "-//WAPFORUM//DTD WML 1.3//EN"
"http://www.wapforum.org/DTD/wml13.dtd">
<wml> <card id="test" title="VIRTUAL
ELECTRO-LAB" newcontext="true"> <do
type="accept" label="Submit"> <go
method="post" href="handlingwml.php">
<postfield name="SP" value="$(SP)"/>
<postfield name="PID" value="$(PID)"/> </go>
</do> <do type="reset" label="Reset"> <refresh>
<setvar name="SP" value="50,0"/> <setvar
name="PID" value="10,0/0,01/0,05"/> </refresh>
</do> <p align="center"> Password: <input
type="password" title="passwd" name="passwd"
value="password"/> <br/> Set Moisture: <input
title="Set Moisture [%]" name="SP"
value="50,0"/> <br/> Kp/Ki/Kd: <input
title="Kp/Ki/Kd" name="PID"
value="10/0,01/0,05"/> <br/> <img
src="images/moist4.wbmp" alt="image"/> </p>
</card> </wml>
```

As it can be seen, the post method is implemented by the handlingwml.php file (writing the values received from the operator in the wmlvalues.txt file into the wap-site main directory .../pros_demo on the WS:

```
<?php header("Content-type: text/vnd.wap.wml"); ?>
<!DOCTYPE wml PUBLIC "-//WAPFORUM//DTD
WML 1.1//EN"
"http://www.wapforum.org/DTD/wml_1.1.xml">
<wml> <card id="process" title="Process">
<onevent type="ontimer"> <go
href="monitor.wml"/>
</onevent> <timer value="25"/> <p align="center">
<big> <b> Wait, processing...</b> </big>
<?php $filename =
/var/www/pros_demo/wmlvalues.txt";
if (!file_exists($filename))
{ touch($filename); // Create blank file
chmod($filename,0666); }
$fp=fopen("/var/www/pros_demo/wmlvalues.txt",
"w");
$strSP="SP=";
$strPID="PID=";
$tmp = fputs($fp, $strSP);
$tmp = fputs($fp, $SP."\n");
$tmp = fputs($fp, $strPID);
$tmp = fputs($fp, $PID);
fclose($fp); ?> </p> </card> </wml>
```

It can be noticed the redirection towards the monitor.wml page, that will only ask confirmation to get the status of the work-bench, providing also a delay long enough for the completion of the automated measurement.

5.2. LabView Implementation of Work-Bench Server

National Instruments LabView [TM] is one of the main alternatives for WBS programming, taking into account its dedicated interfacing and driving of instruments and of data acquisition and distribution boards, together with optimized functions for advanced digital and analog signal analysis and synthesis.

Switching can be controlled as well by direct connection to digital outputs of data acquisition and distribution boards of the relays' command circuits – windings (or LED-s) of (opto-) relays (like in our case) etc. Another possibility, tested by the authors, is simple Centronics – parallel port connection directly to grid/base of field effect-/bipolar junction-transistors (BJT) inserted between ground (connected to emmiter/source) and termination of relays' command circuit (connected to collector/drain); advantages of this insertion include protection of aforementioned digital outputs (one can insert also extra-resistors between these outputs and BJT bases) and also the possibility to command relays fed with any control voltage (less than the maximum one supported by the transistors), not compulsory 5V.

The panel of the Vantaa1.VI that runs continuously on the WBS is presented in figure 6. Besides special stop-buttons for forced interruption of the execution (by any local supervisor of the remote-controlled experiments), the panel includes only indications – the information to be published on the WS but also a chart dedicated to the above-mentioned supervisor;

• the supervisor can grant access by the own web-server of LabView 6.1, in a special procedure that requires the remote-operator to download and plug-in into his browser a Real-Time Engine from National Instruments, but such procedures are beyond the simple and robust approach (centered on mobility, with cheap terminals) that are presented in this paper.

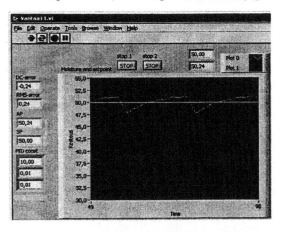

Fig. 6. The panel of the Vantaa1.VI virtual instrument that runs continuously on the WBS, waiting to serve any "new set-point" or "get status" request

The diagram of Vantaa1.VI includes two cycles:

• the shortest cycle, with a period of 500 ms (see figure 7), implements PID control of width (then of frequency) for the pulses that feed the counter driving the air-pump; Basic DC-RMS sub-VI averages the DC and RMS values of errors, in order to offer synthetic benchmarks of the control process, easier to display on the mobile terminal than a chart: e.g., a small DC error, called also Integral Error (normalized to time) in versions of the virtual instruments, compared to a great RMS error (or an Integral Absolute Error, normalized to time) illustrate a a process variable (AP) "vibrating" around SP etc.

• the longest cycle, with a period of 10 s, initialises the above-mentioned counter whose output drives the air-pump; suppose pulses are generated with shorter width (as a result of the PID algorithm) the counter's output toggles (and starts the air-pump) earlier, to compensate a moisture decrease, and so on. Each 10 s, the virtual instrument tests the length of the wmlvalues.txt file (see figure 8), to see if any new setting values have been sent from remote; when length stays 0, nothing runs in the case structure; when it's false that length stayed 0, each new access for setting, the first frame of the sequence structure runs in order to start the elapsed time chronometer and to extract the SP and PID constants from the received string (by "match string" functional blocks that deliver what is before and after the PID= sub-string and before and after the two slashes, / ; the second frame was not represented in figure 8 as it simply deletes the wmlvalues.txt file, preparing the virtual instrument to wait for another access. The length of another similar file, wmlmonitor.txt, created by the 'get status' request (in the same way wmlvalues.txt was created) is also checked each 10 s. If its content is YES, this means a renewal of the results.wml file is ordered and an actualized wap-page is published. This time it's the first frame of the appropriate sequence that wipes out results.wml file (in order to allow its re-publishing) so we detail only the second frame, in figure 9.

The wap-/web- source file results.wml is assembled by simple insertion (string operation "concatenate") of programmed values (SP and PID constants), as well as measured values (elapsed time and actual moisture, AP) and computed values, DC error (or normalized integral error) and RMS error (or normalized integral absolute error), together with direct publishing of the wap-/web- page on the web-server, via a local link.

This simple way of publishing, directly from the WBS to the WS, only in the „Network Neighborhood", could be a robust alternative to invoking web-server capabilities included in LabView or to the use of any other „web deployment" method.

Without any prerequisite skills, one can edit, beforehand, the wap-/web- page with the schematic of the experiment.

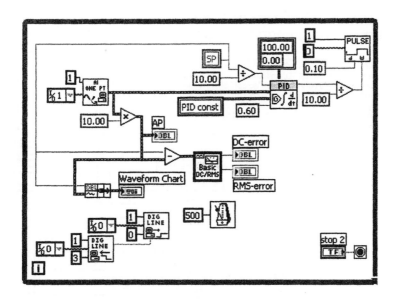

Fig. 7. The shortest cycle (implementing PID control) of the virtual instrument

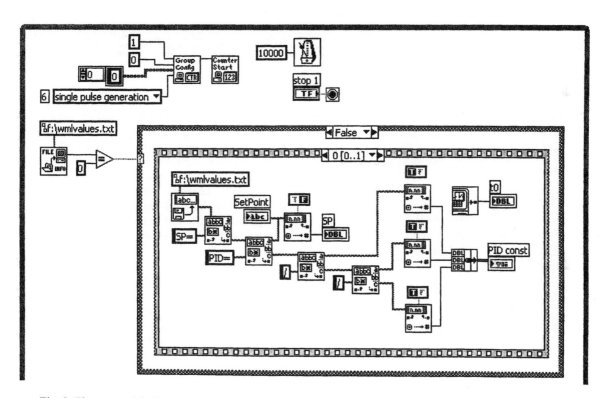

Fig. 8. First part of the longest cycle (implementing initialization and service of the new settings access)

In the proximity of input/output nodes representation, labels can be inserted, after which the above-mentioned stimuli/measured values could be displayed, together with the computed values. With the very friendly "wysiwyg" design tools available, this could be simple for anyone.

The .wml or .html source file can then be split in some alphanumeric constants (practically no problems with their length), immediately after the above-mentioned labels. These pieces can then be easily concatenated with the input, measured or computed values, into a file that will be written locally, directly by LabView, from the WBS (that can "see" the WS in the Intranet „Network Neighborhood") to the WS (the only one that can be "seen" in the Internet), in its appropriate directory (e.g. \\inetpub\wwwroot, established by automatic web-publishing procedure at site creation). This results page (see, in figure 10, how it could look with OPERA browser) could be completed also with a hyperlink taking back to input forms' page.

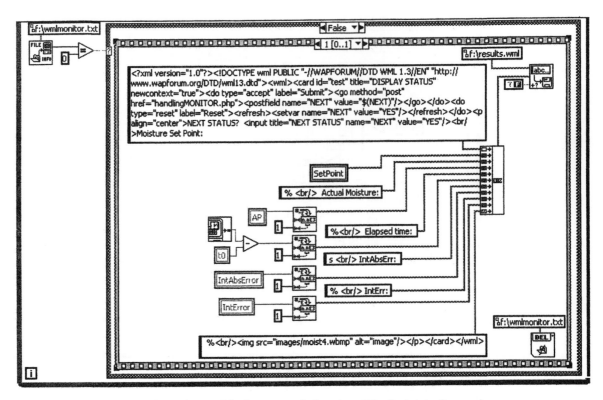

Fig. 9. Second part of the longest cycle (service of the "get status" access)

Fig. 10. Wap-page with results of the experiment, as seen with OPERA universal browser

6. CONCLUSION

The paper presents intuitive, robust and portable solutions for "wysiwyg" remote testing. Their strength is the powerful WBS programming by LabView that can master, in an unified way, multi-disciplinary configurations. Real-time efficiency of LabView processing lets only wap access interval to limit the time constants of controllable systems.
On the other side, simplest and cheapest client software implementation allows easy remote and even mobile access to control processes.

The results of the authors' work can be extended to complex monitoring scenarios, even for remote (re-) configuration by high priority administrators that can restart or adjust important systems from anywhere, at any time (e.g. in emergency cases). Further development becomes possible with the recent release of development tools for Java 2 for Mobile Equipment (J2ME) that is now firmware in the new generation of mobile phones.

REFERENCES

Kayafas, E., F. Sandu, I. Patiniotakis, P.N. Borza (2001). Approaches to programming for tele-measurement. In: *Proc. of the XVII IMEKO World Congress 2001, Lisbon – Portugal.*

Sandu, F., W. Szabo and P.N. Borza, (2000). Automated Measurement Laboratory Accessed by Internet. In: *Proc. of the XVI IMEKO World Congress 2000,* Vienna, Austria.

Sandu, F., D.N. Robu, P.N. Borza and W. Szabo (2002a) Twin- Microcontroller GSM Modem Development System. In *Proceedings of 8th International Conference on Optimization of Electrical and Electronic Equipment "OPTIM 2002",* Brasov - Romania.

Sandu, F., W. Szabo and V. Cazacu. (2002b) Automated Test System with Wireless Access. In: *The Romanian Magazine of Virtual Instrumentation, Mediamira Publishing House,* Cluj-Napoca, Romania

ELSEVIER
IFAC
PUBLICATIONS
www.elsevier.com/locate/ifac

A WEB BASED INDUSTRIAL CONTROL LABORATORY

Wojciech Grega

University of Mining and Metallurgy, Department of Automatics,
30-059 Krakow, Al. Mickiewicza 30, wgr@ia.agh.edu.pl

Abstract: A Web Based Industrial Control Laboratory was developed as a part of university campus real heat distribution control system. The industrial control system designed and implemented for group of heating substation supplying the 18 university buildings have established the foundation of a virtual research and teaching laboratory. The implemented solution have integrated a wide range of state-of-the-art technologies and standards representing modern industrial IT control systems. The virtual laboratory is used throughout the course homework assignments, in which students are required to solve both control system design problems and teleoperation modelling problems. *Copyright © 2003 IFAC*

Keywords: Educational Laboratory, Distributed Control, Supervisory Control, Integrated Plant Control, Energy Control.

1. INTRODUCTION

The control of a real industrial system is an important part in control education, where students can learn to handle imperfection in sensor and actuators or disturbances unavoidable in real systems. Such effects are difficult to simulate but exists in real industrial control system. Moreover, in the operation of a real plant constraints representing device physical limitations, safety and environmental requirements must be satisfied - in contradiction to theoretical simulation models.

Currently, Web-based control laboratories are a hot topic for control education community (Ko, *et al.*, 2001; Overstreet, *et al.*, 1999). The concept is not new. The possibilities of TCP/IP based communication networks and growth of the amount of Internet connections have caused that several universities have developed a distance learning applications that allowed students to remotely conduct experiments in a control laboratory. However, these remote laboratories give the access only to the laboratory process rigs, not to the real industrial process and only partly to the industrial control technology.

The heating network of the University of Mining and Metallurgy (UMM) campus became the focus of the "System for monitoring and optimization of heat energy consumption in UMM" project implemented in the years 2000/01, (Grega, Kolek, 2002). The key objectives of the project were:

- to develop a system allowing remote monitoring and control of a heating substation being a main part of the UMM power and heat supply system,

- to identify new ways to optimise energy consumption and develop adequate control algorithms,

- to establish the foundation of a virtual research and teaching laboratory, allowing practical verification of research results and supporting students projects.

The control system designed for the heating network was implemented for group of heating substation supplying the 18 university buildings in the area of 5 sq. kilometres. During the winter season of

2001/2002 the buildings were consuming up to 20 GJ/h. Such a high power consumption allows significant savings, if a proper control strategy is implemented.

2. AN OVERVIEW OF THE CONTROL SYSTEM STRUCTURE

The developed system of control and monitoring has a multi-layer structure. It consists of the following levels (Fig.1).

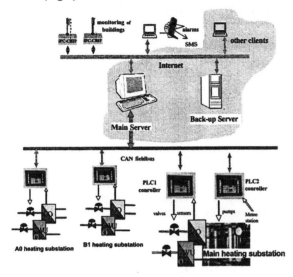

Fig.1. Structure of the heat distribution control and monitoring system.

- Direct control and data acquisition layer collecting information from the heating substations. The key elements of this layer are industrial PLCs equipped with interfaces to the local industrial CAN network (Lawrenz, 1997).

- Data acquisition and monitoring unit (Main Server). It is a high-parameter PC-class computer serving as a platform for the SCADA industrial operating system. The installed iFIX system (Intellution, 2002) performs tasks that are typical for a supervising control unit – acquisition of substation operating data to databases, visualization of the process status as well as detection and reporting of emergency conditions. The main server is also used as a supervisory control platform for tuning of the PLC parameters and changing reference temperature values in particular installations. The iFIX system communicates, via Internet, with the temperature and heat supply monitoring systems in particular buildings, using the modular IPC-Chip network servers (Beck, 2002; Kolek, Bania, 2002). An additional functionality of the main server is the provision of access to the data from outside of the local network. An industrial Internet WEB server has been installed to suport this functionality. This solution allows the authorised operators, researchers and advanced students to access the SCADA system from typical Internet browsers.

- Back-up server – generates and stores back-up copies of databases, performs data analyses and

provides access to heating system operating data to all Internet users. This server is mainly dedicated for teaching applications. A client PC can access the laboratory remotely through a connection to the back-up server that hosts the web site for providing information about system operation and safety access to the process data.

It is worth noticing that the implemented solution integrates a wide range of state-of-the-art technologies and standards representing modern industrial IT control systems. The CAN network applied in this project, supported by the CSMA/CD/AMP protocol (Lawrenz, 1997), is one of the most dynamically developing standards in industrial IT networks. The iFIX software package installed in the main server represents the new generation of SCADA industrial software for Windows NT-based systems, based on the Microsoft object-oriented architecture. The proper data flow between system levels is possible due to implementation of several modern data exchange standards supporting the control system integration, such as OPC (Iwanitz, Lange, 2001), ODBC, TCP/IP or CSMA/CD/AMP.

A significant feature of the proposed solution is the open access structure. The "open" character of the system software allows easy modification, including addition of new modules used in various teaching and research experiments.

3. ACCESS TO THE DATA

Information about the current status of the University heating installations and historical data from the past heating season is available through the Internet via two WWW servers (Fig.2).

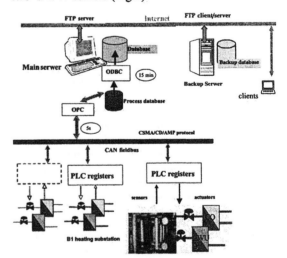

Fig. 2. Data flow between the system components.

The main server (http://wg.agh.edu.pl) contains the process database and copies of the selected synoptic screens in a format allowing browser viewing. Data screens are updated "on-line", every 5 seconds. Due to the licensing limitations of the installed SCADA software the server can simultaneously be used by a maximum of 10 authorised users. The main task of the server is to collect process data and to provide the technical maintenance services with a view of the current substation operating parameters. Only a limited number of advanced supervisory control experiments can performed using this server.

The back-up server (http://wg.ia.agh.edu.pl) contains a copy of the database, and provides access to copies of selected diagnostic screens (updated every 15 seconds) as well as weekly and monthly graphs illustrating the operation of the heating system and customised heating system operating cost analysis reports. The installed software allows unlimited access to the server. The task of the back-up server is to process historical and current data and to provide access to information used for education and research purposes.

4. TEACHING APPLICATIONS

4.1 Access to the teaching materials and process data

They are two access levels provided by the laboratory. After introducing the password students are able to login on the main server and to re-define the supervisory control algorithm for the selected local PLC controller. This opens possibility of conducting a wide range of experiments, like minimal energy control, predictive control or neural control. This level of access is mainly addressed to PhD students or to the students completed their final projects. For safety reasons the experiments must be carefully planned and supervised both by teaching and technical staff. Fortunately, the heat distribution

system components, mathematical models, examples of control algorithms and simulation models. Under this button the students can learn about the control equipment. The data exchange protocols and connection standards are also available under this button. Source process data are accessible under *Data base* button. The user-formated time plots can be obtained using *Data analysis* button. Source data are accessible under the *Data base* button.

It is not possible to influence the controller operation from this level. However, the current controller settings are available under the *Controller* button.

4.2 Experiments

The main motivation behind development of the Web-based laboratory was that the students taking courses in control find it difficult to relate the theory to the real industrial control systems. New possibilities of industrial SCADA systems and Internet tools have changed the situation: students can be provided with a full, on-line access to real industrial process data or even they can desire and conduct his own experiments.

Below, the example list of the Web laboratory-based experiments is given. These are:

- identification of the heat exchanger model,
- neural models of the heating system power consumption,
- prediction of the power consumption,

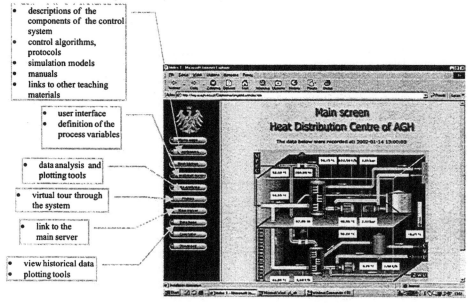

Fig.3. Main user interface of the backup server.

system is very robust to imperfect operation of the control algorithms. Experienced technical staff can easily correct any faulty operation.

Second-level access to the laboratory is provided by the back-up server from any Internet browser. The example of the main user interface is given in Fig.3.

Teaching material is available under *Descriptions* button. This includes description of the control

- optimisation of the supervisory control,
- distributed control via Internet and teleoperation principles, including traffic modelling,
- new technologies for building applications for control and monitoring (ActiveX, SCADA systems).

The experiments corresponds to the advanced level curricula and were carried out as a part of the following courses:

- *Integrated Control* (5th year of Automatics and Robotics specialisation),

- *Identification* (4th year of Automatics and Robotics specialisation),

- *Digital Control* (4th year of Automatics and Robotics specialisation).

The virtual laboratory is mainly used throughout the course homework assignments, in which students are required to solve both control system identification or design problems and teleoperation modelling problems.

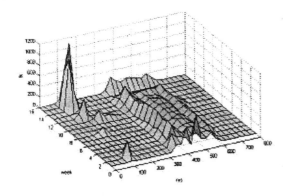

Fig. 4. Transmission delay (*ms*) between substations in the Internet network measured during 16 weeks of operation of the control system.

Fig. 4 shows example results of the student project aimed at the analysis of Internet usability as a part of the remote control system of heating substation. The obtained distribution of the transmission time was used for development and testing of the control methods suitable for variable delay control algorithms or adaptive delay compensators.

5. CONCLUSION

A Web Based Industrial Control Laboratory was developed as a part of real campus heat distribution control system. The laboratory serves students and staff in the Department of Control at the University of Mining and Metallurgy in Krakow.

The laboratory is currently being utilised in teaching of graduate and PhD students. Due to the complexity of the system, the students are carrying out individual project rather than regular laboratory sessions. They are able to collect data and making experiments according to his own time schedule. This is especially useful for the part-time students who can logon and use the system whenever they are free. For safety reasons implementation of the design is supervised by the teachers.

The laboratory offers an excellent platform for researchers to test new algorithms. For example, optimal supervisory control algorithms for the heat distribution system were developed by the author of this paper (Grega, *et al.*, 2002).

REFERENCES

Beck (2002). *http://www.beck-ipc.com.*

Grega W., Kolek K, Bania P. (2002). Internet measurement system, (Internetowy system pomiarowy). *Pomiary, Automatyka, Robotyka,* **vol.10,** no.2, 2002, pp.5-8.

Grega W., Kolek K. (2002): Monitoring and Control of Heat Distribution. In: *Proc. Of International Carpathian Control Conference ICCC'2002,* Malenowice, Czech Republic, May 27-30, 2002, pp. 439-444.

Intellution (2002). *www.intellution.com/products/fix.*

Iwanitz F., Lange J. (2001): *Ole for Process Control,* Huthing GmbH&Co KG, Heidelberg

Ko C.C., Chen Ben M., Zhuang Y., Tan C.K. (2001). Development of a Web-Based Laboratory for Control Experiments on a Coupled Tank Apparatus. *IEEE Transaction on Education,* **vol.44,** pp. 76 - 86.

Kolek K, Bania P. (2002). Optimal Control of the Heating Substation. (Optymalne sterowanie wezlem cieplnym) In: *Proc. of the XIV National Control Conference* (A. Korbicz ed.), vol.2, Zielona Góra 2002.

Lawrenz W.(1997). *CAN System Engineering,* Springer, 1997, ISBN 0-387-94939-9.

Overstreet J.W., Tzes A. (1999). An Internet-Based Real-Time Control Engineering Laboratory, *IEEE Control Systems,* **vol.7,** no. 5/1999, pp. 19-33.

ACKNOWLEDGEMENT

This paper was supported by the SOCRATES 2 - THEIERE Project no. 10063-CP-1-2000-1-PT-ERASMUS-ETNE and the Polish National Research Council grant.

E-LEARNING BY REMOTE LABORATORIES:
A NEW TOOL FOR CONTROL EDUCATION

Marco Casini, Domenico Prattichizzo, Antonio Vicino

Dipartimento di Ingegneria dell'Informazione
Università di Siena, Siena - Italy
Email: {casini,prattichizzo,vicino}@ing.unisi.it

Abstract: This paper deals with the Automatic Control Telelab (ACT), a remote laboratory of automatic control developed at University of Siena. In particular, it focuses on a new chapter of the ACT referred to as "student competition". It is a mechanism through which students can compete to design the controller with best performance for a given remote experiment. Controllers which achieve the given performance requirements are stored according to a ranking criteria. This tool allows students to make control synthesis practice on real remote processes through the Internet, and to compare their favourite controller with competitors designed by other users. *Copyright © 2003 IFAC*

Keywords: Control education, Remote laboratory, Control systems.

1. INTRODUCTION

The Automatic Control Telelab (ACT) is a laboratory developed at the University of Siena. Telelaboratories are instances of the more general distance education problems which is attracting a wide attention in the academic and government communities. Automatic Control is one of the technical areas which most exploited the new technologies to develop new tools for distance learning (Poindexter and Heck, 1999). A thorough treatment about control education by means of web technologies has been recently reported in (Dormido, 2002).

The web–based laboratories are divided in two classes: the virtual labs and the remote labs. The main difference between them is that virtual labs allow to remotely run simulations with possible animations of the controlled system (Merrick and Ponton, 1996; Lee *et al.*, 1998; Schmid, 1998), while remote labs are laboratories where students can remotely interact with real experiments. The ACT at the University of Siena is an example of

remote laboratory and is attracting the interest of many students from our campus and from other national and international institutions (Casini *et al.*, 2001).

In remote labs, users can change control parameters, run the experiment, see the results and download data through a web interface. This is for instance the case of (Knight and DeWeerth, 1996), where a remote lab for testing analog circuits is described; in (Shaheen *et al.*, 1998), a remote chemical control process is implemented and in (Henry, 1998) several laboratory experiments are made available.

In (Exel *et al.*, 2000) a comparison between virtual labs and remote labs is presented. The authors examine a common experiment (ball and beam) from these two points of view, and conclude that virtual labs are good to assimilate theory, but they cannot replace real processes, since a model is only an approximation which cannot reproduce all the aspects of the process, such as for instance unexpected non-linearities. On the other hand remote laboratories allow students to directly act with real processes, which is very important especially for engineering students.

From a pedagogical point of view, remote labs allowing for the design and implementation of the control law are the most exciting. Typically, the price to pay to obtain the controller design feature in many of the existing remote labs is that students must learn and use new control languages which are designed specifically for the remote lab and cannot take advantage of control functions developed in other contexts.

One of the key features of the Automatic Control Telelab (ACT) is that students can choose a control law, change on-line the control parameters and design their own controllers simply through the Matlab/Simulink environment. This feature allows a remote user to synthesize his/her own controller without learning any special language other than the Matlab/Simulink software. It is the authors' opinion that usage of a standard language like Matlab/Simulink will dramatically encourage the exercise with remote labs in control classes.

This paper deals with a recent improvement of the ACT: the "student competition" chapter. In this part of the ACT structure, students, or groups of students, compete to gain the best controller performance for a given remote experiment. This is a very exciting experience and it is attracting a great attention from automatic control classes. A typical competition session starts with defining control system requirements on one of the real experiments of the ACT. Then students access the ACT and design their own controllers which will steer the process during the competition. The ACT server stores the controllers with students data in a database, computes the performance indexes and assigns a score to the controllers. Then a ranking is computed and it is possible to evaluate the ability of students to meet the assigned specifications.

The paper is organized as follows. Section 2 illustrates the main features of ACT. In Section 3 the student competition mechanism is described, whereas in Section 4 a competition example is provided. Section 5 deals with teaching experiences while some implementation aspects are described in Section 6. Conclusions are drawn in Section 7.

2. AUTOMATIC CONTROL TELELAB OVERVIEW

The Automatic Control Telelab is a remote laboratory mainly intended for educational purpose, and since 1999 it has been used in control systems courses (Casini, 1999; Casini et al., 1999; Casini

et al., 2001). The aim of the project was to allow students to put in practice their theoretical knowledge of control theory in an easy way and without restrictions due to laboratory opening time and processes availability. At present, the ACT is accessible 24 hours a day from any computer connected to the Internet by means of a common browser like Netscape Navigator or Microsoft Internet Explorer; no special software or plug-in is required. If a user wants to design his/her own controller, the Matlab/Simulink software is required. A live video window is provided for each remote experiment session. One of the main features of the ACT is the possibility to integrate any user-defined controller in the control loop of the remote process. The controller synthesis is based on the Matlab/Simulink environment which is very popular in the control community. Since Matlab and Simulink packages are standard tools in control systems courses, there are no additional hurdles for a student who wants to design his/her own controller, which simply consists in a Simulink model similar to those commonly used to run a system simulation.

At present, four processes are available for remote control (Fig. 1): a DC motor, a tank for level control, a magnetic levitation system and a two degrees–of–freedom helicopter. The DC motor is used to control the axis angular position and velocity. The level control process has been included because, in spite of its simplicity, it shows nonlinear dynamics, whereas the magnetic levitation process, being nonlinear and unstable, shows very interesting properties to be analyzed in control theory education. Finally, the two degrees–of–freedom helicopter, being a nonlinear and unstable MIMO system, can be used in graduate control system courses.

Since like every remote lab the experiment hardware is controllable by one user at a time, from the web page showing the list of available experiments (Fig. 1) it is possible to know which processes are ready as well as the maximum waiting time needed to access the busy experiments.

3. STUDENT COMPETITION OVERVIEW

A typical remote laboratory allows users to run remote experiments using predefined or user-defined controllers. Students can run an experiment and see the dynamic response, but in general no information on controller performances is provided and it is not possible to know how controllers designed by other people behave on the same process. This is one of the reasons that motivated us to design a student competition mechanism for our remote laboratory. Through this tool a student knows about performance requirements his/her own con-

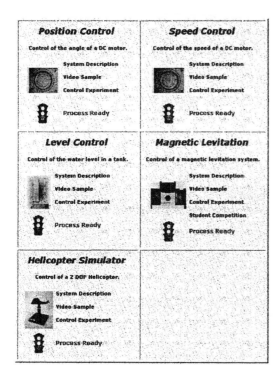

Fig. 1. Automatic Control Telelab's on-line experiments

troller must satisfy. Moreover a final ranking of the best controllers as well the time plots of the relative experiments are provided.

In the following some features of the ACT competition structure are described.

Remote exercises: in addition to standard control synthesis exercises, this tool allows a student to design a controller, which must satisfy some performance requirements, and to test it on remote real processes. At the end of the experiment, the performance indexes are automatically computed and shown to the user; if such indexes fulfil the requirements, the exercise is completed. An overall index is then computed (usually as a weighted sum of the previous indexes) and the controller is included in the ranking list.

Controller comparison: since this tool allows everybody to view the ranking concerning a competition, it is possible to know what kind of controller achieved better results. During the end-competition lesson, students who have designed the best controllers are invited to discuss their projects, while the lecturer shows why some control architectures work better than others.

Many competitions on the same process: it is possible to provide more than one competition benchmark on the same process, thus increasing the number of remote exercises available for students. Due to the software design of the competition structure of the ACT, new benchmarks can be added very efficiently.

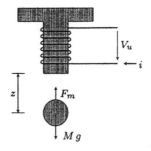

Fig. 2. Sketch of the process

It is the authors' opinion that competition can be considered as a new useful tool for distance learning and, at the same time, a tool which increases the potentiality of remote laboratories.

4. A COMPETITION SESSION DESCRIPTION

In this section, a competition session is described. In particular an example of competition regarding the process of magnetic levitation (see Fig. 1) is reported.

First of all, a student or a group of students who want to compete need to register by filling a form and obtaining a username and a password.

The user can analyze the mathematical model of the process (provided as a *pdf* file) as well as the required performance specifications. In this example it is required that, for a step reference, the settling time (5%) must be less than 1 second and the overshoot must be less than 40%. A more detailed description shows also the working point of the nonlinear benchmark.

The mathematical model of this process, sketched in Fig. 2, is summarized as follows

$$\begin{cases} M\,\ddot{z} = M\,g - F_m \\ F_m = k_m \dfrac{i^2}{z^2} \\ i = k_a\,V_u \end{cases} \quad (1)$$

where z is the absolute distance of the center of the ball from the coil, M is the mass of the ball, F_m is the magnetic force, i is the current in the coil, and V_u is the input voltage of the coil ($0 \leq V_u \leq 5$); k_m and k_a are the magnetic constant and the input conductance respectively. The actual values of these coefficients are reported in Table 1.

Table 1. Magnetic levitation system parameters.

M	Mass of the ball	$20 \cdot 10^{-3}$ Kg
k_m	Magnetic constant	$2.058 \cdot 10^{-4}$ $N(m/A)^2$
k_a	Input conductance	0.5488 $1/\Omega$
g	Gravity acceleration	9.80665 m/s^2
k_y	Unit conversion	100 cm/m

Equation (1) can be rewritten with $x_1 = z$, $x_2 = \dot{z}$, $u = V_u$ (input command) and $y = k_y z$ (output in centimeters).

$$\begin{cases} \dot{x_1} = x_2 \\ \dot{x_2} = g - \dfrac{k_m \, k_a^2}{M} \dfrac{u^2}{x_1^2} = g - k_t \dfrac{u^2}{x_1^2} \\ y = k_y \, x_1 \end{cases}$$

By substituting the actual values of parameters in the above equations, one obtains:

$$\begin{cases} \dot{x_1} = x_2 \\ \dot{x_2} = 9.80665 - 0.0031 \dfrac{u^2}{x_1^2} \\ y = 100 \, x_1 \end{cases}$$

Since the competition is based on an experiment around the state $(x_{10} = 0.05m, \ x_{20} = 0)$, students can choose to linearize dynamics

$$\begin{cases} \Delta \dot{x} = A \, \Delta x + B \, \Delta u \\ \Delta y = C \, \Delta x + D \, \Delta u \end{cases}$$

It follows immediately that $u_0 = \sqrt{\dfrac{g \, x_{10}^2}{k_t}} = 2.811$, thus linearized matrices are:

$$A = \begin{bmatrix} 0 & 1 \\ \dfrac{2 \, k_t \, u_0^2}{x_{10}^3} & 0 \end{bmatrix} = \begin{bmatrix} 0 & 1 \\ 139.4389 & 0 \end{bmatrix}$$

$$B = \begin{bmatrix} 0 \\ -\dfrac{2 \, k_t \, u_0}{x_{10}^2} \end{bmatrix} = \begin{bmatrix} 0 \\ -6.9719 \end{bmatrix}$$

$$C = [k_t \ \ 0] = [100 \ \ 0] \ \ , \quad D = [0]$$

Now a linear controller, such as a PID or a lead-lag compensator, can be synthesized. Of course, advanced students can design controllers with nonlinear techniques.

In order to design the controller, students must run the Simulink environment on their own local computers, then download a template file (*template.mdl*) and connect the signals describing the output, the error and the command to design the desired controller. An example of a PID controller is shown in Fig. 3.

A special interface (Fig. 4) allows a student to describe the structure of his/her own controller (i.e. P.I.D. Controller) and to set the sample time of the experiment; if the controller is continuous time, the sample time is intended as the integration step of the Simulink solver. Moreover, the user has to specify the file containing the controller and, if needed, the Matlab workspace file (*.mat*) containing essential data for that controller. Those files will be uploaded to the server, compiled and, if no error occurs, executed on the real remote process.

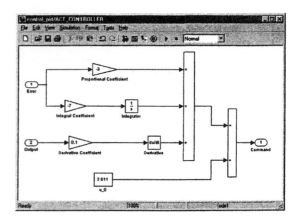

Fig. 3. A Simulink model for a PID controller

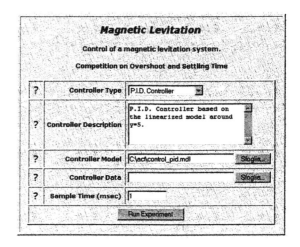

Fig. 4. The interface describing the controller features

A second graphical interface (Fig. 5) allows a user to start the experiment and to observe its behaviour through plots and the live video window.

At the end of the experiment, the performance indexes are computed and are displayed to the user. It is now possible to download a Matlab workspace file containing the full dynamics of the experiment and to view the time plots (Fig. 6). The ranking of the user controller is given as in Fig. 7.

Since several controllers can achieve the requested performance, an overall index is evaluated to build the ranking. This index is obtained by weighting each performance index. If a controller does not satisfy the requirements, the overall index is not computed.

It is possible for every user to view a controller report (Fig. 8) where information on ranking and other data, such as the controller description, the nickname of the user and his/her nationality and institution, are displayed.

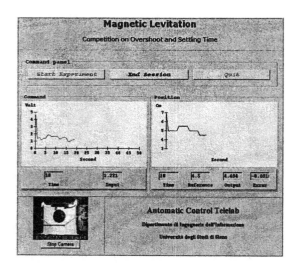

Fig. 5. The interface showing the running experiment

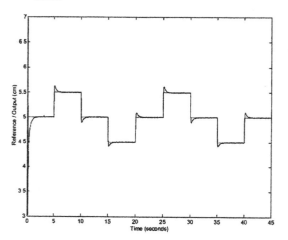

Fig. 6. Time plots of the experiment

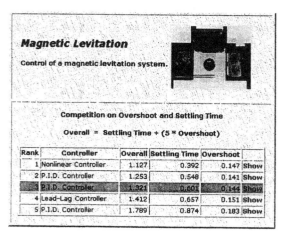

Fig. 7. Rank position of the controller

5. TEACHING EXPERIENCES

In spring 2002 undergraduate control system classes at the University of Siena used the student competition system.

First of all, the lecturer illustrated the physical model of the magnetic levitation system, empha-

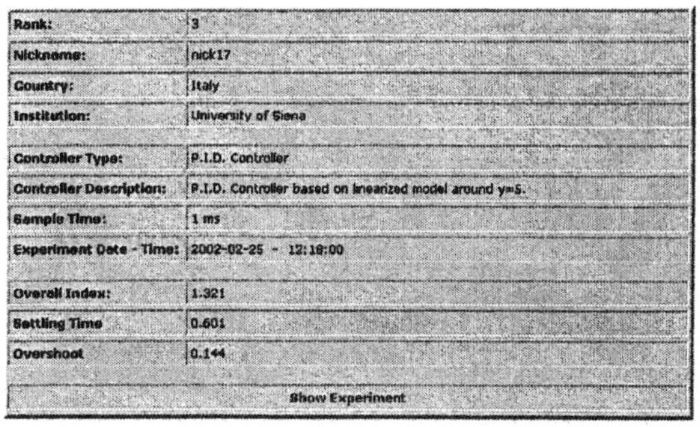

Fig. 8. Controller report

sizing its unstable and nonlinear dynamics, and suggesting the students to linearize dynamics to design the controller.

Since ACT is accessible at any time, students had no problems to analyze the problem and test their own controllers during the days before the second competition class, where the lecturer answered students about their questions and difficulties, and helped them to solve some typical problems. For example, he suggested them to use a pre-filter on the reference to obtain smoother command signals, and, in general, better performances.

At the end of the competition almost all the students were able to design a satisfactory controller, and their feedback was really positive.

After an evaluation process, some conclusions about positive and negative aspects of this experience were drawn:

positive aspects: students seemed to be very interested and excited, and used this tool to put in practice many theoretical notions. Moreover, everyone tried to do his/her best to obtain a good position in the ranking. However, the real motivation for this kind of competition, is not to individuate a winner, but to give students a new tool which can help them to better understand some practical control design issues as well as to increase their interest about control systems and technology.

negative aspects: after a first phase when students were really involved in learning new tools for designing a good controller, many students spent plenty of time to tune controller parameters just to obtain the best controller in the ranking, without any additional educational improvement.

6. IMPLEMENTATION NOTES

In this section a brief description of implementation aspects is provided.

The home page and other descriptive pages are stored in a unique server which is common to every process. Several pages regarding the student

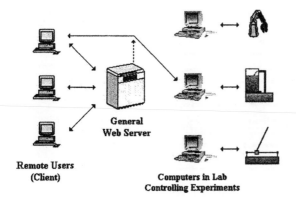

**Remote Users
(Client)**

**General
Web Server**

**Computers in Lab
Controlling Experiments**

Fig. 9. General scheme of ACT connections

competition are dynamically generated by means of the PHP language, and all data about users and controllers are stored in a MySQL database.

When an experiment is chosen, the user host is re-addressed to the machine directly connected to the process (Fig. 9). Data are exchanged through the Internet by a TCP connection between the user (client) and the ACT server. Once the connection has been established, the server sends all the data the client needs, afterwards the process is ready to start.

Server programs run under Microsoft Windows NT/2000 operating system. They are executable files obtained by Real-Time Workshop (RTW), a Matlab toolbox which allows to transform a Simulink model into a C source. To realize all the special features of ACT, specific code is linked with the source code generated by RTW.

To run a controller designed by a user, the controller model is merged with a Simulink model interfacing with the process through a data acquisition board. Thus, the overall file obtained is compiled and executed.

On the client side, Java applets are used, so that platform compatibility is assured. In this way, the user can operate through a very easy to use graphical interface.

To allow video transmission, the software Webcam32 (Kolban, 2002) is used. It is based on a Java applet to display on-line video, so it is not necessary for the user to install special software to perform this task.

7. CONCLUSIONS

The student competition mechanism of the Automatic Control Telelab has been described. The competition session is a natural extension of the remote laboratory at the University of Siena. It is the authors' opinion that competition stimulates students' interest thus improving the learning capabilities. In spring 2002 control system classes at the University of Siena used the ACT competition system with great interest and excitement. The ACT's home page is: http://www.dii.unisi.it/~control/act.

REFERENCES

Casini, M. (1999). Designing a Tele Laboratory for control of dynamic systems through Internet. Master thesis (in italian). Università degli Studi di Siena.

Casini, M., D. Prattichizzo and A. Vicino (1999). The automatic control telelab: a user-friendly interface for distance learning. Technical report. Università di Siena, Italy.

Casini, M., D. Prattichizzo and A. Vicino (2001). The automatic control telelab: a remote control engineering laboratory.. In: *Proc. of 40th IEEE Conference on Decision and Control.* Orlando. pp. 3242–3247.

Dormido, S. (2002). Control learning: present and future. In: 15*th IFAC World Congress b'02.* Barcelona.

Exel, M., S. Gentil, F. Michau and D. Rey (2000). Simulation workshop and remote laboratory: two web-based training approaches for control. In: *Proc. of American Control Conference.* Chicago. pp. 3468–3472.

Henry, J. (1998). *Enginering lab on line.* University of Tennessee at Chattanooga. http://chem.engr.utc.edu.

Knight, C.D. and S.P. DeWeerth (1996). World wide web-based automatic testing of analog circuits. In: *Proc. of 1996 Midwest Symposium Circuits and Systems.* pp. 295–298.

Kolban, N. (2002). Webcam32 - the ultimate webcam software. Technical report. http://surveyorcorp.com/webcam32.

Lee, K-M, W. Daley and T. McKlin (1998). An interactive learning tool for dynamic systems and control. In: *Proc. of International Mechanical Engineering Congress & Exposition.* CA.

Merrick, C.M and J.W. Ponton (1996). The ecosse control hypercourse. *Computers in Chemical Engineering* **20, Supplement**, S1353–S1358.

Poindexter, S. E. and B. S. Heck (1999). Using the web in your courses: What can you do? what should you do?. *IEEE Control System* **19**(1), 83–92.

Schmid, C. (1998). The virtual lab VCLAB for education on the web. In: *Proc. of American Control Conference.* Philadelphia. pp. 1314–1318.

Shaheen, M., K. Loparo and M. Buchner (1998). Remote laboratory experimentation. In: *Proc. of American Control Conference.* Philadelphia. pp. 1314–1318.

ELSEVIER
IFAC
PUBLICATIONS
www.elsevier.com/locate/ifac

REENGINEERING AN EMBEDDED LABORATORY ALLOWING REMOTE EXPRIMENTS THROUGH THE INTERNET

J. M. F. Calado[1], P. M. A. Silva[1], J. M. G. Sá da Costa[2] and V. M. Becerra[3]

[1]*IDMEC/ISEL – Polytechnic Institute of Lisbon*
Instituto Superior de Engenharia de Lisboa, Mechanical Engineering Studies Centre
Rua Conselheiro Emidio Navarro, 1949-014 Lisboa, Portugal
Fax: + 351 21 8317057, Email: {jcalado, psilva}@dem.isel.ipl.pt
[2]*Technical University of Lisbon, Instituto Superior Técnico*
Dept. of Mechanical Engineering, GCAR/IDMEC
Av. Rovisco Pais, 1049-001 Lisboa, Portugal
Fax: + 351 21 8498097, Email: sadacosta@dem.ist.utl.pt
[3]*The University of Reading, Department of Cybernetics*
Whyteknights, Reading RG6 6AY, United Kingdom
Fax: + 44 (0) 1189318220, Email: v.m.becerra@reading.ac.uk

Abstract: The paper presents a methodology for reengineering existing control systems labs allowing remote experiments through the Internet. Such a methodology is based on standard software tools applied to develop and implement a computer gateway. All the devices used in the control system lab considered in the current approach, such as sensors, actuators and controllers, as well as the computer gateway, are connected to a PROFIBUS network. The system implemented allows real-time experiments being performed remotely and could be a powerful way of studying new identification, control or fault detection and isolation algorithms independently the place where the methods are implemented. *Copyright © 2003 IFAC*

Keywords: Remote education, remote laboratory, control education, proxy agent, Internet, World Wide Web and teleoperation.

1. INTRODUCTION

Nowadays control systems are becoming more and more complex and control algorithms more and more sophisticated. As a consequence, industrial requirements for well-prepared control systems engineers are evolving, due to marketplace pressures and progress in technology (Åström, 1994). Thus, as far as the educational activity is concerned, in automatic control education practical experiments have been playing and will sure play an important role (Horacek, 2000). As a matter of fact, one fundamental feature of control engineering education is the laboratory and practical work needed to provide to the engineering students a taste of a real industrial environment with all the associated problems such as control loops configuration, controllers tuning, communications, measurements

and instrumentation problems (Röhrig and Jochheim, 2000).

Therefore, the rapid progress of Internet technology and its popularity has been motivating the development of Internet laboratories allowing the students to be able to perform real time experiments, on real equipment, but through the Internet (Shaheen *et al.*, 1998; Overstreet and Tzes, 1999; Sheng *et al.*, 2000; Sánchez *et al.*, 2001). Thus, the use of remote labs for supporting and integrating the activity of a control engineering course is in fact a widely discussed issue (Bencomo, 2002).

Several developed solutions for remote labs, where different laboratory experiments are performed remotely via a web interface, have been pointed out by Poindexter and Heck (1999). However, a common

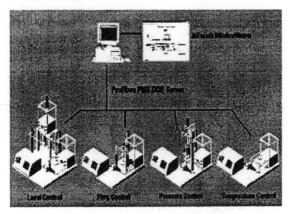

Fig. 1. Process Control System framing

point of all proposed solutions is that the remote user (a client user) is connected by Internet to a dedicated web server that interacts with computers of the lab, which are used for supervisory purposes. The current approach uses a different methodology consisting on a web server that has the advantage of interacting directly with the devices used in the control systems lab through a fieldbus network. The main point of implementing real-time interaction with physical systems through the Internet is to enable the perception at distance of their real dynamics including nonlinearities and noise without need of code development by the user. Therefore, two integrated features need to be achieved. The first includes a graphical representation of current states of the physical system. The second is concerned with the synchronization of that representation with the actual evolution of the real system for animation purposes despite variations in the Internet bandwidth.

This paper presents an architecture for reengineering existing control systems labs allowing web connectivity, based on well known and widely spread software tools applied for embedding an Internet agent. Thus, the paper has the following structure. Section 2 provides a description of the existing control systems lab. In section 3 is described the methodology followed to achieve the web connectivity allowing students to perform real-time experiments considering the Internet constraints. In section 4 some concluding remarks are presented.

2. CONTROL SYSTEMS LAB ARCHITECTURE

The Process Control System (PCS) existing in the lab has been developed for initial vocational training, advanced training and retraining in the field of automatic control and communications. Some experiments on instruments calibration can be performed too. The Process Control System in the Control Systems Lab contains four stations, resembling the following control loops:

- Level Control;

- Flow Control;
- Pressure Control;
- Temperature Control.

Each station is connected to a field network. The networking of the complete PCS is realised via a 'Profibus Fieldbus Messenger Specification Dynamic Data Exchange Server' or Profibus FMS DDE Server. The stations are connected to the network via a Profibus adapter board. The Profibus FMS DDE Server supports the management of the information and data exchange by the high speed DDE mechanism through 'InTouch'. InTouch is the process visualisation software that manages the required input and output signals of the complete system. The processes can be monitored and manipulated with the graphical user interfaces of InTouch. Fig. 1 illustrates the framing of the PCS.

It is possible to combine the stations in various ways. Also, the stations can be individually converted, thus providing the possibility to adapt a station to a particular approach to problem solving and project work. Several configurations could be chosen remotely. The stations are provided with components, which are generally employed in the industry, creating a realistic training environment.

The Level Control station consists of the following main components: an analogue ultrasonic level sensor; two capacitive proximity sensors used to detect the maximum and minimum levels in a tank; several manual valves allowing different system configurations; a pump coupled with a motor allowing speed regulation; two tanks; and a control panel including a industrial digital PID controller and an analogue terminal allowing the connection of external sensors to the PID controller.

The flow Control station can be operated in two different modes:

- Flow control by means of a proportional control valve where a pump is running at constant speed;
- Flow control by means of a pump coupled with a motor allowing speed regulation.

Therefore, the main components of the Flow Control station are the following: a flow sensor – the fluid is guided into a circular motion via a swirl plate in the measuring chamber and directed onto a lightweight triple vane rotor and, then, the speed of the rotor is measured through a built-in optoelectronic infrared system being proportional to the flow rate; a proportional control valve that is a electrical actuated 2/2-way valve with a reset spring and, then, the valve piston is lifted of its seat as a function of the solenoid coil current and it will be closed after the solenoid coil has been de-energised; a tank; two capacitive proximity sensors used for the same purposes mentioned above; several manual valves allowing different system configurations; and a control panel equal to the one described to the Level Control bench.

The Pressure Control bench can also be operated in two different modes as follows:

- Pressure control by means of a proportional control valve where a pump runs with a constant speed;
- Flow control by means of a pump coupled with a motor allowing speed regulation.

Then, the main components of the Pressure Control station are as follows: a piezoresistive analogue pressure sensor with a built-in amplifier and temperature compensator allowing a measuring range between 0 and 100 mbar; a welded steel pressure vessel allowing pressures in the range -0.95 and 16 bar; a tank; several manual valves allowing different system configurations; a pump coupled with a motor allowing speed regulation; and a control panel equal to the ones quoted above.

The Temperature Control station provides the control of the water temperature in a tank allowing a maximum operating temperature of approximately 65 °C. A pump is used to constantly circulate the water in order to achieve a uniform water warming. Therefore, in the Temperature Control station the following components are used: a PT100 temperature sensor having a measuring range between -50 and 150 °C; a stainless steel electric heating unit; and all the remain components are equal to the ones mentioned to the other stations, such as the pump, manual valves and the control panel.

In the next section, it is described how such a control system lab has been reengineering in order to allow remote experiments performed remotely through the Internet.

3. RE-ENGINEERING AN EMBEDDED CONTROL SYSTEMS LAB

Remote education and real-time interaction over the Internet, with the tremendous potential in terms of time and space flexibility, has much to offer to the experimental research and control engineering education. This plays an important role and is special true in fields like the control engineering profession where rapid advancement in technology and innovation need more efficient and flexible means of accumulating latest know-how and updated information (Tan, 1997).

The Internet through the World Wide Web (WWW) provides users with a systematic and convenient means of accessing the wide variety of resources such as pictures, text, data, video, etc. Nowadays, popular software interfaces such as Internet Explorer (IE) and Netscape, facilitate the interaction and use of the WWW (Ertugrul, 1997).

Furthermore, nowadays trend in education (Poindexter and Heck, 1999) and, particularly, in control engineering field (Paproch, 1998), is to incorporate Information Technology tools in the learning process. This is due to the increasing amount of available computers connected to the Internet and decreasing effective teaching hours at University and Polytechnics in new undergraduate programmes. An increase in time and space flexibility of the educational process can be observed when Internet resources are used. Clearly, distance or remote education takes more advantage from new Information Technologies than standard education. However, the traditional approach can make use of Information Technologies as an effective support (Paproch, 1998). Based on Internet, two different options can be found for the laboratories developing aiming to allow distance or teleoperated experiments: virtual laboratories and remote laboratories. A virtual laboratory allows, for example, continuous access to a simulated process in a computer. Halfway between traditional and virtual laboratories are the remote laboratories that can offer real time experiments to remote users. Most equipment needed for setting up a remote laboratory is available in traditional laboratories and only an interface between the local application and a Web server is the additional part to be developed.

The most important aspect of implementing a remote lab is the real-time interaction over the Internet with the physical system enabling the perception at distance of their dynamics. Providing a graphical representation of the physical system is trivial using the available Internet technologies. However, ensuring that these representations are adequately synchronized with the actual evolution of the real system requires dedicated solutions. Furthermore, another issue is the need for the remote operators to get feedback on the actions carried out as quickly as possible. This additional constraint also requires dedicated solutions to avoid erroneous actions due to misperception of the actual system state.

Generally speaking, the implementation of a remote control systems lab will be concerned with the connection of an embedded system to the Internet for the purposes of monitoring and management of the devices used in such a lab. Embedded systems are systems implementing specific functions, such as printers, household devices, communication devices, avionics, medical instruments, and process control systems. The connection of embedded systems to the network enables the replacement of traditional and proprietary user interfaces to standard Internet interfaces, such as the use of familiar Web browsers from any desktop. However, due to hardware and software limitations or legacy compatibility issues, it is generally difficult to reengineer the original embedded product such a control systems lab to add network connectivity. The most popular way to overcome those problems is to implement an interface to translate network requests into direct product access and vice-versa. This interface is called proxy agent because it works like a proxy between clients' requests and responses from Internet and the control systems lab.

There are two ways to implement proxy agents: a computer gateway or an embedded proxy agent. The computer gateway is a desktop computer connected to the embedded product that receives the requests from the Internet clients and translates the requests directly to the product. The communication between the computer gateway and the embedded product is implemented by traditional embedded system access, like serial interface or a Fieldbus network such as PROFIBUS.

Embedded proxy agents implement the same functionality as computer gateways in an integrated way. However, the main disadvantage of an embedded proxy agent is the need of specific hardware and non-standard available software packages. Therefore, the first approach to implement a proxy agent has been followed in the current approach.

Thus, in the current approach a computer gateway has been implemented to achieve a remote control systems lab. All the devices used in the control systems lab, such as sensors, actuators, controllers and the computer gateway, are connected to a PROFIBUS network as previously mentioned. The current design has followed a three steps procedure:

- First a web server has been installed, which will serve as the gateway mentioned earlier.
- Second, a DDE connection has been established between a database on the computer gateway and the PROFIBUS network responsible for the connection with all the devices used in the lab;
- Third, a communication channel between the database and the Internet has been developed and implemented.

In the first step the web server was installed and configured in a computer that is connected to the Profibus network and is using the PROFIBUS FMS DDE Server. The web server chosen was the one present in the Internet Information Server 5.0 (IIS), fundamentally because of three aspects:

- Compatibility with the operating system used by the DDE Server.
- Very easy to use and to configure.
- Easy to program and to interact with other applications of the operating system.

The IIS as also some other servers that can be very useful later: an electronic mail server and a file transport protocol (FTP) server. These two capabilities can be used to send some electronic mail (mail server) and exchange files between the server and users (FTP server). Later on this two aspects will be discussed.

In order to perform the second step a DDE connection with a database has been developed. Thus, with knowledge of the DDE principle, the PROFIBUS FMS DDE Server and the basics of Visual Basic, a connection between the PROFIBUS FMS DDE Server and a Microsoft Access database has been implemented. The choice of MS Access was based on its obvious relationship with the Windows operating system and the massive usage of this kind of databases.

The aim of such a connection was to retrieve the DDE output of the devices used in the lab in a table of the database. Once this is accomplished, the information of the table can be requested in an Active Server Page (ASP).

However, the main point is to create a link that dynamically updates the table in the database. Thus, to create such a link a Visual Basic Application (VBAP) that connects and manages the DDE outputs with the table has been developed. Hence, each time a DDE output is activated the VBAP sends the new information to the database. The database contains a number of sufficient tables to keep all the information received. Let's now consider for instance the situation where the system has to deal with two measurement variables: ActualValue and SetpointValue. Thus, with the activation of a specific event (the arrival of these variables to the DEE Server) of the VBAP, a sub procedure is activated and will be responsible for the following mechanisms:

- In the first part of the event procedure, the variables are declared, followed by an error handling;
- A link is created between the VBAP and the database.
- The value of the variable will be assigned to the field of the corresponding table, and evaluated by comparison statements.
- When the information is processed the link to the database is closed.

A similar procedure is followed to the other process measurement variables. Of course, that the table where the data is stored will grows with the addition of new records. Thus, in order to avoid memory problems the table should be cleared time to time. At the current stage, the attempt to create a DDE connection between the PROFIBUS Server and a database was successful and the second step is actually completed.

The third step is to create means of communication between the database on the computer gateway and the Internet. Communications with Internet involves building web pages, which involves HTML scripting and Program language scripting. Thus, in order to perform such a task it has been chosen the Microsoft Visual InterDev as a development environment, which is based on the implementation of Active Server Pages (ASP) facilities. ASP is an object model for Microsoft Internet Information Severs that is compatible with web servers. ASP technology provides the ability to dynamically generate browser-neutral contents using server-side scripting. The code for this scripting can be written in any of several languages and is embedded in special tags inside the HTML coding for page content. ASP is the answer to the possible complex problems posed by the development of Common Gateway Interface (CGI). The CGI provides a mechanism by which a web browser can communicate a request for the execution of a specific application on the web server. The result of such an application is converted into a browser readable form (HTML) and sent to the requesting browser. CGI is the basis for information processing on the web.

In order to display the retrieved data from the devices in the control systems lab into the Internet environment, a connection between the database and an ASP page, has been established. To achieve such a task, a 'Default.asp" page has been designed. That page is an active server page containing server objects that can be used to create pages that can read and write to a database. The "Default.asp" page should contain the following objects:

- An Activex Data Object – To open a connection to a existing database.
- A Recordset Object – To navigate in a chosen table of the database.

An Activex Data Object is an object that permits a data connection. Thus is possible, with this object, to specify a database and display the corresponding data on the Internet page. The connections created like this can be stored in a special file, "global.asa", which manages all the events of the Internet Application, and can be used to change some parameters of the Internet Information Service. The Recordset Object creates a "navigation system" in a chosen table of the database, making possible to: navigate in the records of the table; delete records; edit records; add new records.

When all the connections between the ASP page and the database are completed, is necessary a way to dynamically update the Internet page with all the most recent values of the process. One-way of solving the problem is creating an open loop in the ASP. This ASP must have a connection to the database that is being updated. This open loop will create a bridge between the database and the Internet user, making possible the continuous reading of the database new values by the ASP page and the continuous updating of these values in the Internet user environment. Using this strategy it is possible to have on-line all the recent values of the process, independently of the speed of the Internet connection. When the connections are complete it is available an Internet Laboratory that can be easily used by students with different kinds of knowledge.

Making the system more intuitive and easier to use is one of the main tasks as well. Some work has been done in this area and some tools have been developed:

- Graphical tools that can give to the user a much easier understanding of the process dynamics.
- Using the possibilities of the IIS 5.0, an FTP server and a mail server were created.
- Web Cameras are available through the Laboratory.

The graphical tools can be developed using tools like Direct Animation. Some simple animations were developed using these tools, making possible to have a correct perception of the process dynamics. Using on-line animated charts the user can have a real feeling of the process behaviour.

The Internet user has at his disposition a FTP server. It's possible to download all the tests done by a certain user. Thus, a student can collect data, build a model, build a controller and then download a file with all the results. This file can be in MatLab format or in simple text format. Another way to receive the files is by email. Using the mail server the files are sent to the users' email. Therefore, the information acquired during the user registration task will be very useful to allow the information exchange between the system and the user.

Furthermore, when the Internet Lab is remotely accessed the users don't know what is really going on in the lab. Thus, the use of web cameras enables the users to have available real time images of the lab allowing establishing an easier relationship between the data that has been received and the experiments that are currently running. This is fundamental to an Internet user that is far from the real process.

Fig. 2 depicts the interface achieved when the lab is accessed remotely through the Internet. From such a figure, it can be seen a graphical tool where it is shown a dynamic chart represent the results achieved during a process identification task; the possibility of using a FTP server and the mail server to download and send results; images from one of the web cams focused on the bench working at the moment; and finally, there is a field with the user details consisting of his name and email allowing future contacts with him.

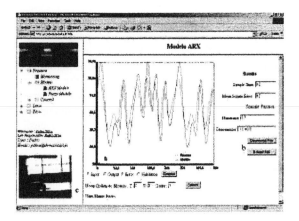

Fig 2. The Web Interface.

4. CONCLUSIONS

An existing embedded control systems lab has been successful reengineering allowing real-time experiments to be performed remotely through the Internet. A proxy agent based on a computer gateway has been developed and implemented. Such a computer gateway, which is connected to the embedded lab through a PROFIBUS FMS DDE server, receives the requests from the Internet clients and translates them directly to the devices used in the lab. Active Server Pages technology has been used to develop and implement a Common Gateway Interface allowing the communication between any Internet client and the web server. It has been demonstrated that most equipment needed to implement a remote lab is available in the traditional labs and only an interface between the local physical system and a web server is the additional part to be developed. In the current approach such a task has been achieved by using well known and widely spread software tools. Furthermore, as far as experimental research and control engineering are concerned, it has been demonstrated that Information Technology tools could provide a tremendous potential in terms of time and space flexibility.

On the other hand, the methodology used to reengineering an existing control system lab could be used in the modern industrial plants allowing process teleoperation or at least remote process supervising with all the inherent advantages.

It has been observed that the most important aspect of implementing a remote lab is the real-time interaction over the internet with the physical system enabling the perception at distance of their dynamics. Therefore, further research is needed in order to find solutions to keep a seamless degree of interactivity despite the Internet bandwidth variations. Other open issues to the connection of embedded systems to the Internet include the development of automatic tools for the generation of embedded proxy agents, and the inclusion of security mechanisms compatible with the CPU power and memory available to these systems, since any Internet-based device is vulnerable to clandestine invasion.

ACNKNOWLEDGEMENT

Work partially supported by The British Council under the Treaty of Windsor Programme and FCT under program POCTI – U46-L3 subsidized by FEDER.

REFERENCES

Åström, K.J. (1994). The future of control. *Modelling, Identification and Control*, **15**, No. 3, pp. 127-134.

Bencomo, S.D. (2002). Control Learning: Present and Future. *Preprints of the 15th IFAC World Congress on Automatic Control*, July, Barcelona, Spain.

Ertugrul, N. (1997). *Towards virtual laboratories: a survey of LabView-based teaching/learning tools and future trends*. University of Adelaide, Department of Electrical and Electronic Engineering, Adelaide, Australia, 5005.

Horacek, P. (2000). Laboratory experiments for control theory courses: A survey. *Annual Reviews in Control*, **24**, pp. 151-162.

Overstreet, J.W. and A. Tzes (1999). An Internet based real-time control engineering laboratory. *IEEE Control Systems Magazine*, **19**, No. 5, pp. 19-34.

Paproch, K., (1998). *Distance Learning: The Ultimate Guide*. Sage Publications, London.

Poindexter, S.E. and B.S. Heck, (1999). Using the Web in your Courses: What can you do? What should you do?. *IEEE Control Systems Magazine*, **19**, No. 1, pp. 83-92.

Röhrig, C. and A. Jochheim (2000). Java-based framework for remote access to laboratory experiments. *Preprints of the IFAC/IEEE Symposium on Advances in Control Education*, Gold Cost, Australia.

Sánchez, J., F. Morilla and S. Dormido (2001). Teleoperation on an inverted pendulum through the world wide web. *IFAC Workshop Internet Based Education*, Madrid, Spain.

Shaheen, M., K.A. Loparo and M.R. Buchner (1988). Remote laboratory experimentation. *Proc. of the ACC*, pp. 1326-1329, Philadelphia, United States.

Sheng, W., L. Choo-Min and L. Khiang-Wee (2000). An integrated Internet based control laboratory. *Preprints of the IFAC/IEEE Symposium on Advances in Control Education*, Gold Cost, Australia.

Tan, K.K., T.H. Lee and F.M. Leu. (1997). *Development of a distant laboratory using Labview*. Department of Electrical Engineering, National University of Singapore, 10, Kent Ridge Crescent, Singapore 119260.

ELSEVIER

IFAC

PUBLICATIONS
www.elsevier.com/locate/ifac

INTEGRATING A WEB-BASED LABORATORY INTO A REUSABILTY-ORIENTED FRAMEWORK

R. Pastor*, J. Sánchez*, S. Dormido*, C. Salzman**, D. Gillet**

* *Dpto. de Informática y Automática, UNED, Avda. Senda del Rey n° 9, 28040 Madrid. Spain.*
E-mail: {rpastor, jsanchez,sdormido}@dia.uned.es

***Swiss Federal Institute of Technology in Lausanne, ME Ecublens, 1015 Lausanne. Switzerland.*
E-mail: {christophe.salzmann,denis.gillet}@epfl.ch

Abstract: *RELATED* is a XML-based framework for the development, integration and reusing of Internet-based laboratories. In this approach, a XML definition file is used to connect a set of laboratory elements (servo motor, PID controller, experiments) and conduct the remote system access. A full example of integrating and reusing of a web-based laboratory and its remote on-line control is presented in this paper. *Copyright © 2003 IFAC.*

Keywords: Remote operations, PID control, virtual laboratories, distance learning, Web-based experimentation.

1. INTRODUCTION

At the moment, the practical education (Dormido, 2002) in control topics faces different problems that can be summarized in two ones: no room for didactical setups and lack of financial resources. To solve these deficiencies, there are many works focused to the development of virtual (simulation-based) and remote (real plants) laboratories conducted through Internet (Schmid, 2000; Gillet, *et al.*, 2001a; Gillet, *et al.*, 2001b; Ling, *et al.*, 2000; Ko, *et al.*, 2001; Salzmann, *et al.*, 2001; Sánchez, *et al.*, 2002).

But all these remote and virtual laboratories are punctual efforts of different research groups. The use of the software and hardware of different universities is not contemplated in any case, not taking advantage of the work previously carried out by others. That is to say, until now, a methodology or a standard for the construction of networks of virtual/remote laboratories based on previous developments does not exist.

In order to create these networks of virtual/remote labs (Figure 1), new tools and languages are needed for the definition and integration of the elements of different control labs (plants, controllers, user interfaces, experiments, access permissions, etc). Once the components are declared by means of these new definition languages, the tools carry out the integration and release in the WWW of a new virtual/remote laboratory with independence of the component location. These components (plants, control code, models, etc.) could belong to other previous developments but the user experimentation interface will hide this point.

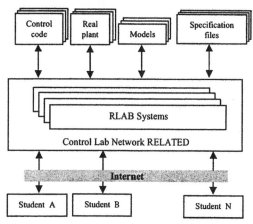

Fig. 1. Architecture of the control lab network.

2. WHAT IS RELATED?

RELATED (**RE**mote **L**aboratory **Ext**ended) is a XML-based framework for the development of new paradigms on Internet-based laboratories. **RELATED** has been defined for the teleoperation of academic control systems as simulation problems or real time plants (pilot or didactical setups). The main idea is to define an abstract entity called *RLAB System* by a **XML DTD** and to publish it on a *RLAB Control Web Server* in order to let students the remote access. Therefore, it will be possible to work with any RLAB System plugged to the Web Server and to share different resources as plants, simulation engines, or, simply, control code. The main RELATED features are:

1. Internet access via a RLAB Control Web Server to the services provided for different RLAB Systems.

2. Teleoperation for any RLAB System defined on the Control Systems Network. Using digital certificates and signatures, which are generated for each RLAB System user, will provide security.

3. Remote system specification via XML language. Different XML files define different types of systems on different computers or platforms.

4. Platform independency and portability. The RELATED framework is fully developed in Java language. According to the Java developer claim "write once and run everywhere", the RLAB System could run in any Java platform as Wintel boxes, Unix/Linux systems, or even Java Cards.

5. Local code reusability. Any control code developed for the local operation of a RLAB System can adapt for the implementation of other new RLAB Systems.

6. Control code repository. The code of other remote RLAB Systems can be used to build other remote RLAB System.

7. Open architecture. Changes are easily done in the system. For example, the addition of new system features just consists in *"to add new XML lines to the definition files"*.

3. SYSTEM ARCHITECTURE

As shown in Figure 2, RELATED system architecture is quite simple. There is several RLAB Systems (formally *RLAB Component Server*)

connected to the *RLAB Control Web Server* that acts like an information system. First, it allows users to connect to a particular RLAB System, i.e., it works like a reference server (connections are possible only with sufficient permission). Also this main web server can point to the RLAB systems web pages of other associated HTTP servers, i.e., it works too like a publication system. In this sense, explanatory text about local system behavior can publish, as available hands-on assignments or information pages. Hence, users conduct the RLAB Systems using an Internet navigator in order to browse the RLAB Control Web Server pages and find the RLAB System to manipulate.

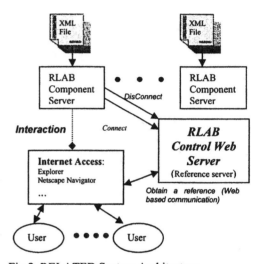

Fig.2. RELATED System Architecture.

4. BASIC XML DEFINITIONS

XML is used in the definition of each RLAB System. All the objects included in a RLAB system are "written" using an XML file in terms of tags and attributes. These elements represent formally the different laboratory elements that can be present in a virtual/remote setup. So a **RLAB System** is a set of different objects (later on they are explained): variables, parameters, modules, experiments, views, and references. It is a black box that can be manipulated across their variables as it happens in block-oriented applications (e.g., Simulink or LabVIEW). The **<system>** XML tag defines a system and **only** one tag is allowed on a definition file.

A **module** is an entity with variables and parameters, and it gives access to the user to get and set values. This entity runs code inside a local thread while a user is conducting a remote experiment. At present **RELATED** supports two types of implementation: Native code and Java code.

The development of a module implementation is the biggest effort that a RLAB System administrator must do, but many advantages are clearly got. The **<module>** XML tag defines a module and the **<implementation>** tag is used to indicate what code will be run.

A **variable** is an object that stores a value (*double, float, integer, long, boolean,* or *string*) that can be changed while the system is running an experiment. A variable is always associated with a module since the module implements the **get** and **set** functions to access to these variables. A variable is defined by the **<var>** XML tag.

A **parameter** is like a variable but it can not be changed while the system is running an experiment. Parameters are useful to initialize the internal variables of a module implementation.

A **view** defines a GUI form composed of graphical components and multimedia capabilities (video, animation, sound). These components allow users to view and manipulate the remote data of a RLAB System. The **<view>** XML tag defines a view and the syntax looks like a module definition.

A **reference** is a URL direction that allows to publish information using HTML pages. Any information about local RLAB Systems can add via references. The **<reference>** XML tag defines this object.

An **experiment** describes the dynamic behavior of a RLAB System, as for example, a controlled system, the manual operation, etc. An experiment is built describing what modules are going to be used and how the module variables can be setted and getted from other remote or local modules and also if these variables can be changed by the student (i.e., they are interactive variables). Of course, the experiment duration must be indicated using the **<duration>** tag.

A RLAB System could include so many experiments as the system administrator adds to the definition file. Another possibility consists in allowing users to add their own experiments to the definition file.

5. A CASE STUDY: EASY IMPLEMENTATION OF A REMOTE LAB

5.1 The Servo-Motor Setup

A remote lab for conducting experiences with a servo-motor has been developed by the Swiss Federal Institute of Technology in Lausanne (EPFL) team. Some of the main features of this lab are the following:

- Control of the velocity or the position with a PID controller.
- Different input signals (square, step, zero, triangle, random).
- Automatic and manual control.

Up to now, there are two approaches to teleoperate the servo-motor: a LabVIEW client (Figure 3) and a Java-based client (Gillet *et al.*, 2002). In the next paragraphs, the RELATED framework is used to reuse and integrate the Java code developed in (Gillet *et al.*, 2002), publishing a new lab and endowing it with different experiments and views.

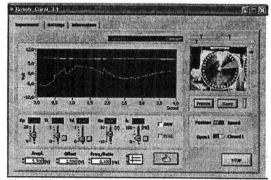

Fig. 3. The LabVIEW client.

5.2 RELATED XML Definitions.

As a first step, the different variables and parameters necesary for the remote conducting of experiments must be classified. In this example the name of the variables for ruling the didactical setup are:
- SPEED. It represents the speed measurement.
- POSITION. It represents the position.
- KP, TI, and TD. Proportional gain, integral time, and derivative time of the built-in PID controller.
- SETPOINT. It is used in the closed-loop operation to set either the speed or the position control.
- SPEED_OR_POSITION. Type of control.

- CLOSED_LOOP. Boolean variable to switch to automatic or manual mode.
- GENERATOR_FREQUENCY, GENERATOR_AMPLITUDE, GENERATOR_SYGNALTYPE. Three variables to set the excitation signal.

Once the variables set is defined, a software module must be developed. In this case, the programming task is short since the Java code of the eMersion project is reused (Nguyen, *et al.*, 2002). Briefly, this task just adapts the existing code to the RELATED guidelines. This adaptation consists of endowing the code with an interface to access to the values of the variables. More details about the guidelines to develop software modules can be found in (Pastor, *et al.*, 2001) and (Pastor, *et al.*, 2002).

After that, the XML specification of the software module is created using RELATED tags. This specification can be appreciated in Figure 4. The name of the module is "SERVO module" and the variables are defined by **<var>** tags. The **<implementation>** tag for setting the type, the name, and the location of the module is located at the end of the declaration. In this case, the module is a Jar file since the software source is Java code.

```
<module name="SERVO module">
  Manual operation module
  <param name="ExecutionTime" type="long" value="100">Thread time</param>
  <param name="MOTOR_NUMBER" type="int" value="3">Number of Motor used</param>
  <var name="SPEED" type="double" vector="yes" initial="0" max="10" min="0" units="V">Speed of motor</var>
  <var name="POSITION" type="double" vector="yes" initial="0" max="10" min="-10" units="V">Position of motor</var>
  <var name="ACTUATOR" type="double" vector="yes" initial="0" max="10" min="0" units="V">Speed of motor</var>
  <var name="SETPOINT_VECTOR" type="double" vector="yes" initial="0" max="10" min="-10" units="V">Vector for position or
  speed setpoint for servo motor</var>
  <var name="SETPOINT" type="double" initial="0" max="10" min="-10" units="V">Position or speed setpoint for servo
  motor</var>
  <var name="MOTOR_IMAGE" type="rawbytes" initial="0" max="0" min="0" units="N/A">Image of motor</var>
  <var name="KP" type="double" initial="1.0" max="2" min="0" units="%">Proportional gain</var>
  <var name="TI" type="double" initial="0.0" max="2" min="0" units="seconds">Integral time</var>
  <var name="TD" type="double" initial="0.0" max="1" min="0" units="N/A">Derivative time</var>
  <var name="SPEED_OR_POSITION" type="boolean" initial="false" max="1" min="0" units="N/A">Type of control</var>
  <var name="CLOSED_LOOP" type="boolean" initial="true" max="1" min="0" units="N/A">Closed or Opened loop for
  control</var>
  <var name="GENERATOR_FREQUENCY" type="double" initial="0.3" max="10" min="0" units="Hz">Frequency of signal
  generator</var>
  <var name="GENERATOR_AMPLITUDE" type="double" initial="0" max="10" min="0" units="%">Amplitud of signal
  generated</var>
  <var name="GENERATOR_SIGNALTYPE" type="int" initial="1" max="4" min="0" units="N/A">Reference signal: 0-zero,1-
  manual,2-Steps,3-ramps,4-Sin</var>
  <implementation type="JAVA" jarfile="file:///home/rafa/RELATED/epfl/Module/servomodule.jar"
  classname="servo">Access to servo</implementation>
</module>
```

Fig. 4. RELATED module for servo motor remote access.

The next step is to develop graphical views to help the students to understand what they are doing with the motor. In this case study an unique view is developed since the manipulation of variables (get and set actions) is done by the RELATED applet features. The motor image is shown in this view, thus a visual feedback is presented to the student.

Again, the Java code is reused to implement the "Image panel" view (Figure 5). The RELATED XML specification of the view uses the **<view>** tag to declare the properties of the implementation:

code location, file name, accesible variables. In this example, the MOTOR_IMAGE variable of the "SERVO module" (Figure 4) is used with the internal name "image".

```
<?xml version="1.0" encoding="utf-8" ?>
- <system name="EPFL_SERVOMOTOR" type="0">
  EPFL Servo Motor Remote System Definition
  + <module name="SERVO module">
  - <view name="Image Panel" jarfile="file:///home/rafa/RELATED/epfl/View/imageview.jar" classname="JpegImagePanel">
    JPEG Image Panel view
    <use name="MOTOR_IMAGE" module="SERVO module" as="image" />
  </view>
  + <experiment name="Manual operation" sampleTime="100">
  + <experiment name="Open loop velocity" sampleTime="100">
  + <experiment name="Open loop position" sampleTime="100">
  + <experiment name="Closed loop velocity" sampleTime="100">
  + <experiment name="Closed loop position" sampleTime="100">
  </system>
```

Fig. 5.- XML definition of the "Image panel" view.

Finally some experiments are defined. For example the "Manual Operation" experiment definition (Figure 6) lets users interact with all the variables of the "SERVO module" and open the "Image Panel" view during an unspecified experimentation time.

```
<experiment name="Manual operation" sampleTime="100">
  Manual Operation
  <duration type="User">Infinite time</duration>
  <run module="SERVO module">This module is necessary</run>
  <interactives
    names="KP,TI,TD,SETPOINT,SPEED_OR_POSITION,CLOSED_LOOP,GENERATOR_FREQUENCY,
    show="true,true,true,true,true,true,true,true,true" />
  <open view="Image Panel" UpdateTime="1000">You'll be able to open this view</open>
</experiment>
```

Fig. 6. "Manual operation" experiment.

5.3 Running Experiments.

Once the XML definition file is completed, the Component Server (RLAB Cserver) loads the definition file, releasing all the features to the students. From this moment, the didactical setup can be accesed from the RELATED applet (Figure 7). Thus a student can run/stop the experiments and open the "Image panel" to view the real actions he/she is doing with the remote laboratory. For example, a student could start the "Manual operation" experiment and watch in the "Image panel" (Figure 8) what is going on.

Normally, the control actions a student can do are changes of the interactive variables values previously declared in the XML experiment definitions (see **<interactive names>** tag in Figure 6). These changes are made using a GUI, termed "Interactive Variables panel", that is configurated after reading the tags of the experiment definitions related to interactive variables.

In Figure 9 the "Manual operation" experiment, defined in Figure 6, is running and its associated "Interactive Variables panel" is presented to the user.

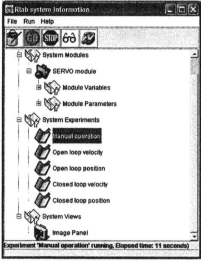

Fig. 7. Main windows of the RELATED applet.

Fig. 8.- Image panel view during the "Manual operation" running.

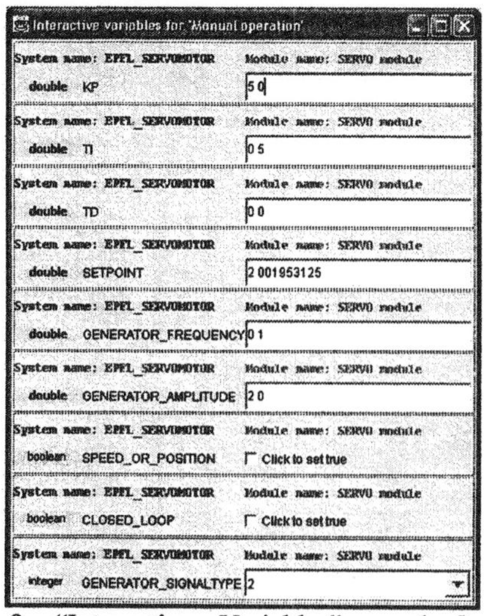

Fig. 9. "Interactive Variables" panel for the "Manual operation" experiment.

Also, the RELATED applet gives user the posibbility of selecting the most important data (Figure 10) to show the temporal evolution in different scopes.

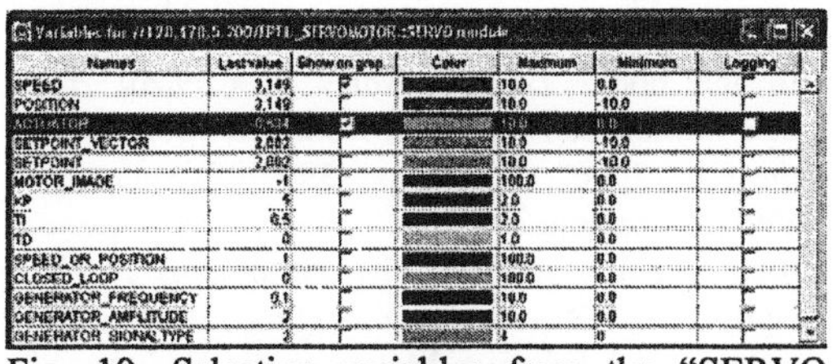

Fig. 10. Selecting variables from the "SERVO module" for the painting window.

6. CONCLUSIONS

The RELATED framework is being designed as a new generation tool oriented to the construction of educational control lab networks. Three words can resume the main features: reusing, simplicity and flexibility.

In this paper, a case study shows how the behavior of a real remote laboratory can be represented using the basic definitions that RELATED provides. Also it has been demonstrated that it is possible to reuse common code in order to get a fast and easy integration of different control systems.

With this aim, the behavior of a servo motor has been modeled with XML tags and the RELATED tools has been used for the remote execution of experiments. Also, the most important characteristics of the RELATED applet has been presented here.

ACKNOWLEDGEMENTS

This work has been supported by the Spanish CICYT under grant DPI2001-1012.

REFERENCES

Dormido, S. (2002). Control Learning: Present and Future. Plenary Lecture, *15th Triennial World Congress of IFAC*, Barcelona, Spain, Plenary Papers, Survey Papers and Milestones pp 81-103

Gillet, D., H.A. Latchman, Ch. Salzmann, and O.D. Crisalle (2001a). Hands-On Laboratory Experiments in Flexible and Distance Learning.

Journal of Engineering Education, April, pp.187-191.

Gillet, D., Ch. Salzmann, and P. Huguenin (2001b). A Distributed Architecture for Teleoperation over the Internet with Application to the Remote Control of an Inverted Pendulum. In the book: *Lecture Notes in Control and Information Sciences 258: Nonlinear Control in the year 2000*", Vol. 1, pp. 399-407, Springer-Verlag, London, 2001.

Gillet, D., F. Geoffroy, K. Zeramdini, A. V. Nguyen, Y. Rekik and Y. Piguet (2002). The copkpit: An Efective Metaphor for Web-based Experimentation in Engineering Education. *International Journal of Engineering Education "Special Issue on the Remote Access/Distance Learning Laboratories"*, Publication in August 2003.

Ko, C.C., B.M. Chen, J. Chen, Y. Zhuang, and K. C. Tan (2001). Development of a Web-based laboratory for control experiments on a coupled tank apparatus. *IEEE Transactions on Education*, **44**, no. 1, pp. 76-86.

Ling, K., Y. Lai, and K. Chew (2000). An online Internet laboratory for control experiments. In: *IFAC/IEEE Symposium on Advances in Control Education*, Gold Coast (Australia).

Nguyen A. V., Y. Rekik and D. Gillet (2002). Integrated Environment for Web-based Experimentation in Engineering Education. In: *ED-MEDIA 2002*, Denver, Colorado (USA).

Pastor, R., J. Sánchez, and S. Dormido (2001). RELATED: A Framework to Publish Web-based Laboratory Control Systems. In: *IFAC Workshop on Internet Based Control Education*, Madrid (Spain).

Pastor, R., J. Sánchez, and S. Dormido (2002). A Xml-based Framework for the development of web-based laboratories focused on control systems education. *International Journal of Engineering Education "Special Issue on the Remote Access/Distance Learning Laboratories"*, Publication in August 2003.

Salzmann Ch., D. Gillet, and P. Huguenin (2000). Introduction to Real-Time Control using LabVIEW™ with an Application to Distance Learning. In: *International Journal of Engineering Education* "Special Issue: LabVIEW Applications in Engineering Education", Vol. 16, No. 3, pp. 255-272, 2000.

Sánchez, J., F. Morilla, S. Dormido, J. Aranda, and P. Ruipérez (2002). Virtual Control Lab using Java and Matlab: A Qualitative Approach. *IEEE Control Systems* Magazine, vol. 22, n° 2. pp. 8-20.

Schmid, C. (2000). Remote Experimentation Techniques for Teaching Control Engineering. In: *4th International Scientific - Technical Conference PROCESS CONTROL 2000*, Kouty nad Desnou, Czech.

ELSEVIER

IFAC

PUBLICATIONS
www.elsevier.com/locate/ifac

A WEB BASED LECTURE NOTE ON FEEDBACK CONTROL WITH REMOTE LAB

Yasuhiro Ohyama, Jin-Hua She, Hiroshi Hashimoto,
Tomio Yamaura, Kunio Oishi

School of Engineering, Tokyo University of Technology
1404-1 Katakura, Hachioji, Tokyo 192-0982, Japan
E-mail: {ohyama,she,hasimoto,yamaura,kohishi}@cc.teu.ac.jp

Abstract: This paper describes a course on Feedback Control for beginners and its Web-based lecture notes. First, an outline of the lecture, which concerns a simple arm robot, is presented. Second, Web-based lecture notes on simulations of the control system, animations for the robot, and videos of experiments are discussed. Finally, a remote-controlled arm robot system accessible via the Internet is described. *Copyright © 2003 IFAC*

Keywords: Control Education, Educational aids, Web Based Training, Remote Laboratory

1. INTRODUCTION

This paper describes a set of Web-based lecture notes that are used by students before and after class. A common complaint about lectures on control theory is that they are too theoretical and too mathematical. "Learning by doing" is a much more effective pedagogical approach; and many classes employ demonstrations and videos of experiments. However, the students need to be individually motivated and to do their own laboratory work to acquire a basic grounding in the field. In a large class, both demonstrations and laboratory exercises are very difficult.

Many computer- and web-based training programs are available nowadays (Jochheim,1999; Schmid, *et al.*,2000; Ubell, 2000; Wittenmark, *et al.*, 1998). The authors also have developed a set of web-based lecture notes that include the lecture material for the class, simulations of various control systems, and a remote lab. Students access the notes with a Web browser. One theme running through the course is the control of an arm robot, and the material on the Web site stresses this theme.

This paper describes the course outline and web-based training programs of CONTROL 1, which

is opened to students of mechatronics course in the 5th semester. Finally, a remote lab system is presented.

2. GOAL OF THE COURSE

The main goals of the course CONTROL I are to ensure that students know
1) how to derive the transfer function of a dynamic system;
2) how to analyze the response and stability of a system; and
3) how to analyze the frequency response of a system.

The use not only of textbooks but also of laboratory equipment is very effective in the study and teaching

Photograph 1.   Arm robot and amplifier

of control technique (Ohyama, et al., 1994). In this course, a simple single-arm robot is used as the target control object and the following ideas are explained:

- what a control system is;
- what the basic components of a control system are;
- how to control a system with a PC; and
- how to design the parameters of a controller.

The control of this equipment is demonstrated in class, and Web-based lecture notes provide exercises employing simulations.

The structure of the system is shown in Fig. 1. A small motor (1 Watt) with gears drives a small bar called the arm. The angle of the arm is determined with a hand-made encoder. The amplifier shown in Fig. 2 is an interface circuit between the robot and a PC. An 8-bit signal from the PC drives the motor via a D/A converter with a power amplifier. The encoder signal is counted by a PIC microcontroller, and the data are transferred to the PC. Photograph 1 shows the arm robot and the amplifier.

In a classroom, topics are lectured by using both a black board and a PC. Web-based lecture notes are used before and after class.

3. COURSE OUTLINE

The outline of the course is shown as follows.

3.1 Getting started

In the first stage of the course, a controller is designed by trial and error with the goal of driving the arm to a reference position as quickly as possible.

First, the concept of a block diagram and an empirical ON/OFF type controller (Fig. 3) are explained. The following operations are necessary to achieve the control objective:

- Subtract the output from the reference.
- When the error is positive, move the motor forward (positive voltage).
- When the error is negative, move the motor backward (negative voltage).
- When the error is zero, stop the motor (zero voltage).

This type of control is demonstrated by using a simple manual switch, but the results are always unsatisfactory.

Second, the controller in Fig. 4 is explained. The error, which is obtained by subtracting the output from the reference, drives the input (motor). However, the following statement is incorrect:

Reference [rad.] - Output [rad.] ==> Input [volts].

So, the units have to be changed. The parameter proportional gain, g, is introduced as shown in Fig. 5.

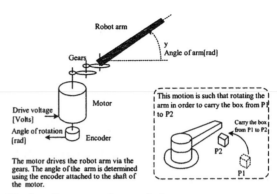

Fig. 1. Structure of arm robot

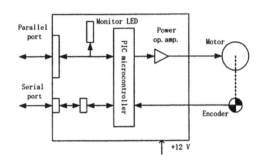

Fig.2. Amplifier

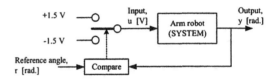

Fig. 3. Block diagram of ON/OFF control

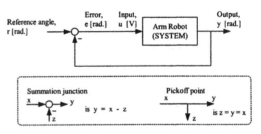

Fig. 4. Proportional control 1

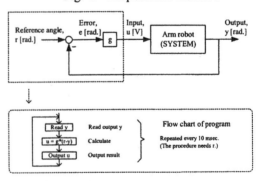

Fig. 5. Proportional control 2 and programming algorithm

The operation in the dotted box is carried out on a PC. The following control algorithm is also taught:

1. Read the reference and the output, and calculate the error;
2. Multiply the error by g; and
3. Drive the motor based on the result.

An experiment using the arm robot and a PC is demonstrated and a response like that in Fig. 6 is obtained. One of the most important concepts for students is the following relations:

small	<==	g	==>	large
long		rise time		short
none		overshoot		large
large		error		small

The last steps in the procedure for designing the controller, which have the purpose of making the angle of the arm track the reference angle, are:

1) Determine the control algorithm shown in Fig. 5 (Structure of the control system)
2) Determine the parameters (Parameter g)
3) Execute control (Carried out by PC)

3.2 Analysis of Dynamics

The second stage of the course concerns the derivations of dynamic equations and transfer functions. A dynamic equation is derived by using Newton's law of dynamics:

$$J\frac{d^2}{dt^2}y(t) + c\frac{d}{dt}y(t) = ku(t) \qquad (1)$$

where J: moment of inertia
        c: coefficient of viscous friction
        k: coefficient of motor torque.

The Laplace transform is given simply as

$$\frac{d}{dt} \Rightarrow s \, 、 \; \frac{d^2}{dt^2} \Rightarrow s^2 \, .$$

Making the substitutions $y(t) \rightarrow Y(s)$, $u(t) \rightarrow U(s)$ converts Eq. (1), in which y and u are functions of t, into the following equation, in which Y and U are functions of s:

$$Js^2Y(s) + csY(s) = kU(s). \qquad (2)$$

Then, the transfer function is obtained:

$$Y(s) = \frac{k}{Js^2 + cs}U(s) = \frac{b}{s^2 + as}U(s) \qquad (3)$$

The transfer function from $R(s)$ to $Y(s)$ is derived as follows:

$$Y(s) = \frac{b}{s^2 + as}U(s) \qquad (4)$$

$$E(s) = R(s) - Y(s) \qquad (5)$$

$$U(s) = gE(s) \qquad (6)$$

$$Y(s) = \frac{bg}{s^2 + as + bg}R(s). \qquad (7)$$

The graphical derivation is also explained.

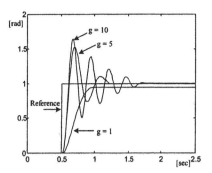

Fig. 6.   Step response

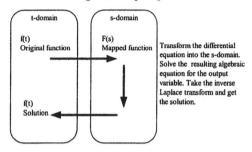

Fig.7. Laplace Transform

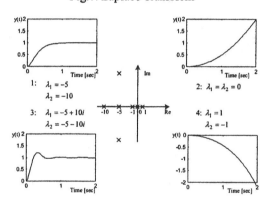

Fig. 8.   Poles and step response

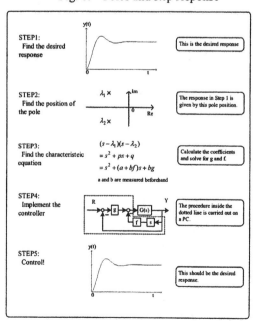

Fig. 9.   Design procedure

3.3 Laplace Transform

In the third stage, the details of the Laplace transform technique illustrated in Fig. 7 are explained. The following problems on solving ordinary linear differential equations are given:

1. Transform a differential equation into the s-domain using a Laplace transform.
2. Solve the resulting algebraic equation for the output variable.
3. Perform a partial-fraction expansion so that the inverse Laplace transform can be obtained from the Laplace transform table.
4. Perform the inverse Laplace transform.

Finally, the relationship between the step response of a 2nd-order system and its pole locations is explained, as in Fig. 8. One control design technique called the pole assignment method is explained, as in Fig. 9.

3.4 Frequency Response

In the fourth and final stage of the course, the concept of frequency response is explained, with a special focus on Bode diagrams. The responses are demonstrated when various reference inputs, $r(t) = \sin(\omega t)$, are applied to the arm robot control system. A typical response is shown in Fig. 10. The students do exercises on drawing Bode diagrams based on hand calculations using data like that in Fig. 10.

4. WEB BASED TRAINING

The Web-based lecture notes are used to support the classroom lectures. In a large class, it is very difficult for all the students to see demonstrations. These lecture notes have the following features.

- The lecture notes are text-based and are synchronized with the class material.
- Students can carry out the simulations explained in class with just a Web browser.
- The operating principle of encoders and D/A converters is shown through animations.
- Videos of some experiments are available for viewing.

MATLAB$^{(TM)}$ is widely used in control engineering to simulate control systems. However, it is too expensive for most students to buy and has too many functions for beginners. In these Web-based lecture notes, students can carry out simulations with just a Web browser (Netscape™ or Internet Explorer™).

Figure 11 shows a note explaining the structure of the arm robot, and Fig. 12 shows the simulated 2nd-order step response of the system. All the algorithms used in the simulations are coded in JavaScript, and

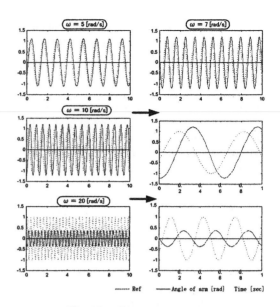

Fig. 10. Frequency response

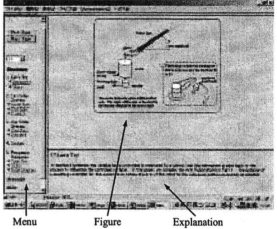

Menu   Figure   Explanation

Fig.11. Client window of the lecture note (text)

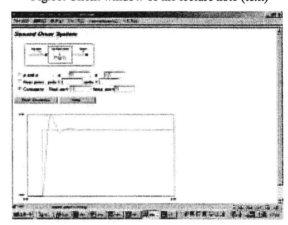

Fig.12 Client window of the lecture note (simulation)

graphics are displayed by means of Java applets.

Figure 13 shows an access record for the Web site. About 200 students took the course. Lectures were held once a week on Monday afternoon. The record shows that most of them accessed the notes on Sunday or Monday. The peaks indicate exercises, tests, and the submission of homework. Most of the students accessed the notes from the computer room at the university, but some of them accessed them from their homes via the Internet.

5. REMOTE LAB

The simulations mentioned above do a fairly good job of giving the students confidence that the theory actually works and is practical. However, since the control results are strongly influenced by disturbances and sensor noise, the accuracy of the measurement instruments, and the unmodeled dynamics of the plant, experiments are indispensable to making students aware of these factors and making them mindful of the gap between theory and practice; and they are also one of the best ways to provide students with a deeper understanding of the theory. In addition, the Web-based lecture notes provide a great deal of information on how an experiment system should be set up and used, and allow us to examine these issues in a practical setting.

Several methods of employing a remote lab over the Internet have been reported (Fitzpatrick, 1999; Johansson, *et al.*, 1999; Junge, *et al.*, 2000; Overstreet, *et al.*, 1999; Ramakrishnan, *et al.*, 2000; Rohrig, *et al.*, 2000; Sanchez, *et al.*, 2000). The authors previously built a prototype Internet experiment system for control engineering (Ohyama, *et al.*, 1999; Ohyama *et al.*, 2001); and the same ideas have been applied to the arm robot equipment for the course. The experiment system was set up in a laboratory and connected to the Internet. Students can run the experiment just by accessing the Web site for the course. The experiment system is used at different stages of the course to help students gain a deeper understanding of the theory.

The structure of the experiment system is shown in Fig. 14 and the inside of the arm robot system is shown in Photograph 2. The client is a personal computer used by a student, and the provider for the course is comprised of three parts: the plant (arm robot), a control machine, and a server. Three programs are executed on these machines to perform an experiment. The client program, which is stored on the server, is run from a Web browser on a client machine to set an experiment up. The server program, which is stored and executed on the server machine, processes commands from clients, supervises the control machine, and sends experimental results back to clients. The control program, which is stored and executed on the control machine according to instructions from the server, runs the plant. The client displays the experiment, which is shown by a camera as Fig.16.

This system will be opened to students in the next term.

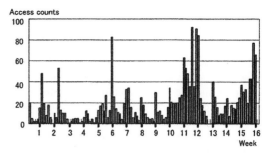

Fig. 13. Access record for the Web site.

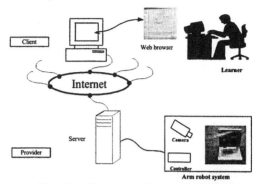

Fig. 14. Structure of remote-lab system

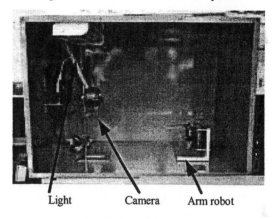

Photograph 2. Inside of the arm robot system (The controller is behind this box.)

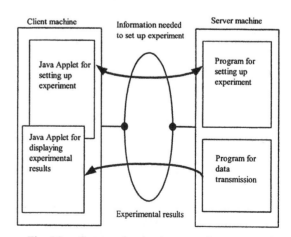

Fig. 15. Communication between the server and a client

6. CONCLUSION

This paper describes a course on Feedback Control for beginners and its Web-based lecture notes. The subject of the classroom lectures is a simple arm robot. The purpose of the course is to provide an understanding of

1) the derivation of the transfer function of a dynamic system;
2) the response and stability of a system; and
3) the frequency response of a system.

The equipment is very simple and students can easily understand its motion. The Web-based lecture notes are also designed to be convenient and helpful to students by providing the following features:

1) Students can access them with just a Web browser.
2) The classroom lecture is posted after each session.
3) Several simulations are available.
4) An actual experiment system can be remotely controlled.

The access record for the Web site showed that most of the students accessed the site before and after a class. Many students pointed out the usefulness of the simulations. The benefits of incorporating a remote lab into the course will not appear until the next term.

REFERENCES

Fitzpatrick, T. (1999). "Live Remote Control of a Robot via the Internet", *IEEE Robotics & Automation Magazine*, Vol. 6, No. 3, pp. 7-8

Jochheim, A. (1999) "Real Systems in the Virtual Lab",
http://prt.fernuni-hagen.de/virtlab/KurzInfo_e.html

Johansson, M., M. Gafvert and K.J. Astrom, (1998). "Interactive Tools for Education in Automatic Control", *IEEE Control Systems*, pp. 33-40

Junge, T.F. and C. Schmid (2000). "WEB-BASED REMOTE EXPERIMENT USING A LABORATORY-SCALE OPTICAL TRACKER", *American Control Conference 2000*,

Ohyama, Y. and J. Ikebe (1994). "Exercise in Control Technique Using Simple Handmade Equipment", *The 3rd Symposium on Advances in Control Education*, pp. 215-218

Ohyama, Y., J.-H. She and K. Watanabe (1999). "A Prototype Internet Experiment System for Control Engineering", *Proc. of 1999 IEEE Int. Conf. on Systems, Man, and Cybernetics*, Vol. 1, pp. 894-898

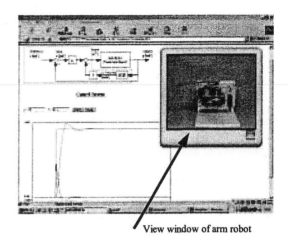

View window of arm robot

Fig.16   A sample of browser display including view window of arm robot

Ohyama, Y., J.-H. She, Y. Yoshizawa and M. Kimura (2001). "Construction of an On-line Education Course Using an Experiment System", *Proc. Int. The 5th World Multiconference on Systemics,Cybernetics and informatics*, Vol. 8, pp. 210-213

Overstreet J. W. and A. Tzes (1999). "An Internet-Based Real-Time Control Engineering Laboratory", *IEEE Control System*, Vol. 19, No. 5, pp. 19-33

Ramakrishnan, V., Y. Zhuang, S. Y. Hu, J. P. Chen, C. C. Ko, Ben M. Chen and K. C. Tan (2000), "Development of a Web-Based Control Experiment for a Coupled Tank Apparatus", *Proc. of the American Control Conf.*, pp. 4409-4413

Rohrig, C. and A. Jochheim (2000). "JAVA-BASED FRAMEWORK FOR REMOTE ACCESS TO LABORATORY EXPERIMENTS", *The 5th Symposium on Advances in Control Education*,

Sanchez, J., F. Morilla, S. Dormido, J. Aranda and P. Ruiperez (2002). "Virtual and Remote Control Labs Using Java:A Qualitative Approach", *IEEE Control Systems Magazine*, April, pp. 8-20

Schmid, C. and A. Ali (2000). "A WEB-BASED SYSTME FOR CONTROL ENGINEERING EDUCATION", *ACC 2000*, pp. 1-5

Ubell, R. (2000). "Engineers turn to e-learning", *IEEE Spectrum*, Vol. 37, No. 10, pp. 59-63

Wittenmark,B., H. Haglund and M. Johansson (1998). "Dynamic Pictures and Interactive Learning", *IEEE Control Systems*, pp. 26-32

eLEARNING FOR BATCH CONTROL

Outi Rask and Seppo Kuikka

Tampere University of Technology, Institute of Automation and Control,
P.O. Box 692, FIN-33101 Tampere, Finland
Email: Outi.Rask@tut.fi, Seppo.Kuikka@tut.fi

Abstract: Learning in the Internet is a new style of distance learning. When planning a web-based learning environment, or eLearning environment, you have to consider its objectives on various organizational levels. For example, developing a learning environment for batch control is needed – in addition to individual and company wise objectives - also because there is no other course in any Finnish university for batch control only. *Copyright © 2003 IFAC*

Keywords: batch control, learning environment, web-based, education

1. INTRODUCTION

Use of the Internet in learning has increased significantly in recent years. As late as in the late eighties, the Internet was fairly unfamiliar and computers and networks were still rarely used on basic education. At present, however, elements of computing are taught already in primary schools and computer networks are utilized in teaching both on lower secondary and vocational as well as on senior secondary and university level education.

The Finnish ministry of education developed in 1999 a project plan "Koulutuksen ja tutkimuksen tietostrategia 2000-2004" (The Finnish strategy for education and research in 2000-2004), (The Finnish ministry of education, 2002). The strategy covers a wide educational area from primary schools to universities and research institutions. It also includes a Finnish initiative for Virtual University. According to its guidelines, Tampere University of Technology

(TUT) has a virtual university of its own, the purpose of which is to further network-based learning within TUT.

Due to its importance and lack of university level education, batch control is a suitable subject for network-based learning. The subject is also suitable for eLearning because cases to be learnt deal with upper level control systems and thus it is possible to use a simulator to represent controls of lower level. Also the simulator is possible to be accessed via the Internet.

The novel standardization efforts and research in the field have achieved results, which are worth delivering to wider audiences than TUT alone. Therefore our intention is to offer batch control education for other Finnish universities and enterprises too. The learning environment for Batch control, reported in this paper, is completely web browser based and thus usable from everywhere, universities and industrial companies alike.

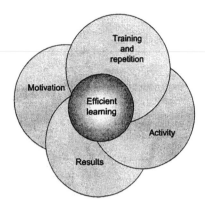

Figure 1. Efficient learning according to (Rask, 2002).

2. THE INTERNET BASED LEARNING ENVIRONMENTS

2.1 On learning in general

Good learning results depend on a student's motivation, activity, thirst for learning and training as well as repetition. When sufficiently good fundamentals are formed, begins orientation. After orientation the student has to internalize the issue. Only issues which are internalized can be used in practice.

Good learning includes evaluating the learning results as an essential part. A student can estimate his (her) own results by himself (herself). Evaluating can also be done by someone else, like the teacher or other students.

Figure 1 presents efficient learning with all its essential constituents. This picture indicates the fact that also training and repetition are very important parts of efficient learning.

Cognitive psychology is one of the key issues when talking about learning. A cognitive function as a concept is very wide. It usually includes observation, remembering, argumentation, reasoning, conception, parlance and thinking. In the learning environment, various cognitive tools are used for conceptualization, analysing and understanding the issues to be learned. (Rask, 2002)

Table 1. The general levels of value added in learning (Kaufman, et. al. , 2001).

Level of Planning	Primary Client and Beneficiary	Name of Results	Defining Statement
Mega	Society	Outcomes	Ideal Vision
Macro	Organization	Outputs	Mission objective
Micro	Individual /Small group	Products	Individual's objectives

2.2 Learning in the Internet

Distance learning offers learning opportunities to people at a time and location that is best suited to them. However, in order to add value for learners and their organizations, the offered opportunities have to be useful in a well defined manner.

One way to define usefulness is to measure, how well the distance learning has achieved its objectives at different organizational levels (Kaufman, et. al., 2001). Table 1 states three levels of learning and respective beneficiaries, results and definitions for success.

When developing an Internet-based learning environment for batch control we have focused mainly on Macro and Micro levels of learning, trying to reach the objectives on both organizational and individual levels. However, the fact that there is no other course in any Finnish university dedicated entirely to batch control, seems to make the learning environment interesting on so called societal Mega level, too.

There are many phases in developing an Internet-based learning environment. There are several models for a course planner to make this work easier. Examples of these models can be found e.g. in the Internet and in (Rask, 2002).

Learning in the Internet proceeds in practice as illustrated in figure 2. On the lower level we learn basic skills like how to use a modem and a browser.

On the second level we get used to the learning environment itself and its services. On that level we also get acquainted with other students.

Figure 2. Learning in the Internet according to (Ojaniemi et al., 1999).

On the third and fourth level our courage is increased and groups are formed. Our goal is on the top of the pyramid. There the using of environment is fluent and flexible.

2.3 eLearning environments

Learning environments are physical or conceptual places where single student's or group of students' learning process occurs. A learning environment consists of many factors like students, teachers, a technical environment itself, different kinds of views of learning and forms of activity, sources of learning and different tools.

eLearning environments are learning environments where students and teachers communicate with each other using computers in a way differing from traditional learning occurring in a classroom.

eLearning environments can be used in many kinds of situations. They can be used, for example in a case in which the whole course is lectured and organized on a network.

The eLearning environment can also be used for supporting traditional teaching. In this case all materials and current announcements can be located in the environment.

A third way to use a web-based environment is to use it only for communication between parties. In this case e.g. a material used in the

course can be constructed during web discussions.

More information about learning environments can be found e.g. in the Internet, (Rask, 2002) and (Meisalo, V. et al., 2000).

3. BATCH CONTROL IN CONTROL EDUCATION

Currently, about 50% of the industrial processes include batch processing. The main industries are the manufacturing of pharmaceuticals, food and beverages industry, metallurgical industry and chemical industry. The increasing emphasis on specialized high-technology products, customer service, and product quality all highlight the benefits of batch processing.

The equipment used within batch processes is usually more complex and more expensive than the equipment used within respective continuous processes. Moreover, continuous processes are usually designed to produce products of one kind or grade only. Batch processes are in this respect more flexible. A plant of an enterprise may have one or several process cells, the equipment of which can be used to produce several products or product variants.

The product-specific recipes define requirements for pieces of equipment needed to produce the product. More importantly, they define the production sequence and parameters (set points, raw material and product quantities and quality targets) needed for the production.

Batch processes have been defined in the standard ISA-S88.01 as follows:

"The batch processes lead to the production of finite quantities of material (batches) by subjecting quantities of input materials to a defined order of processing actions using one or more pieces of equipment. Batch processes are discontinuous processes. Batch processes are neither discrete nor continuous; however, they have characteristics of both."

Batch control is very interesting also as to its educational and didactic aspects:
- due to the nature of batch processes, batch control contains both continuous control, binary and sequence controls as well as high

demands for exeption handling and production quality

- interfaces to and interoperability with product design (via recipes) and production control (via production schedules) are important in fulfilling the overall automation goals (Johnsson, 1997)
- there exist new, software component based batch control systems, which can be flexibly configured or assembled for plant specific batch control needs
- so called "intelligent" batch control is an important, emerging research topic (Kuikka, 1999)

Thus batch processes both cover several important control topics and offer interesting challenges for utilizing the newest control implementation technologies.

4. A NEW WEB-BASED LEARNING ENVIRONMENT FOR BATCH CONTROL

In the Institute of Automation and Control of the Tampere University of Technology there began last spring a new course called Batch Process Control (2 cu). The course is meant for third and forth year students studying automation as their main or secondary subject.

The course consists of 14 hours lectures during seven weeks, seven exercises (one for each week), one project assignment at the end and examination.

Because the course was meant to be attended from outside of our university, web-based learning environment was designed and implemented for it. The environment contains the whole training material, weekly exercises, the project assignment, and all material belonging to the course.

4.1 Designing contents

Designing a web-course can be divided into two separate entities: designing the contents and designing the learning environment. Some good tips for designing an environment (and web pages in general) can be found from (Nielsen, J. 1993 and 2000).

In our case we started with planning the contents. When the goals were clear, we designed and

wrote training materials. The materials are based on the standard ANSI/ISA-S88. The topics of the course deal mainly with standard's first part, S88.01.

First the material was written with a regular word processor. After that the text was delivered into seven parts comprising the materials for each lecture and week. These parts were converted into HTML format and the HTML pages were inserted into the learning environment.

After writing the material, weekly exercises and project assignment were designed. In all the exercises, Intellution's well known batch control software iBatch was used. Students were allowed to install the iBatch demo version on their own computers.

There were two types of weekly exercises. The first type had very detailed instructions including all necessary commands of iBatch and explanations for them. The second type was more like a description of the needed development task. In this type of exercise the student had to apply the knowledge he/she got from the exercise in the previous week.

There was also a tight connection between lectures and exercises. Exercises strived to apply the theories taught during previous lectures.

4.2 Designing the learning environment using WebCT

WebCT was chosen for the platform of the learning environment, because of its flexible structure and variety of tools. WebCT included nearly all tools we needed.

The Batch Process Control environment was to include all materials and also some self-tests for students. Opportunity for virtual discussions with other students and with staff was also deemed beneficial.

WebCT offers over 40 tools for web-course designer. There are also some completed platforms for environment which a designer can use. These platforms were introduced in (University of Turku, 2001).

4.3 Structure of the environment

The user interface of the environment was to be kept as simple and clear as possible. The course was quite short and thus it was very important that time for getting to know the environment (levels 1 and 2 in figure 2) was minimized.

In the figure 3 the final structure of the environment is illustrated. There are also mentioned all tools of WebCT which are used in the environment.

4.4 Building up the environment using WebCT

When constructing learning environment with WebCT and its tools, we used so called Designer Map. Designer Map includes all tools which are available. More detailed information on how to use this map can be found in (Rask, 2002) or (University of Turku, 2001).

In the figure 4 there are illustrated essential parts of Designer map (Tools and Utilities).

5. CONCLUSIONS

Batch control is a vital part of industrial control. Currently, about 50% of the industrial processes include batch processing. In spite of that there is hardly any education of this area of automation in Finnish universities.

Learning in the Internet differs from traditional learning due to human social aspects. E.g. communication in the Internet might turn out to be very difficult. Still, learning in the Internet has so many positive aspects that improving it is valuable. For example, with web-based environments it is possible to share the information with a larger audience.

A web-based learning environment for batch control has been developed in the Institute of Automation and Control (ACI) of Tampere University of Technology (TUT). The environment is used and further developed

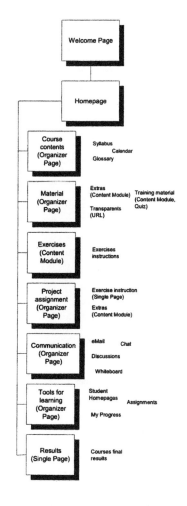

Figure 3. Structure of the learning environment according (Rask, 2002).

within a course called Batch Process Control which is lectured in ACI.

The course Batch Process Control was lectured first time in spring 2002. During the year 2002 the course was further developed mostly according to the feedback we collected from our students. For this development work we received financial help from the Virtual University of Tampere University of Technology. Batch Process Control was one of its pilot projects of year 2002.

Figure 4. The part of designer map.

REFERENCES

ANSI/ISA-S88.01 (1995). Batch Control Part 1: Models and Terminology. ANSI, ISA.

The Finnish ministry of education. Koulutuksen ja tutkimuksen tietostrategia 2000 – 2004. (online) http://www.minedu.fi/toim/koul_tutk_tietost rat/index.html (24.6.2002). (in Finnish)

Johnsson, C. (1997). Recipe-Based Batch Control Using High-Level Grafchart. Sweden, Lund, Lund Institute of Technology, Department of Automatic Control.

Kaufman, R., Watkins, R. and Guerra, I. (2001). The Future of Distance Learning: Defining and Sustaining Useful Results. Educational Technology.

Kuikka, S. (1999). A batch process management framework: Domain-specific, design pattern and software component based approach. Finland, Helsinki, Technical Research Centre of Finland (VTT) (doctoral thesis).

Meisalo, V., Sutinen, E. & Tarhio, J. (2000). Modernit oppimisympäristöt (Modern learning environments). Helsinki, Tietosanoma Oy. (in Finnish)

Nielsen, J. (2000). WWW-suunnittelu (Designing Web Usability). Oy Edita Ab, Edita.

Nielsen, J. (1993). Usability Engineering. USA, Academic Press.

Ojaniemi, K., Nurmela, S., Suvanto, J. & Bruun, P. (1999). Opiskelijana tietoverkossa – työkaluna WebCT (As a student in a network using WebCT). University of Turku, The Centre for Extension Studies. (in Finnish)

Rask, O. (2002). Panosautomaation oppimisympäristö (Learning Environment of Batch Control). Tampere University of Technology (master thesis). (in Finnish)

University of Turku, The Centre for Extension Studies (2001). Tee verkkokurssi! – WebCT versiolla 3 (Make a web-based course using WebCT 3.0). (in Finnish)

THE SYSTEM OF THE TRAINING IN CONTROL TECHNIQUES FOR ENGINEER STUDENT

Ádám, T., Czekkel, J., Dalmi, I.

University of Miskolc, Department of Automation
H-3515 Miskolc, Egyetemváros

Abstract: The history of University of Miskolc and the Department of Automation are introduced. An overview is given about the subjects. "Industrial process control" and "Automation" are the most important subjects of the department. The theoretical education is complemented by laboratory exercises. Three laboratories are dedicated for these purposes, where conventional controllers, modern control theories and PLC technologies are available. A "Totally Integrated Automation System" has been realized. It consists of different technology models, PLC-s, industrial communication and process visualization systems. There is a distance connection to the laboratory of the Logistic Department through the university's LAN network. *Copyright© 2003 IFAC*

Keywords: Industrial process control, Totally Integrated Automation System, Industrial Communication System

1. INTRODUCTION

The University of Miskolc has a two-and-half century long history. In 1735 the Court Chamber of Vienna founded a school of mining and metallurgy in Selmecbánya (today Slovakia, Banska Stiavnica) in order to train specialists according to the requirements of the industrial revolution and to upgrade precious metals and copper mining of Hungary that was playing important role in Europe in this time. In 1919 the school was moved to Sopron to the Hungarian west border. After the Second World War two faculties – Faculty of Mining and Metallurgy – were moved again to another place, to Miskolc, to the new industrial centre, where a new university was established in 1949. A third faculty - Faculty of Mechanical Engineering - was also founded at the university. Though the heavy industry companies - metallurgy and machine tool industry-came to a chrisis, the university has been developing

continuously. In 1981 The Faculty of lawyers, in 1987 the Faculty of the Economy, in 1992 the Faculty of Arts started with new training programs. A Music Conservatory, a Teacher Training College and a Collage for Health Workers are also integrated into the structure of the University today. Now the number of students is more than fifteen thousands. There are five years MSc. and three years BSc. programs on the technical faculties. It is to be noted that there is no electrical engineering MSc. training program in Miskolc at present. The Faculty of Mechanical Engineering has Mechanical Engineering (BSc. and MSc.), Information Engineering (MSc.), and Electrical Engineering (BSc.) education programs. The Electrical Engineering Institute of the Mechanical Engineering Faculty is responsible for the Electrical Engineering BSc. program. The institute consists of two departments, as Department of Electrical and Electronic Engineering and Department of Automation.

2. EDUCATION PROGRAMS OF INDUSTRIAL PROCESS CONTROL AND AUTOMATION

2.1. Lectures for Different Engineering Students.

The Department of Automation is responsible for the education of several different subjects for the non electrical engineering students of the three engineering faculties. Both MSc. and BSc. level training programs are on the engineering faculties. For example the subject Automation is educated for mechanical engineering BSc. and MSc., information engineering MSc., and metallurgy engineering MSc. students. It has to be taken into consideration, that these subjects, among them the Automation and Process control are only subsidiary subjects for these students. Therefore this training programs are only one semester long that are usually divided into about 30 hours theoretical lectures and 30 hours laboratory practical exercises.

2.2. Education programs for Electrical Engineering Students.

The education of the BSc. electrical engineering students is the main task of the department. After the third semester the students can change among different specializations as follows:
industrial process control technology,
informatics of signals and systems,
application of digital systems,
telecommunication,
vehicle electronics,
Power electronics and energetics.

The main subjects of the Department of Automation are:
Digital Systems,
Process Control,
Intelligent control systems,
industrial communication systems,
telecommunication systems,
Programmable Logical Controllers
Microcontrollers
Image Processing and Multimedia Systems
Industrial Measurement Systems,
Vehicle automation, vehicle information systems and vehicle diagnostics.

According to the specialization possibilities, the department has different kind of laboratories where the students come acquainted with the practical problems of the different fields. The laboratories and the equipments are:
Digital Systems Laboratory, basic logical circuits, microcontrollers, microcontroller application models, programmable logic circuits with application models

(FPGA, PAL), software and hardware development tools.

Telecommunication Systems Laboratory: basic telecommunication circuit measurement tools, digital telephone switchboard, mobile communication circuits, digital signal processors, digital filters, encoders, decoders, software simulators.

Multimedia laboratory: digital video camera, infrared camera, voice analysis, recognition, pattern recognition, video conference , distance learning tools. Width band distance connection is to our partner university laboratory and research institute.

Measuring Systems Laboratory: measurement tools for non-electric quantities, pc based data acquisition systems, special measuring instruments.

Process Simulation Laboratory: MATLAB and SIMULINK simulation tools.

Microprocessor and Digital signal processor laboratory: DSP and general-purpose eight-bit microprocessor development tools.

Process Control Laboratory: conventional industrial controllers, DCS with technology, PC-based control education kit with three-phase and H bridge inverter, ac and dc servo drives, fuzzy demonstration systems.

Fig. 1 Tank system

Industrial Control and Communication Systems Laboratory, PLC systems, technology models, industrial communication system.

SCADA and Process Visualization Laboratory: process visualization workplaces connected to the LAN. Process control, industrial communication and control and SCADA and process visualization laboratories are all connected to the LAN network, as it is discussed later.

After graduating, the students who change industrial process control specification, get job in the field of industrial control. They need deeper and more detailed knowledge of control engineering than the others, therefore this education program gives more theoretical and practical knowledge. The fundamentals of this training are the subjects studied during the earlier semesters, as theoretical electrical engineering, digital techniques, mathematics, mathematical modeling, etc.

3. LABORATORIES OF THE TOTALLY INTEGRATED AUTOMATION SYSTEMS

3.1 Process Control Laboratory

Automation, process control, control system simulation and process visualization are the most important subjects that need laboratory background. The laboratory exercises connecting the above mentioned subjects are made in the process control laboratory. Conventional controllers, for example PI temperature controller, modern control theories such fuzzy controllers can be demonstrated here. There is also a tank system that consists of four tanks, connection to the cold and hot water pipe systems and is used for realize water level and temperature controller. PLC and DCS system is being installed to this purpose. The tank system is shown in Fig. 1.

3.2 PLC laboratory

Programmable logic controllers are important tools of industrial automation systems. The PLC laboratory is equipped with different devices that are controlled by PLC. There is a manufacturing system model that consists of six assembling parts as it can be seen in Fig. 2. Work pieces have to be moved from one part to the others. In this way automated assembling systems can be modeled. Each segments are controlled by it's own PLC, but each PLC's are able to communicate with the others. One induction motor drive with PWM inverter is also in the laboratory. It is also connected to the industrial communication system. One pneumatic actuator demonstration system is used to show the basics of pneumatic actuator's applications in a control system. A PLC controlled lift model is also in the laboratory.

3.3 Process visualization laboratory.

The third laboratory of the department connecting to the integrated automation system is the process visualization laboratory. Ten workplaces based on IBM PCs are available for the students to get experiences in the process visualization methods. The computers are connected to the industrial information systems of the process control and the PLC laboratories on the Ethernet LAN.

Fig. 2 Technology model in the PLC laboratory

The simplified scheme of the laboratories with the industrial network is shown in Fig. 3.

4. THE COMPONENTS OF THE INDUSTRIAL INFORMATION SYSTEM

4.1 Actuator/Sensor Interface Bus

The Actuator/Sensor Interface bus works in a master slave mode. In our system the master function is built in PLC's. Pneumatic actuators, sensors and small PLC's are connected to the ASI master with a single two-wire cable The master controls the communication between the slaves and master, and in the same time it communicates with the higher level field bus. Different type of sensors is assembled to the technology model. Some of them are simple, conventional ones; some are intelligent devices that have communication interface. These intelligent actuators and sensors are connected via ASI bus. The conventional sensors and actuators are connected to the I/O modules. ASI bus represents the lower level of the industrial information system.

4.2 Multipoint interface

The serial communication ports of the PLC's are used as a multipoint communication interface. The PLC programs can be downloaded from the programmer PC to the PLC's on the multipoint interface. The programmer server computer connects the ETHERNET LAN and the multipoint interface.

Process vizualization laboratory

Process vizualization and PLC
program developing

WEB server

Programmer workplaces

PLC laboratory

SCADA terminals

Ethernet LAN

Programmer server

PLC   PLC

Fieldbus

PLC   PLC   PLC

MPI

ASI bus

Actuators   Sensors   PLC

Logistic laboratory

Konveyor

Truck

RF board

PLC

Ethernet LAN

Fig. 3 The Integrated Automation System

5. DISTANCE CONNECTION TO THE LOGISTIC LABORATORY

The material handling laboratory of the Logistic Department has different material handling devices. There is a storage system where the article stored on the shelf system can be selected, picked up and moved to a conveyor automatically. The conveyor can transfer the article to another conveyor system, or to a remote controlled truck. There are special robots as the part of the system. The robots are equipped with CCD camera that is used to recognize patterns, or for position control of the robot arm. The devices are all connected to the university LAN by PC. The stable devices, as the conveyors, the storage system, the robots are controlled by PLC's. The controller PLC's have communication interface to their PC. The mobile truck has different communication interface. One PC has a LAN interface and also a RF interface board. The mobile truck is equipped with PC, that also has RF interface board. In this way wireless communication is possible between the truck and the other parts of the material handling system. The truck is able to follow optical signal or RF wire laid on the floor. Other positioning strategies are also studied and tested.

Fig. 3 Pneumatic actuators

4.3 Fieldbus

The technologies in the process control and in the PLC laboratories are controlled by different PLC configurations. A PROcess FIeld BUS, or PROFIBUS is used as the next level of the industrial information system. The communications of intelligent components-PLC's, PC's, programming terminals – are carried out on the PROFIBUS. A PLC is used as the interface between the Ethernet LAN and the PROFIBUS. The PROFIBUS master block and the Ethernet interface is built into this PLC. The initialization of the data transfer is always done by the master.

REFERENCES

Ádám T., Czekkel J., Dalmi I. Laboratory for Education of Integrated Automation Systems. International Carpathian Contorl Conference ICCC' 2002 Conference Proceedings, pp 637-640. Malenovce, Czeh Republic.

J. WEIGMANN, G. KILIAN: Decentralization with PROFIBUS-DP. Siemens, 2000.

FACULTY EXCHANGE IN CONTROL EDUCATION, ONE ASPECT OF INTERNATIONAL CO- OPERATION

Omar Zia and Steven Liu

Southern Polytechnic State University, Marietta, Georgia , USA
Hochschule Harz (Harz University of Applied Sciences), Wernigerode, Germany

Abstract: This paper is an outcome of a faculty exchange program between an American and a German university, in the summer semester of 2001. The goal is to present and promote, the faculty exchange program as one of the many components of international cooperation in engineering education. While no single recipe for a successful approach can be provided, an attempt has been made to create a list of crucial guidelines. The objective is to encourage and share author's experience with fellow educators. It is hoped that international offices of universities will be able to use findings from this paper as a tool to help faculty members in the development of faculty exchange. *Copyright* © *2003 IFAC*

Keywords: Control, Education, Faculty, Exchange

INTRODUCTION

A recurring theme and frequent topic of discussion in education, business, and media, in the last decade has been global economy, and international competition. Most businesses, industrial and financial firms have had an international perspective. Many employers have been looking for some international knowledge in addition to technical skills in their future employees. As a consequence, engineering practice in many European countries is becoming more and more international. Engineering education throughout the world, especially in the developed countries, is beginning to have much in common, requiring more and more emphasis on globalization of engineering education. Engineering curricula are being redesigned to incorporate elements of international cooperation. Engineering educators in those countries are more knowledgeable in the international aspects of engineering education. They are more actively involved in the international co-operation activities, such as faculty and student exchange programs. This has not been quite true in the United States. In the past, as studies have shown[1], United States educational systems have not delivered either the level or the nature of education required for success in a competitive global economy

GLOBALIZATION OF ENGINEERING EDUCATION IS A NECESSITY

In the past two decades international trade has expanded five percent per year. Today 22% of our GNP comes from international trade. Foreign business has become an increasingly critical element of America's corporations. A substantial part of the United State's corporate profits flow from abroad. More and more firms, both large and small, are looking overseas for opportunities. No American firm can afford to assume that it is impervious to foreign competition. Failure to understand and adapt to the overseas environment is a cause of executive failure in international operations. To be successful a global perspective must be maintained at all times.

While the list of arguments supporting globalizations of engineering education is long, their common premise however, can be summarized as follows:

- There is an increasingly obvious need for business competing in a global economy to employ technical staff with an international perspective[3]
- In global industrialization, it is imperative that engineers and technologists have an international perspective. Most employers who plan on placing employees in international service prefer their employees have international knowledge in addition to their technical skills[2]

ENGINEERING EDUCATION FOR INTERNATIONAL PRACTICE[4]

To adequately prepare new engineering graduates for careers in the international arena, It takes more than just adding a foreign language course to the existing curriculum, The minimum requirement must include:

- Foreign language proficiency, written and spoken fluency
- Understanding culture of peoples in regions of the world where graduates may practice
- Understanding of international business issues, such as competitiveness, free market development, multinational companies, varying ethical norms, and varying consumer protection mechanisms.
- Familiarity with measurement systems, varying standards and codes, environmental concerns

PARTICIPATING INSTITUTIONS

Southern Polytechnic State University, an American University located in Marietta, Georgia. The mission of the university is to provide the residents of the state of Georgia with university-level education in technology, engineering, arts and sciences, architecture, management, and related fields. In Georgia, which is one of the fastest growing states, the realization that we are a part of "global economy" is very strong. Therefore, the mission of Southern Polytechnic State University (SPSU) is unambiguous about seeking international opportunities to participate in the teaching and transfer of technology. Both the faculty and administration are keenly aware of the importance of educating engineers able to practice internationally. SPSU has participated in foreign exchange activities for a number of years. The International Programs Office has worked with students and faculty members to establish international exchange programs.

Hochscule Harz (FH), a German University of applied sciences, located in Wernigerode, Germany.

Engineering education in Europe is a complex issue to deal with because of the very diverse engineering education systems within it. In Germany, there are two kinds of university:

- Classical universities offering a more research oriented education.
- Fachhochschulen (university of applied sciences) offering application oriented education.

The curricula of the Fachhochschulen have a bias towards a practical approach, whereas the classical universities have a more theoretical (research oriented) approach.

The Dipl.-Ing. degree from the Hochschule Harz is similar to BEng (Hor.) offered in U.K. and levels between a BEng. and MEng. offered by an US university . The Hochschule Harz serves students studying automation, information technology and business. It features extensive teaching laboratories providing the students with a "hands-on" learning experience.

WHEN AND HOW DID THE AGREEMENT BETWEEN THE INSTITUTIONS BEGIN?

A formal agreement of cooperation between the two universities has been in effect since 1994. Faculty and students from both institutions have visited each other universities on different occasions and for different academic purposes. The faculty exchange, however, is the first of its kind between the two institutions and it began as follows. During the fall semester of 2000, initial contacts between two faculty members from Southern Polytechnic State University and Hochschule Harz were made. Several areas of interest for ongoing exchange activities in the area of Electrical and Computer Engineering technology were identified. Control systems seemed to be the most convenient area at the time, mainly due to the availability of volunteers to take part in the exchange. The authors of this paper were selected as the first participants of Summer 2001 Faculty Exchange program. Most of the negotiations and communication were made via electronic mail. Both sides offered solutions for issues confronting them and made the following decisions with regard to the scope and implementation of the exchange.

1. THE MEDIUM OF TEACHING, WILL THE COURSES BE TAUGHT IN ENGLISH OR GERMAN?

The medium of teaching for classes offered under Faculty Exchange program will be English. The obvious reason for this decision was the fact that the majority of students at Hochschule Harz do speak and understand English. As a matter of fact, students enrolled in summer of 2001, in the class taught by the authors, had asked for English to be the medium of teaching. A less important factor in choosing English as medium of teaching was the availability and method of selection of textbooks. In the US the scope, content and course outline is determined and the textbooks are selected by either the instructor or a committee and the

students have to purchase it in advance. In contrast, the professor in Germany is the one who decides about the form and the content of the course without using any textbook

2. HOW WILL THE STUDENTS BE TESTED AND BY WHOM

Participating faculty will make the evaluation in accordance to the procedures and rules effective at the host institution. It is important to mention that in Germany, teaching methods and procedure differ greatly from what is generally applied in the US institutions. To mention just a few, in general there are no quizzes in Germany; very often there is only one test (final exam) that can be either oral or written. . Also, it is unusual for the students in Germany to be assigned homework problems every week. In contrast in the US unlike Germany homework and quizzes can make an significant portion of final grade.

3. WHO IS RESPONSIBLE FOR HOUSING AND TRANSPORTATION?

Travel from US to Germany and from Germany to US by the participants would be expected. The time period from mid-May through August was required for this particular exchange to be completed. Housing was and remained a concern. The question of housing availability and cost in both countries seemed to be the biggest obstacle. It was clear from the outset that neither institution was in a position to directly provide housing abilities, however, they offered important assistance in housing arrangement.

4. ISSUES THAT NEED TO BE GIVEN MORE CONSIDERATION IN THE FUTURE

FUNDING

Funding for international initiatives is a significant issue for institutions in the US. It is important to note that housing and transportation is very expensive both in the US and Germany. Therefore, funding and the availability of limited resources remains a critical issue for American side. This is not the case for German institutions. In Germany extensive efforts are made by the government to make sure that universities have international standards and are closer to international system of education. Therefore, certain amount of funds and resources are at the disposal of university authorities for international activities.

COMMUNICATION

Due to time zone differences the use of direct phone calls is not always an easy way. Therefore attacking the communication issues early on is essential. Business letters from Germany are printed on paper that is longer than our 11-inch sheets which causes fax problems. It turned out that electronic mail provided significant help in reducing communication problems. In Germany, especially in a small town, it could be a problem to find enough people speaking English in city administration and communication offices, causing sometimes problems in managing the everyday life.

REFERENCES

Kozak Michael, Engineering Technology Curriculum Revitalization. *Proceeding, 1993 ASEE A annual Conference*

Kraebery W.Henry, Student Exchange Program Critical Issues- A case Study, *Proceedings 1993 ASEE Annual Conference.*

ELSEVIER

IFAC
PUBLICATIONS
www.elsevier.com/locate/ifac

THE BENEFITS AND EXPERIENCE OF STUDENTS EXCHANGE BETWEEN UNIVERSITY OF SKOPJE AND UNIVERSITY OF ROSTOCK

Prof. Dr. Vangel Fustik

Institute of Power Plants and Systems, Faculty of Electrical Engineering,
University St. Cyril and Methodius, P.O. Box 574, 1000- Skopje, Republic of Macedonia
Fax: ++389 2 36 42 62; E-mail: vfustic@etf.ukim.edu.mk

Abstract: The paper presents an overview of the student's exchange in the framework of the Project "DYSIMAC" and few examples of students work and their results. The paper also identified benefits and experience from the research studies performed with study stay by students from University of Skopje at modern control laboratory at University of Rostock.
Copyright © 2003 IFAC

Keywords: Education, Control applications, Power systems, modeling, Simulation

1. INTRODUCTION

Since January 2000, in the framework of Stability Pact in South Eastern Europe, DYSIMAC Project (Dynamic Simulation of Macedonian Power Plants) has been realized and has been sponsored by DAAD (German Exchange Service). The Project was organized and performed between Institute of Power Engineering, University of Rostock (Germany) and Institute of Power Plants and Systems, University "St. Cyril and Methodius" (Macedonia). The project consists of several tasks such as modeling and simulation of the Macedonian power plants in a new technological and market environment, measurements of the most important hydro unit parameters in operation and practical application of the performed investigations. As one of the most important issue and experience was students exchange between institutions.

The scientific research that has been performed has analyze the foreseen new technological and market demands and contribute in establishing real and useful rules in control of units in hydro power plants that are in rehabilitation phase. These new rules have acquire the technical and market operation of the power plants in Macedonia and the currently isolated interconnected system of the south eastern part of Europe. The results of the project became a useful tool that can be used and directly incorporated into the electrical power system strategy of today.

2. "DYSIMAC" PROJECT IN BRIEF

The Project established the following goals:

- Establishing a close co-operation between two Institutes and their laboratories in Macedonia and Germany.
- Exchange of the background research activities.
- Developing of models for dynamic analysis of the power plants and power systems not available yet.
- Searching for and including original contributions in control of the plants concerning the further development of the Macedonian power system regarding the deregulation process.
- Improvement of the dynamic stability behavior of the Macedonian power system in the new technological environment (new Control philosophy, new Energy Management System).
- Student and researcher exchange.

Project has been based on the following resources:

- Professors, assistants and students from both sides have organized the research team. The approx. number was roughly 20 persons including 14 students.
- The available power plant models from the previous scientific research were quite sufficient only for starting phase of the project. During the project implementation the researchers have directly created the appropriate software.

The methodology that has been used in the project:

Since the project topics have to be implemented in a new technological and market environment, the most sophisticated approach and a modern vision of the researches was needed. The used methodology acquires the proven know-how of the power system technique and has a step forward in implementation of the new demands both from the utilities within one country and also between the countries with interconnected power systems. Especially, the central region of Balkans – Macedonia, shall be considered.

3. SOCIAL AND POLITICAL ENVIRONMENT

The project has been performed in a very difficult and sensitive social and political moment and environment. Republic of Macedonia is a post socialistic country with significant steps towards the market economy and reforms in all fields of the social life. However, difficult economic situation creates constraints that should be overcome. The war in 2001 that Macedonia has been faced with made also uncomfortable environment for performing all visits and researcher's exchange with stay in Macedonia. With great engagement of the project leaders all scheduled tasks were realized.

Moreover, the rehabilitation of the six biggest hydro power plants in Macedonian power system is underway. In that context the rehabilitation of the control system is also foreseen as one of the most important tasks. In that sense the main unit components in hydro power plants should be prepared for modern control system and full automatic operation with remote control.

That was also a very nice opportunity to include in models and investigations complex rehabilitation of the plants. The project has established a useful database for the main components and control system for further usage in the rehabilitation activities.

4. EDUCATIONAL ENVIRONMENT

4.1 University of Rostock, Institute of Power Engineering

In the start-up moment only the Institute in Rostock had an appropriate laboratory under leadership of Prof. Dr. Harald Weber (http://www.e-technik.uni-rostock.de/ee/). Laboratory had capacity for all research activity and has been equipped with the most necessary measurement equipment and modern computer hardware for the main project phases. The available research places for the students (Laboratories, cabinets) in Rostock are sufficient for study stay of approximately 3-4 students from Macedonia.

Educational background, hardware and software recourses and available staff have created educational environment for successful research work and student's exchange.

4.2 University "St. Cyril and Methodius", Institute of Power Plants and Systems

The researchers in Skopje have a fundamental knowledge and experience in control systems analysis. However, the Institute of Power Plants and Systems in Skopje itself has no appropriate resources to establish a modern laboratory with all necessary equipment for student's education in control of power plants (http://www.etf.ukim.edu.mk). Concerning the offices and rooms there was enough conditions for training sessions that were performed from the Uni Rostock staff during their stay in Skopje.

5. TIMING, DURATION OF STAY AND OTHER ORGANIZATIONAL ACTIVITIES

The student exchange has been realized with a following program that has been agreed between the project leaders:
  1. Students from Skopje to realize one semester stay at Institutions in Rostock and
  2. The staff from Uni Rostock (professors and assistants) to have short visits in Skopje for training sessions (hardware and software development).

In particular the exchange has been realized in the following steps:

- In the period February - April 2001, 3 students from Skopje realized 3 months stay in Rostock,
- In the period April - June 2002, 4 students from Skopje realized 3 months stay in Rostock,
- In the period October - December 2002, 4 students from Skopje shall realized 3 months stay in Rostock,
- In the period 2000 - 2002, overall 7 visits of 4 persons (1 professor, 1 assistant and 2 post-

graduated students from Skopje to realize one week stay in Rostock.

- In the period 2000 - 2002, overall 8 visits of 6 persons (1 professor, 1 assistant and 4 PhD students from Uni Rostock to realize one week stay in Skopje.

6. SCOPE OF STUDENTS RESEARCH AND DIPLOMA THESIS

In the period-indicated students and staff from both universities have undertaken the specific tasks and steps for the following scope (Report 5/2002):

- Overview of the Macedonian power systems and the interconnections.
- Data acquisition of the technical parameters of the plants, substations and systems.
- Experience and advantages of the west European countries concerning the project subject.
- Establishing an optimal methodology for the Macedonian system.
- Establishing a technological and market model of the Macedonian System.
- Optimization of this new methodology of the system.
- Modeling and Simulation of Hydro Power Plants
- Estimation of the usefulness of the optimized new methodology.
- New software tools for power system control
- Preparing project reports.

The main efforts were directed towards working on diploma thesis and contributions in already started master thesis in the field of control of power plants and systems.

7. RESEARCH RESULTS AND OUTCOMES

As illustration of the research work performed during students' stay in Rostock the following results are outlined:

- **Construction and programming virtual instruments useful in the power engineering**

Each virtual instrument (Kolondzoski, 2001) is structured of five elements:

- Transducers;
- Signal conditioning;
- DAQ hardware;
- Personal computer;
- Software.

The most important element of the virtual instrument is the DAQ board because it acquires the values that should be measured. Its characteristics are:

- sampling rate;
- multiplexing;
- resolution;
- range;
- relative accuracy.

The capabilities of the personal computer used for data acquisition system can significantly affect the maximum speeds of data acquisition. As the most popular application software for programming virtual instruments is used LabVIEW (Laboratory Virtual Instrument Engineering Workbench). It is a development environment based on the graphical programming language G. The LabVIEW could be used with little programming experience.

By using LabVIEW two virtual instruments were programmed:

1. Universal virtual instrument with 16 channels

2. Virtual instrument for frequency measurements

With the universal virtual instrument 16 physical values could be measured in the same time and also to save them in certain file. First active channels are used by pushing their buttons. When the program started, measured values could be seen in p.u. ones on the upper graph. Also, there are presented measured values from several channels and their difference is in the different colors. Also the user can choose how many lines (channels) could be seen in the upper graph. On the graph user can measure physical values from only one channel but in real values. User can choose which channel to see and also there is an indicator for showing the value in the measured moment. User can control the measuring speed (sampling rate) for each channel. The information about each channel (physical and electrical minimum and maximum, nominal value etc.) is based in a cluster.

The process of recording the values begins by switching ON the RECORD button and stops by switching OFF. User can see all the information about the channels and measuring values exactly in the file. The name of the file presents the date and the time when it was created. Furthermore, use can set the path of the file by the path of the file control. In the front panel there is a place for writing comments. It might be a normal text that the user needs as useful for the program information and it is also automatically written in the file. The measurement stops by pushing the STOP button.

With the virtual instrument for frequency measurements user can measure the frequency of the grid.

This program also gives an alert when the frequency is low. On the graph user can see measured values of the frequency and also the low limit of the frequency. First the user has to choose the channel for the frequency measurement in the CHANNEL control. Than the low limit control is defined for frequency in the LOW LIMIT control. User can see the value of the frequency in the real moment in the MOMENTARY VALUE indicator. If the value in the moment is less then the low limit, the LOW FREQUENCY indicator blinks in red color and also in that moment there is a sound signal.

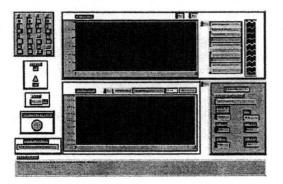

Fig. 1. Universal virtual instrument with 16 channels

- **Virtual Instrument for Data Acquisition in Hydro Power Plants**

The Virtual Instrument for Data Acquisition in Hydro Power Plants (Dimova, 2002), is created by the following components:

> -front panel (graphic user interface-GUI),
> - block-diagram.

The front panel consists of control buttons and indicators that are I/O for the instrument. The block-diagram consists of graphic source code of the instrument and defines its functionality.

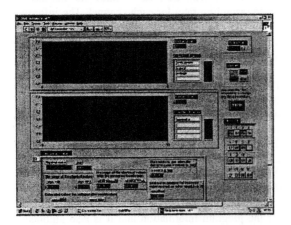

Fig. 2. Front panel of the virtual instrument for Data Acquisition

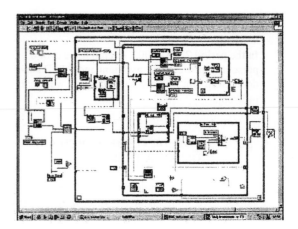

Fig. 3. Block diagram of Universal virtual instrument with 16 channels

The instrument has 16 channels, and the values of the measured parameters are presented on two plots on the front panel. Al values are recorded and stored in appropriate database. The measurements and obtained data by the virtual instrument could be used for modeling and simulation of the hydro power plant with the Matlab software.

- **Model for Automatic Control of Power Transformer with DigSilent Software**

With DigSilent software the students have studied power system analysis. One work that has been done (Mladenovski, 2002), was creation of the model for automatic control of power transformer 110/20 kV. As input variables have been treated: the voltage of the high voltage bus bars and the position of the regulating switch.
The following parameters have been treated as known: nominal voltage, maximal and minimal regulating range, etc.

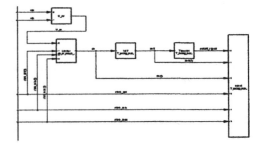

Fig. 4. Model for automatic control of power transformer 110/20 kV

Another task was creation of the model for automatic control of capacitor with 10*50 kVAr. As input data was the voltage of the bus bars where the battery is connected, the reactive power by consumer's etc. As known variables were treated nominal voltage of the bus bars, maximal and minimal range of the capacity regulating, etc.

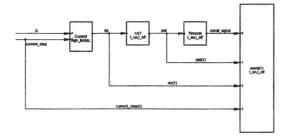

Fig. 5. Model for automatic control of the Capacitors

Also the students created a model for automatic control of the wind power plant with the output up to 1 MW. This model consists of a number of programs and subprograms unified in one completed main program. As input data are treated: greed voltage in complex form and frequency, and as output parameters are the active and reactive power generated by the wind-powered generator.

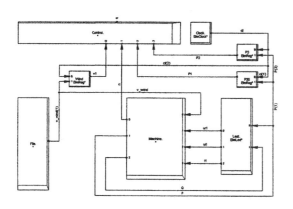

Fig. 6. Model for automatic control of the wind power plant with output up to 1 MW

- **Dynamic Model for Power Control in Hydro Power Plant Tikves in Macedonia,**

The main task in student's research work (Markovik and Krstevski, 2002) was designing Dynamic Model of the Tikves Hydro Power Plant - as a part of the Macedonian Power System. It includes the hydraulic system and one Unit of the hydro power plant.
The complete modeling and simulation process of the mathematical model of HPP Tikves was realized in the software package MATLAB and MATLAB integrated SIMULINK.

The actual model was a design and integration of several technological connected models of the components: the hydraulic system, the Turbine Regulator and the Generator.

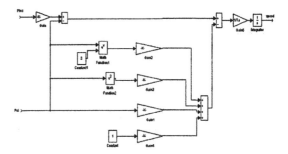

Fig. 7. Model of the hydraulic part of the generator

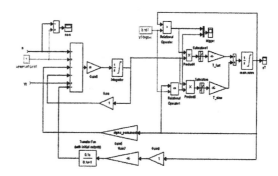

Fig. 8. Turbine regulator model

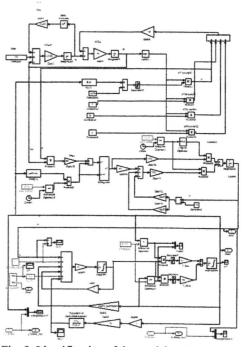

Fig. 9. Identification of the model parameters

Each of the above mentioned models have a corresponding behavior and performance as the real physical system in terms of the relevant physical values (such as water flow, turbine gate opening, active power etc.).

The model parts were created in their initial parameters and then they are integrated in the model and while all the parameters of the model were exactly identified so that the model outputs are as close as possible to the real measured values for certain simulated scenarios. The same scenarios were measured on the real object-HPP Tikves with the LabView software and hardware equipment. The input in the model was active power but the active power measurements were not completely available. During identification a special function was used as input, depending on main valve openings $Pel = f\ (Yt)$. The output of this model was the speed of the turbine wheel n The practical use of this model together with the voltage regulator model is that various scenarios and operational modes of the unit, or the whole power system can be simulated.

8. THE BENEFITS AND EXPERIENCE

The complex analysis of the student exchange has showed that the overall benefits of the project are as follow:

- Control education tasks become more popular among the students
- Creation of an environment for involving control paradigms in student's curricula
- The students have work and studied on the real and practically oriented needs from the industry
- Significant contribution in education process especially in automatic control of power plants and power systems for both institutions.
- Creation of independent and released thinking in solving specific tasks in the field of automation and control of power system.
- Successful establishing a joint team for research activities including education and utility.
- Development of creative and solution oriented students studying.

Our experience from students' exchange is:

- We established an active link between both universities that create environment for future joint applications and projects.
- The students were very satisfied with the research in one university laboratory from developed countries.
- They can now really understand their role as future engineers and to apply and work really needed tasks.

9. CONCLUSION

The students' exchange and acquiring new and modern know-how is very important strategy in education of today. However, the most relevant, useful and helpful results are for the countries that are in specific social and political environment. The overall benefits and a very positive experience are the guaranties that such strategy should be supported much more from the universities and companies from developed countries.

ACKNOWLEDGEMENT

The author would like to express his gratitude to Prof. Dr. Harald Weber and his research team from University of Rostock and DAAD (Germany) for their engagement and financial support.

REFERENCES

Report 5/2002 (2002), Dysimac project, Faculty of electrical engineering, Skopje.

Dimova, I., (2002) Diploma thesis: Virtual Instrument for Data Acquisition in Hydro Power Plants, Faculty of Electrical Engineering, Skopje.

Kolondzoski, Z, (2001), Diploma thesis: Construction and programming virtual instruments useful in the power engineering, Faculty of Electrical Engineering, Skopje.

Mladenovski, Lj., (2002) Diploma thesis: The possibilities of the DigSilent Software for Power System Analysis, Faculty of Electrical Engineering, Skopje.

Markovik A., and Krstevski V., (2002) Diploma thesis: Dynamic Model for Power Control in Hydro Power Plant Tikves in Macedonia, Faculty of Electrical Engineering, Skopje.

Copyright © IFAC Advances in Control Education
Oulu, Finland, 2003

www.elsevier.com/locate/ifac

POSITIVISTIC TENDENCIES DUE TO ENGINEERING EDUCATION

Mustafa Suphi Erden

Department of Electrical and Electronics Engineering,
Computer Vision and Intelligent Systems Laboratory,
Middle East Technical University, Ankara, Turkey.
e-mail: suphi@metu.edu.tr

Abstract: In this research it is claimed that engineering education increases the positivistic tendencies of students regarding to the social problems. A theoretical background is given in the introduction and literature review. Two hypotheses are constructed and a micro-questionnaire survey is performed to test the hypotheses. In the survey positivistic tendencies of engineering students and social science students are compared. The results are depicted and discussed in detail. One of the hypotheses was approved by the results while the other was rejected. It is concluded that the rejected hypothesis was not in accordance with the claim since it overlooked some factors effecting positivism. The results of the survey support the claim by stating that 'engineering students are more positivist than social science students'.
Copyright © 2003 IFAC

Keywords: Education, Engineering Education, Social Science Education, Positivism, Ideology, Questionnaire, Survey.

1. INTRODUCTION

Many factors affect the thoughts of people about the social problems. Among these are the social status, cultural domination, family, religion, age, and education. The ideas or thoughts about social problems can be named as 'ideologies' or 'ideological thoughts' when they are characteristic of a specific social group. The ideologies are results of the mentioned factors affecting the human thought. Among the factors, that is intended to be the scope of this research is university education. More precisely 'the effect of technological education in formation of positivistic thoughts' is searched in this research.

1.1 Ideologies and Education

The term 'ideology' has many different conceptions. The conception taken here, as 'the group of thoughts that have a common characteristic and are peculiar to a specific social group' is only one of the many. This conception here is not directly related to political means in the sense of a class struggle, as many other conceptions of the term would. Rather it is related with the technical approach to social problems. Eagleton's book of *Ideology* is a very good one to introduce the various conceptions of the term in the literature. Among those, the one most close to the

conception here is Raymond Gauss' 'descriptive' definition of the term. Eagleton (1991) writes, "Raymond Gauss has suggested a useful distinction between 'descriptive', 'pejorative' and 'positive' definitions of the term ideology...In the descriptive sense, ideologies are belief systems characteristic of certain social groups or classes, composed of both discursive and non-discursive elements. It is close to the notion of a 'world-view', in the sense of a relatively well-systematized set of categories which provide a 'frame' for the belief, perception and conduct of a body of individuals"[1].

When people are commenting on the solutions of social problems the dominating factor is the ideological thoughts in people's minds. Education has a spurious effect in formation of those thoughts. This is a natural outcome, since education is nothing else but the process of shaping and reshaping the thoughts, and reconstructing the dynamics underlying formation of thoughts. Then the question arises: "What kind of educations impose what kind of ideologies?" However, it seems that it is not so easy to establish a direct relation between a particular form of education and a particular form of ideology. Therefore at this moment it is better to put the question in another way: "What kind of educations

---

[1] Eagleton, T., *Ideology* (London-New York: VERSO, 1991). p.43.

impose what kind of thinking forms?" This report argues that technological educations impose *positivistic ideas* on students. Of particular interest here are the engineering students at universities.

Universities are the places where the education given to students differs the first time. This division is a consequence of the labour division in the society. What is aimed in universities is to give education in different disciplines in accordance with the different characteristics of the problems in social life. Different people are shaped to think in different ways according to the problems they are dealing with. Therefore, one should focus on the university education in order to scope the relation between education and forms of thoughts.

In technological science educations like engineering, students are always dealing with practical problems and applicable solutions of those. Therefore, pragmatism is always dominant. An idea or technique that does not help to gain anything practical is always disregarded. A four-year of education dominated by pragmatism is inevitable to shape the way an engineering student thinks. This is expected to be so not only about engineering problems, but also in approaching to social problems. Therefore it is not surprising to be faced with pragmatism in the solutions commended or supported by the engineers about social problems. This points to the possibility that the education given to engineers is one of the major factors for the pragmatic approaches of the engineers to social problems, hence for the positivism of engineers.

On the other hand, the education of social science students does pay attention to ideas and views that are not necessarily practically applicable. What dominates social sciences is 'understanding', not changing or constructing in a particular manner. Therefore, there seems to be no imposing of pragmatism in social sciences. This fact provides us a frame to search for the effect of engineering education to impose positivism, and a possibility of making a comparison between engineering education and social science education. A comparison of the ideas of the students in engineering and in social sciences will reveal that the two education forms have dispute effects on formation of the thoughts of students.

1.2 Positivism: Technocratism, Elitism, and Rationalism

The concept of *positivism* is a huge subject for which one should refer to mainly Comte and Durkheim for its significance in sociology. The concern of this paper is not to give a discussion of positivism as a sociological methodology of inquiry, or as a methodology to develop solutions to social problems. Rather, the term is considered in its most general meaning as 'taking the society as a thing out there' regarding to the social problems, which means social

life can be understood by scientific methods and changed in some manner using this knowledge directly. The impact of positivism in generating solutions to social problems is of interest here. In this sense, positivism is closely related with pragmatism to the extent that both seek an *immediate applicable solution* to the problem. The concern of this research is the forms of pragmatic solutions for the social problems imposed by positivism. In this respect Göle(1998)[2] has made her research paying attention to three notions: elitism, technocratism, and rationalism. Göle argues that these three are dominating the engineers' approach to social problems as a consequence of positivism. This research fallows Göle in this respect.

It may be proper to clarify the meanings for which the terms technocratism, elitism, and rationalism are used in this context. "Technocratism" is used for the behaviours and ideas that support and approve the intervention of technically educated and experienced (at least believed to be so by most of the people) people for social problems. These people are generally regarded as having the necessary knowledge to solve problems and their situation is termed with 'know-how' in common sense usage. The term "elitism" is used to refer to the idea that some group of people should be respected more than other groups in the society. Therefore everybody should approve that these groups of people would benefit more in social life even at the expense of others' benefit. In this research this specific group of people is considered to be the educated people in the society. As a result the term 'elitism' in this report is used to refer to the ideas, which approve the higher status of educated people in the society. The last term "rationalism" is used in a quite different meaning from its general usage. In its general usage 'rationalism' refers to reliance on human mind, human logic, and the inferences of human thought. Here, on the other hand, the term is used considering three tendencies in producing solutions to social problems: reliance on calculations, aiming a quick solution, and lacking the pursuit of a deep understanding. The three terms, technocratism, elitism, and rationalism, with the specific meaning loaded on each are referred to as "positivism". Therefore the term positivist is used to mean that the person is more technocratist, more elitist, and more rationalist in the meaning described above.

Considering the effect of engineering education in formation of positivist ideas it is expected that the engineering students are more positivist compared to social science students. If the different aspects of positivism are considered, following Göle, the following hypothesis may be put forward:

1. Engineering students approve *elitism* more than social science students.
2. Engineering students approve *technocratism* more than social science students.

---

[2] Göle, N., *Mühendisler ve İdeoloji ~Engineers and Ideologies*, (İstanbul: Metis Yayınları, 1998).

3. Engineering students prefer immediately applicable solutions for social problems instead of understanding them in detail (more *rationalist*), whereas the social science students prefer a deep understanding to an immediate solution.

Besides comparing engineering and social science students, comparison of first-class and fourth-class engineering students may also reveal some information. In the report this comparison is also performed. However, the result was not to support an *immediate* hypothesis like 'fourth-class engineering students would be more positivistic than first-class students'. The discussion of this result is given in the results section with the other outcomes of the research.

2. LITERATURE REVIEW

Eagleton's book (1991) 'Ideology' is the most referred one for an introduction of different conceptions of the term ideology. There are many books about statistical analysis of social science researches. Hamilton's (1995) book is a general one that mentions about survey techniques and evaluating the statistical results.

In her book, 'Mühendisler ve İdeoloji', Göle(1998) argues that the ideology of engineers is marked as being positivist and rationalist in early 80's in Turkey. Elitism, technocratism, reliance on measurement and calculation are dominant in the approach of engineers to the social problems. Despite mentioning these factors, she does not discuss the underlying reasons in formation of these thoughts in detail, or at least does not point to the education specifically. The only underlying reason derived intuitively from the book is that engineers are to gain an upper status as a new class of elites in early 80's. Being in agreement with Göle, that status offered by the occupation and education is effective in formation of one's ideology for pragmatic reasons, it seems that there exists more than this related to education. The form of education imposes particular forms of thinking.

Johnson, discusses the relation between engineering and social problems via the term *discourse*, regarding its usage by Foucault. Johnson argues that engineering needs a wider discourse to be able to consider social and environmental problems in its applications. Furthermore he points to the fact that "...their education has prepared engineers to identify with the interests of their employers, and that they simply do not feel an urge to be freed to protect the public interest".

Abraham (1994), makes a discussion of two theories, structuration theory and differentiation-polarisation theory, in sociology of education. Although the main discussion on these theories is out of concern of this research, many of the topics mentioned throughout the text are relevant with regarding to the positivist approaches in education. In the text a quotation from

Shilling is made to mention his concern about "bringing together micro and macro-approaches in the sociology of education" "...to facilitate a more sophisticated understanding of the degree to which structures are reproduced in and through social interaction in schools and classrooms...".

Poser, makes a discussion of the similarities and differences of the concepts of science and engineering. He compares those regarding to their being pure or applied, being creative, dealing with nature or artifact, seeking aims, means, or functions, using laws or rules, seeking for know-how or know-why, and having aims and values. In his discussion he confronts no clear-cut distinction between science and engineering, and one can derive from this text that engineering is a science with its special features with regarding to the points mentioned above. It is noted in the text that pure sciences aim to understand things better, whereas applied sciences like engineering aim to improve our mastery of them, therefore "...engineers...do not want to get better and deeper knowledge, but better ends...". "Engineers do not aim at truth but efficiency".

Seltzer's article is on the concept of technological infrastructure of science. Joseph Pitt's definition of technological infrastructure as "sets of mutually supporting artifacts and structures which enable human activity [including scientific activity] and provide the means for its development" is of concern. He points to the fact that this definition includes human interactions besides technological tools. Seltzer raises the problem of determinism and autonomy in Pitt's definition, and he argues that the technological infrastructure cannot develop autonomy from the surrounding culture, opposing what Pitt would argue. And he goes on his discussion on the terms of 'agency', 'natural world', 'reality', and 'structure and function of a technological instrument'.

3. THE QUESTIONNAIRE

In preparation of the questionnaire it is planned to be consist of four parts. In the first part the introductory information about the respondent is gathered. These are the questions about the age, department, class, accommodation, monthly-expense, and gender of the respondent. In the remaining three parts it is aimed to measure the three positivistic tendencies, namely technocratism, elitism, and rationalism of the respondent. Three questions are prepared to measure the technocratic tendencies, four question for elitist tendencies and four questions for rationalist tendencies. Each question has about four answers, which reflect the amount of the corresponding tendency. The answers to the questions are graded with points between one and five. For example, in a question to measure the elitist tendency of the respondent, the answer with point five reveals that the respondent has a great tendency of elitism, while the answer with point one reveals that he/she has a rather little tendency of elitism. The total of the points for the three groups of questions gives a grading of the respondent's tendencies of technocratism, elitism, and rationalism. And these grades make it possible to compare different respondents according to their three tendencies. The sum of these three grades reveals the 'positivistic' tendency of the respondent, and this makes it possible to compare the respondents based on their positivistic tendencies.

After the planning of the questionnaire the actual questionnaire is prepared. In this actual form the questions in the last three groups are mixed randomly. Furthermore, the answers to each question are also mixed randomly. This is done so not to reveal the purpose of the questions and their answers to the respondents. Otherwise the respondents would be inclined to specific answers considering their ethical attitudes. What we are searching is not consequences of their ethical ideas, but their ideas that are formed regardless of their ethical attitudes. In this actual questionnaire form it seems to exist only two groups of questions: introductory questions, and questions related to some social problems. For the questions of the second part the respondents are strongly warned to mark only one of the answers, which they approve much. This is to force the respondent to reveal his/her most dominant attitude.

The questionnaire is performed on twenty-four engineering students, and sixteen social science students. Of the twenty-four engineering students, eleven are first class students, twelve are fourth-class students, and two are doctorate students. The sixteen social science students are from any class.

4. EVALUATION OF THE DATA

The evaluation of the data is performed using the analysis tools of the SPSS program, which is a well-known statistical analysis environment. Since the questions for the three different tendencies and the answers to each question are mixed, they lack an order. An order is necessary to put the answers of the respondents to the SPSS environment. For this reason the answers of the respondents needed a preprocessing to evaluate the grades for the three tendencies. This preprocessing is a matter of associating each question with its corresponding group, and the answers of the questions with their points for the tendency they are representing. After these, the grades of the respondent for the three tendencies are calculated by adding up the points of the answers to the questions of each group. This preprocessing needs a small programming and some matrix calculations. These are performed in MATLAB environment. The thing this preprocessing provided is the grades of the respondents for the technocratic, elitist, and rationalist tendencies.

After that, the grades of the respondents are recorded in the SPSS environment with their introductory information. In this environment it was possible to analyse the relation between being engineering or social science student and positivistic tendencies.

5. RESULTS

The claim of the research is that 'engineering education has an increasing effect on positivistic tendencies'. The results presented here are based on the bar graphs obtained by the analysis tools of the SPSS program. The following two main hypotheses, which have their sub-hypotheses, are used to test the main claim:

H_1: Engineering students are *more positivist* than social science students.

  H_{11}: Engineering students are *more technocratist* than social science students.

  H_{12}: Engineering students are *more elitist* than social science students.

  H_{13}: Engineering students are *more rationalist* than social science students.

H_2: 4th class engineering students are *more positivist* than 1st class engineering students.

  H_{21}: 4th year engineering students are *more technocratist* than 1st year engineering students.

  H_{22}: 4th year engineering students are *more elitist* than 1st year engineering students.

  H_{23}: 4th year engineering students are *more rationalist* than 1st year engineering students.

For testing the first hypothesis the mean of the grades of the engineering and social science students are compared. This comparison is done for the three tendencies, and for their sum as being the main positivistic tendency. Figure I presents the bar graphs for the mean values of the grades of engineering and social science students for technocratist, elitist, and rationalist tendencies. In Table-1 are given the frequencies of the positivistic grades in the two groups.

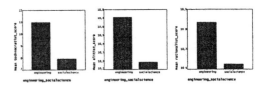

Fig. I. : Bar graphs for the mean values of the grades of engineering and social science students for technocratist, elitist, and rationalist tendencies.

In Figure I it is observed that in all the three cases the mean of the scores of engineering students are higher than those of social science students. These graphs reveal that the three tendencies are apparent in engineering students more than the social science students; hence the approval of the three sub hypothesis of the first hypothesis (H_{11}, H_{12}, H_{13}). Considering the sum of the scores of the three tendencies above, the following graph in Figure II is obtained. Figure II shows the relation between positivistic tendency and being either an engineering or social science student. This graph is an evidence for *acceptance of the first hypothesis*, H_1.

Table 1. Cross-tab showing the relation between positivistic tendency and being either engineering or social science student

positivist_score * engineering_socialscience Crosstabulation				
Count		engineering_socialscience		Total
		engineering	socials cience	
positivist_score	19	1		1
	23	1	3	4
	24		1	1
	25		1	1
	26		1	1
	27	2	2	4
	28	1	2	3
	29	3		3
	30	3	3	6
	31	2		2
	32	1		1
	33	2		2
	34	1		1
	35	3		3
	36	1	1	2
	38		1	1
	39	1		1
	42	1	1	2
	43	1		1
Total		24	16	40

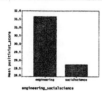

Fig. II: Bar graph for the mean values of the grades of engineering and social science students for the positivistic tendency.

For testing the second hypothesis the means of the grades of the fourth year engineering students, first year engineering students, and social science students are compared. According to the second hypothesis (H_2) and its sub hypotheses, the mean values of the fourth year engineering students were expected to be the highest, the mean values of the first year engineering students to be the middle, and the mean values of the social science students to be the lowest one. This was expected since the engineering education would have strengthened the positivistic tendencies in the four year of engineering education. However, the results obtained are not in accordance with this idea. The mean values of the three tendencies are shown in Figure III for fourth year engineering students, first year engineering students, and social science students.

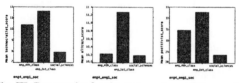

Fig. III. Bar graphs for the mean values of the grades of fourth year and first year engineering students, and social science students for technocratist, elitist, and rationalist tendencies.

In the graphs of Figure III it is observed that for all the three tendencies the first year engineering students have the highest mean grade, the fourth year engineering students have the middle mean grade, and the social science students have the lowest mean grade. The social science students having the lowest mean grade of the three was expected as a consequence of the first hypothesis. However, the mean values for the first year engineering students being higher than the mean values for the fourth year

engineering students was something not expected. According to the second hypothesis the four-year engineering education should have reinforced the positivistic tendencies of the first class engineering students. However, it seems that the positivistic tendencies of the first year engineering students have decreased after the period of four-year engineering education. The graphs in Figure III are enough evidences for *rejection of the sub hypothesis of the second hypothesis* (H_{11}, H_{12}, H_{13}). And the mean positivistic tendency grade, being the mean of the sum of the scores of three tendencies, is depicted in Figure IV for the three groups of students. This graph is also enough for *rejection of the second hypothesis*, H_2.

Fig. IV. Bar graph for the mean values of the grades of fourth year and first year engineering students and social science students for the positivistic tendency.

6. DISCUSSION AND COMMENTS

The results presented above, apparently approve the acceptance of the first hypothesis, and rejection of the second hypothesis. Accordingly, what the results approve may be stated as the following:

- H_1: Engineering students are *more positivist (more technocratist, more elitist, more rationalist)* than social science students.
- $-H_2$: 4th class engineering students are *LESS positivist (LESS technocratist, LESS elitist, LESS rationalist)* than 1st class engineering students.

Approval of the first hypothesis states that engineering students are more positivist than social science students. At a first attempt, rejection of the second hypothesis seems to reject that engineering education increases the positivistic tendency. However such a conclusion would be an *immediate* one lacking the deep understanding. The rejection of the second hypothesis states only that he fourth year engineering students are less positivist than the first-year engineering students. With this result, one should *hesitate* to *reject* that engineering education *increases* the positivistic tendencies. Figure IV states that fourth year engineering students are *less* positivist than first year engineering students but says nothing about the reason of this. And this reason is not something that may be revealed from this research. The things effective in decreasing the positivistic tendency in the four-year engineering education duration may be so variant and strong that the increasing effect of the engineering education may be less than those decreasing effects. For example age may be a significant factor. It is probable that age has a decreasing effect on positivistic tendencies. The results presented here only say that 'engineering students are more positivist than social science students', and 'fourth year engineering students are less positivist than first year engineering students, whereas they are still more positivist than the social science students'. Comparing the tendencies of engineering and social science students in corresponding classes would give more detailed ideas about the effect of age. Unfortunately, such an observation is lacking in this research.

It is evident that the results of a questionnaire are strongly affected by the questions prepared to reveal the specific relation. Besides, the scaling used is also effective. The questions prepared for this research and their scaling seem to be successful to reveal the relation between positivistic tendencies and being either an engineering student or social science student. The questionnaire is successful to determine that the engineering students are more positivist than the social science students, but it does not say much about the reason of this determination. The rejection of the second hypothesis may not be considered to be an outcome of the un-success of the questions or scaling. Rather it should be due to the hypothesis' own failure to support the main claim. Otherwise there wouldn't be a consistency of the relation between the three tendencies and the three groups of students (Figure III). Also the consistent relation observed between engineering and social science students for all three tendencies (Figure I) points to the success of both questions and scaling. It should not be forgotten that the aim of this research is not to make a scaling of the respondents according to their tendencies; rather the aim is to determine that the positivistic tendencies are *stronger* in engineering students than in social science students, without giving a degree of the strength.

7. CONCLUSION

The incredible improvement in technology day by day increases the importance of engineers in the society. The effect of engineers in social life become so significant that the attitudes of the engineers regarding to the social problems draws attention of many researchers. Engineering education has a peculiarity to form the characteristic structures of the thoughts of engineers regarding to the social problems. This research claims that engineering education has positivistic impacts on the students. Namely, it increases the 'technocratic', 'elitist', and 'rationalist' tendencies of the students. To test the claim a micro-questionnaire research is performed on the two hypotheses that 'engineering students are more positivist than social science students' and 'fourth year engineering students are more positivist than first year engineering students'. The first hypothesis is approved by the results of the research while the second one is rejected. Considering the results it is concluded that the engineering education increases the positivistic tendencies compared to

social science educations. However there may be other significant factors to decrease the positivistic tendency, like age.

As the claim is supported by the acceptance of the first hypothesis, and at least is not rejected by the rejection of the second hypothesis, it may be concluded that this research strengthens the claim; engineering education increases the positivistic tendencies. What is the importance of this result? The importance of the result ought to be revealed with a discussion of positivism on social problems. Since the aim here is not to go in this discussion, it will be content to point a few things about understanding social problems. Social life is so complicated and still not much understood by human that, sticking to some specific tools in solving social problems may result in undesired effects. Understanding the social life necessitates a wide range of world-view. Being inclined to computational and technological methods in understanding the social life may be misleading. Since the object of engineering applications and social life experiences are totally different the understanding of the two may necessitate totally different approaches. Engineers may make significant contributions to understand the social life with their analytical understanding, but this may be lacking and even misleading when it is not fed with different forms of thoughts and different forms of world-views. In this respect engineering students should be encouraged to gain different perspectives regarding to social life. It should be noted that life is not technology only, but a compound of many variant dimensions of human. Technology can only make the life better if it is in harmony with this compound.

REFERENCES

Eagleton, T. (1991) *Ideology,* VERSO, London-New York.

Abraham, J. (1994). Positivism, Structurationism and the Differentiation-Polarization Theory: A Reconsideration of Shilling's Novelty and Primacy Thesis. *British Journal of Sociology of Education,* **Vol. 15, No 2,** 231-242.

Göle, N. (1998).*Mühendisler ve İdeoloji,* Metis Yayınları, İstanbul.

Johnston, S., A. Lee, , and H. McGregor. Engineering as Captive Discourse. *Society for Philosophy & Technology,* **Vol. 1, Nos 3-4.**

Poser, H. On Structural Differences Between Science and Engineering. *Society for Philosophy & Technology,* **Vol. 4, No 2.**

Seltzer, M. The Technological Infrastructure of Science: Comments on Baird, Fitzpatrick, Kroes and Pitt, *Society for Philosophy & Technology,* **Vol. 3, No 3.**

Hamilton, L. (1995) *Data Analysis for Social Sciences,* Duxbury Press.

A NEW SELF-STUDY COURSE ON THE WEB: BASIC MATHEMATICS OF CONTROL

P. Riihimäki, T. Ylöstalo, K. Zenger, V. Maasalo

*Helsinki University of Technology, Control Engineering
Laboratory, Konemiehentie 2, FIN-02150 Espoo*

Abstract: The possibility to use the WEB for teaching purposes is utilized more and more. It is anticipated worldwide that almost all university courses will be available on the Internet in the near future. In the paper the results of a pilot project have been described, in which a program package for teaching basic mathematics of control in the web has been developed. *Copyright © 2003 IFAC*

Keywords: Education, Educational aids, Teaching, Web-course

1. INTRODUCTION

In Helsinki University of Technology (HUT) a lot of emphasis has been given to the development of education by utilizing modern information and communication technology. The goal is to improve the flexibility of studying and to make it possible to teach and learn independent of time and place. HUT co-operates with national and international universities in order to develop on-line education and training. HUT is, for example, a member in the Finnish Virtual University (FVU), in the International Network of Universities (INU), and in the European University Network for Information Technology in Education (EUNITE). Virtual HUT offers services to the university's teachers and students, for example technology support, exchange of experiences, counseling with experts, information of seminars and other educational occasions, electronic collections of the library, funding, and reports.

HUT offers courses to the teachers to improve their skills in information and communication technology. One alternative in this respect is to participate in the national teacher training program 'TieVie - Information and Communication Technology in Higher Education'. The training program is carried out in five universities: the University of Oulu (co-ordinator), the University of Helsinki, the University of Jyväskylä, HUT, and the University of Turku.

In 2001 -2002 the Control Engineering Laboratory at HUT participated in the TieVie-program in order to develop laboratory's education. A pilot project for the on-line teaching was started. The aim of the project was to develop self-study material into the net for the course AS-74.102 Basic Mathematics for Control Theory. During the TieVie training program the action plan for the project was developed and guidelines determined. One decision was that the self-study material is presented in www-pages instead of some www-based learning environment (for example WebCT, Discendum Optima). There were several reasons for this selection: price, simple to create, easy modification, ease of use, versatility, and custom (all information of the laboratory's courses are presented in www-pages). The project organization included the teaching staff, which provided the core material of the contents of the course, and summer trainees, who created the www-pages and programmed interactive demos.

2. BASIC MATHEMATICS FOR CONTROL THEORY

The course *Basic Mathematics for Control Theory* is intended for undergraduate students with not so strong background in mathematics. The course was founded several years ago, when it was noticed that some students met difficulties in basic courses of Control Engineering, because they had to concentrate too much on learning basic facts about matrix calculus, differential equations, Laplace transformations etc. The new course was intended to make the students mathematically mature to meet the "hard" courses of Automation, Process Control and Control Engineering.

The course has been lectured for several years now. There are altogether 5 lectures (two hours each) and 5 exercise hours (three hours each) and an examination, in which the students have 10 problems to solve in 3 hours. Passing the course gives 1 credit, which corresponds to one week's work.

The new self-learning package covering all the topics of the course has just been established on the Internet. The idea is that the net course will in the future substitute the normal lectures and exercises, so that the whole course could be studied totally as a self-study. For the time being, the course is still lectured in the normal way, and the net course is an additional aid to the students.

The main idea in this course has been gradually evolving during a number of years and has now reached some kind of a maturity stage. The following facts give the background:

- the students are not very "mathematically oriented"; usually they find the applications of mathematics more motivating,
- the students are interested in automation technology and practical process control.

From these principles it is relatively easy to plan a suitable course, which teaches the mathematical preliminaries in a way, which the students find motivating. The course starts from the description of the general control problem by using a simple example. The concepts open loop and closed-loop control become clear to them in a tentative way. They understand the basic block diagram representation and somehow the concept of a dynamical system. The key question is then: "How to control the system, for example a valve in a liquid level control problem, such that the reference point is reached without too much oscillations? What if the oscillations start to grow? How can we study these phenomena; the basic courses of engineering seem not to give much help for these kinds of problems"?

The answer is that the course will teach the necessary tools. The dynamics of the process, the controller and the actuator etc. can be described by differential equations. The combination of all information in a system is carried out e.g. by using the Laplace transformations and transfer functions, the stability can be studied by the location of roots of the characteristic equation or by the eigenvalues of a certain matrix in the state-space representation etc. It has been noticed that the students being in the first lecture of the course appreciate the knowledge that eigenvalues of a (certain) matrix really determine the stability in the described control problem. The students understand the problem fully, but they do not have any idea how to analyse it. Also, they cannot understand what matrix calculus or linear algebra in general have to do for these kinds of practical problems. This gives the necessary motivation to learn more.

So the basic pedagogical idea in the course is to learn the mathematical preliminaries together with some basics in control engineering. The mathematical tools are used only to the extent that is needed. It is emphasized that this course cannot replace more extensive mathematics courses. Also, it must be noticed that this is not a course in Control Engineering. Basics of control are learned as a "by-product" without even noticing it.

3. THE SELF-STUDY ENVIRONMENT ON THE WEB

The aim of the project was to develop a clear learning environment. It is possible to find several web sites in which control theory is comprised. The sites may be disorganized. There are several links in the text and the readers often gets off the subject. The aim was to avoid the problem.

Material was planned to be on the WEB in a form which is easily readable and understandable to the students, whose mathematical skills are not very good. The material was grouped in six sections:

- Introduction
- Matrix calculus
- Laplace transformation
- Dynamical systems
- Discrete systems
- Non-linear systems

Nowadays it is not uncommon to find a learning environment on the web, where you can study a topic of your interest. However, the problem is how a student can make sure what he/she has really learned. Furthermore, it is not always easy to read pure mathematical text. We have developed a learning environment which takes account of the problem. This is one way to teach a course on the WEB.

3.1 The Structure of the Web Sites

The structure of each sites are similar. The navigation bar, where the site is controlled, is on the top of the site. There are links to theory, exercises and demos in the bar, and the colour of the bar is changed when moving in different section.

Fig. 1. The starting site of the WEB course.

3.2 The Theory Sites

The structure of the theory sites are presented in Figs. 2 and 3. The names of the sections are on the left part of the theory sites, and by clicking the name it is possible to go to another section. The content of the selected section is also seen on the left bar. Further, each section is grouped to lessons, and there are also run-through questions after each lesson. When the reader has answered these questions he/she can make sure whether he/she has understood the section. More on this point can be found in Section 3.4.

It is important that the reader knows, what the subject is, where he is now and what the next subject is. Because of this, the colour of the selected subject is changed in the left content bar. As an example, consider Fig. 3. Because the selected subject is "What is a matrix?" this title is bold in the left navigation bar. When the next subject is chosen then the next title "The basic operations" is changed into bold.

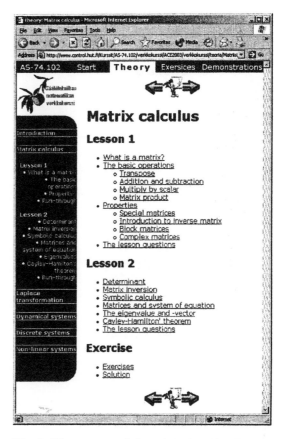

Fig. 2. The content of the selected section is seen in the first site of each section.

Fig. 3. The content of the section is in the left part of the site.

When the section is changed, the contents of the left bar is also changed. For example, if the section "Matrix calculus" has been studied and the student goes to the next section "Laplace transforma-

125

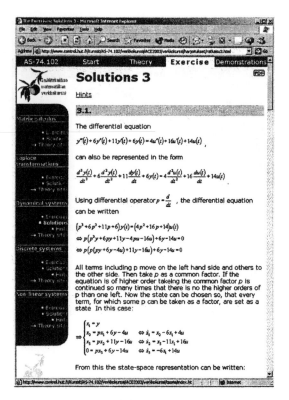

Fig. 4. The structure of exercise sites is similar to the theory sites. Compare with Figure 1.

tion" the content of the Matrix calculation site is hidden and the content of Laplace transformation is showed. The student can concentrate on the section he/she is currently learning and the other sections do not interfere.

3.3 The Exercise Sites

These sites contain the exercises, which are traditionally done by paper and pencil. Now the exercises and solutions were converted into HTML code. The structure of the exercise sites are similar to the theory sites. There are the same names of the sections on the left part of the site as on the theory sites. See Fig. 4.

3.4 The Lesson Questions

One target of the project was to investigate, what would be the best way to add interactivity, because the reading of pure text is not necessarily very educational. To add pictures is already a step towards better but it does not really use the possibilities of WEB. One way is to add Java Applets and it has been used very much although it is very laborious. Another problem is to make sure that it works in all browsers.

Another alternative is to use Matlab WebServer or WebMathematica. These are very good products

but the way you must implement an exercise may be too complicated.

One way to add interactivity to a WEB site is to use ASP (Active Server Pages). ASP is a server-slide scripting environment, which you can use to create dynamic and interactive web-sites and which are easy to develop and modify. You can combine HTML and script commands using ASP. For example, it is possible to connect to a database by using ASP. Traditionally, to process user input on the server you had to use a common CGI (Common Gateway Interface) application. With APS it is possible to collect HTML form information. APS is trademark of Microsoft.

An ASP script, which generates HTML code, was developed in the project. Using the script you can easily define lesson questions making it possible to the students to immediately check what they have learn. An example is shown in Fig. 5.

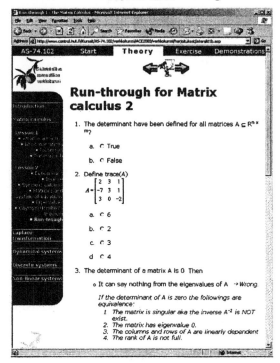

Fig. 5. There are lesson questions after each lesson, by which the students can make sure whether they have understood the main ideas. For example, the answer to question three is wrong in this case.

The lesson questions are defined in the following form.

> with(question.addNew())
> text = "The question. It can also contain html tags";
> newAnswer("The first answer. This alternative is wrong", false);
> newAnswer("This alternative is true", true);

126

:

rightAnswerText = "This contain what you want to write out if answer is right";
wrongAnswerText = "What you want to write out if answer is wrong";

The number of questions or alternatives have not been limited. The pages are modified by using an editor, for example Microsoft Notepad. The questions and all alternatives can contain HTML tags. For example you can insert an image or a list.

Answering the lesson questions is also an educational situation. The students can imediately see if their answers are wrong and, in this case, why they are.

3.5 The demo site

There are three Java Applets on the demo sites. One of them is a matrix calculator. The idea was to illustrate how matrix calculation works.

The second Applet illustrates the behaviour of first and second order systems. The user can change the system parameters with the sliders on the right side and the response of system updates immediately. The idea is that students grasp an intuitive understanding what the meaning of the system parameters is. See Fig. 6.

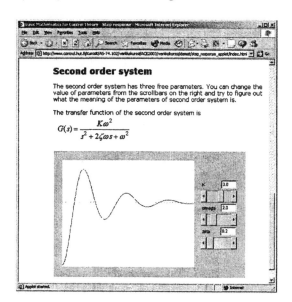

Fig. 6. The Applet illustrates the behaviour of the system.

4. CONCLUSION

A self-study course operating on the WEB was developed in the project. The course is a basic course of mathematics in control theory. The pedagogical idea was to teach the necessary mathematical tools together with some basics in control theory. There are a lot of examples and drill problems in the course.

The new idea in the course has been to write a "web-book" with easily editable lesson questions after each section. By answering the lesson questions the students can make sure whether they have understood the main ideas. These questions were implemented by using ASP script.

One of the goals was to improve the flexibility of studying. The possibility to study independent of time and place is more and more important. Therefore, the material was also converted to pdf-files which are possible to download or print.

The feedback has been positive. According to the students the material has encouraged them to study, and they have passed the course easily.

The Web course is not protected, so it is also available for you.

For further information

WEB-course
http://www.control.hut.fi/Kurssit/As-74.102/verkkokurssi

Finnish Virtual University
http://www.virtuaaliyliopisto.fi/index.php?language=eng

International Network of Universities
http://www.flinders.edu.au/international/links/inu.html

European University Network for Information Technology in Education
http://www.eunite-online.org/

Virtual HUT
http://virtuaali.tkk.fi/index.html

TieVie
http://tievie.oulu.fi/index.html

AN ON LINE COURSE FOR SUPERVISORY CONTROL TEACHING

Jana Flochová, Ronald Lipták, Peter Bachratý

Department of Automatic Control Systems,
Faculty of Electrical Engineering and Information Technology
Slovak University of Technology, Ilkovičova 3, 812 19 Bratislava, Slovak Republic
++421 2 60291 667, ++421 2 60291 341
emails: flochova@kasr.elf.stuba.sk, bachraty@kasr.elf.stuba.sk

Abstract: Discrete event systems (DES) describe the behavior of a plant as it evolves over time in accordance with the abrupt and asynchronous occurrence of events. Such systems are encountered in a variety of fields, for example sensor or actuator faults diagnosis, robotics, computers, communication networks and traffic. The control objectives of the supervisory control of DES are to prevent a set of forbidden states from being reach, to preserve the reachability of some non forbidden states, while at the same time enabling a maximal set of achievable state sequences. The aim of this paper is to introduce several approaches for the supervisors synthesis, a Matlab tool SUPCON and an on line multimedia Petri nets and DES course. *Copyright © 2003 IFAC*

Keywords: Discrete-event systems, Supervisory control, Control education, Petri-nets, Modeling and Simulation, Distance Education

1. INTRODUCTION

Today's plants and their control systems become more complicated and intensively interconnected. They are subjected to many constraints concerning safety and reliability conditions, energy consumption, environment protection, next to ever-increasing demands on economical production and trading-results. As complexity of systems increases rapidly, their forbidden state avoidance is becoming an importance issue. The supervisors of DES have been used to assure that the plant evolution will not violate a set of constraints imposed on its operation. Several approaches for the synthesis of supervisor's have been studied lately. One of the most powerful approaches developed to solve this problem - the supervisory control theory proposed by Ramadge and Wonham in 1987-1989 is based on untimed finite-state automata and formal languages.

This paper presents methodologies taught in a graduate course for modeling and implementation of discrete event supervisory control systems using a Petri net formalism. The outline of the paper is organized as follows: in the second chapter we explain supervisory control principles, in the third chapter we give a brief survey of Petri net approaches to the supervisory control theory. Chapter four presents algorithms of three methods for constructing Petri net supervisors. In the fifth chapter we introduce an interactive multimedia course and tests that have been designed for our master students and finally in the sixth chapter we describe a Matlab tool SUPCON. The on line course pages, some Matlab and C++ program algorithms used in supervisor's synthesis, and the design procedure of several Petri net-based supervisors will be included in the presentation.

2. SUPERVISORY CONTROL

Ho, Cassandras and Lafortune (1982, 1999) described the dynamic behavior of discrete event dynamic systems with a possibly infinite set of states and a finite set of events. The problem of supervisory control of

DES was presented by Ramadge and Wonham (1987, 1989) and since then has been studied extensively. For a given DES, it is of interest to synthesize a supervisor that prevents the occurrence of undesirable states and that guarantees that certain termination states are reached. Disabled events are certainly prevented from occurring and enabled events are not forced to occur.

A DES (plant to be controlled) is modeled as an automaton that generates a formal language over a finite alphabet, say Σ, whose elements label the automaton's transitions, or events. Events labeled by an element in a fixed subset Σ_c of Σ are declared to be controllable, that means they can be disabled by an external controller or supervisor. Events labeled by an element in a fixed subset Σ_u of Σ are uncontrollable, they can't be disabled e.g. machine breakdown. The problem is to design a controller that prevents the occurrence of states that violate the behavioral constraints directly or may lead to a violation of the constraints through the action of uncontrollable transitions. The supervisory controllers specification does not contradict the behavioral specifications of the plant model, i.e. the closed loop system is nonblocking, and achieves maximal permissiveness of the closed loop system within the specification (all events which do not contradict the specification are allowed to happen). Under suitable conditions the control law can be optimized, in the sense of minimally restricting plant behavior. A major goal in the field of supervisory control is the synthesis of supervisor under conditions where certain state-to-state transitions can neither be prevented by any action of the supervisor nor observed in the supervisory control system. The key properties of this theory are: controllability, nonconflicting, and observability. These properties are stated as language properties, i.e. independently of any particular DES modeling formalism.

3. PETRI NET-BASED SUPERVISORS

Petri nets possess many assets as models for DES. Concurrent processes and events can be easily modeled within the framework. The nets provide for systems with a large reachable state space a more compact representation than finite automata, they are better suited for representation of systems with repetitive structures and flows, but large reachable state-spaces. In addition they have an appealing graphical representation that makes it possible to visualize the state-flow of a system and to quickly see the dependencies of one part on another. Many reduction and decomposition techniques have been developed for these models, procedures exist for verifying their behavioral and structural properties.

A Petri net is a directed bipartite graph. The structure of a Petri net is described by (P, T, D^+, D^-), where P and T are representing the vertices of the graph, known as places and transitions. And D^+ and D^- are integer matrices with nonnegative elements representing the flow relation between the two vertex types. Places in a Petri nets hold tokens, whose distribution indicates the net's states or it's marking. Transitions direct the flow of tokens between places, thus the firing of a transition is a state-changing event in a DES model. A Petri nets incidence matrix represents the weighted connections of directed arcs between its places and transitions. It is composed of two matrices, D^+ representing arcs from places to transitions, D^- representing arcs from transitions to places. In a *PN* without self-loops we can write $D = D^+ - D^-$.

Formal definition of a Petri net: (Murata 1989, Češka 1994, David and Alla 1994).
A Petri net PN is the quintuple

$$PN = (P,T,F,W,M_0) \qquad (1)$$

Where
(1) $P = \{p_1, p_2, ..., p_n\}$ is a finite non-empty set of places.
(2) $T = \{t_1, t_2, ..., t_m\}$ is a finite non-empty set of transitions.
(3) $F \subset (P \times T) \cup (T \times P)$ is the set of directed arcs connecting places and transitions.
(4) $W : F \rightarrow \{1,2,3...\}$ is the weight function of the arcs.
(5) $M_0 : P \rightarrow \{0,1,2,3...\}$ is the initial marking.

The state evolution of Petri net *PN* is given by:

$$M_{i+1} = M_i + (D^+ - D^-)n_i$$
$$M_i = M_0 + (D^+ - D^-)x_i \qquad (2)$$

Where the vector n_i has exactly one non zero entry with value 1 indicating which transition fires, $x_i = \sum_{j=1}^{i} n_j$ is the firing count vector.

A *PN* is said *k-bounded* if $m_i \leq k$ for all $p_i \in P$ and all $M \in R(M_0)$, k is an integer constant. $R(M_0)$ denotes the set of states reachable from M_0, $R(M)$ the set of states reachable from M. A net is called live if every transition is live (potentially firable) in M_0, and it is said reversible if $M_0 \in R(M)$ for any $M \in R(M_0)$.

The controlled Petri nets is septuplet:

$$S = (P,T,F,W,C,B,M_0) \qquad (3)$$

Where:
(1) P, T are the non-empty sets of state places and transitions.
(2) F the set of directed arcs connecting state places and transitions.
(3) C is the finite set of control places, most one per transition.
(4) $B \subset (C \times T)$ is the set of oriented arcs associated the control places with transitions.
(5) M_0 is the initial marking.

A general PN supervisory control problem can be stated as follows. Given the specifications for a closed loop plant, and a Petri net model of the open loop DES design a feedback control law that disables as few controllable transitions as possible but that guarantees that the extra safety properties expressed by the specifications (such as boundedness, liveness and reversibility) are satisfied (Zhou et. al. 1992ab). In a manufacturing example these three properties imply the absence of overflows and deadlocks, and guarantee repeated execution of critical tasks and successful completion of production cycles.

Holloway and Krogh (1990) specified a class of Petri nets called cyclic controlled marked graphs with an easy control design algorithm. The specification of the behavior of the controlled system is determined with help of the forbidden markings. The forbidden markings are described by so-called place conditions, set conditions or class conditions.

Boel et al. (1995) treated the forbidden state problem for the class of DES modeled by controlled state machines - a dual class to the marked graph class. The authors characterized forbidden markings through general constraint sets obtained from unions and/or intersections of simpler constraint sets expressing that some places of the net cannot contain more than a certain number of tokens. In Stremersch and Boel (1998-2000) this approach has been extended to general Petri net models, and to more general forbidden sets.

Yamalidou et al. (1996) described a method for effectively constructing a Petri net supervisor given by places and arcs attached to the process Petri net model. The method is computed based on the concept of P-invariants. The supervisory control is specified using a single matrix multiplication without any state enumeration. The methods become computationally more cumbersome, and the maximally permissive control law cannot be found when one takes into account that some transitions are uncontrollable or unobservable. In Moody and Antsaklis (2000) the concept of place invariants has been used for enforcing linear constraints on a plant with uncontrollable and unobservable transitions.

Guia and DiCesare (1994) defined a class of Petri nets, called elementary composed state machine net (ECSM). To describe concurrent systems the model is extended by composing the state machine modules through concurrent composition, an operator that requires the merging of common transitions. The final model can represent both choice and concurrent behavior. The reachability problem for this class can be solved by a modification of the classical incidence matrix. The set of reachable markings is given by the integer solution of a set of linear inequalities.

Li and Wonham (1993) introduced the supervisory control of DES modeled by vector addition systems. Sreenivas (1997) presented a necessary and sufficient condition for the existence of a supervisory policy that enforces liveness in arbitrary controlled Petri nets.

Hrúz et al. (1996a, 1996b) described a class of interpreted conflict free Petri net models for real-time control of DES and their use in the supervisory control theory.

The severe complexity issues that arise in the design of supervisory controllers of DES due to combinatorial explosion, motivated decentralized, centralized hierarchical, and adaptive approaches to the supervisory control theory (Lin and Wonham 1988, Rudie and Wonham 1992, Barret, Yoo and Lafortune 1996-2001, van Schuppen 1996, Boel 2002). The proposed controller's architectures have been either independent of particular models or exploited their structures and properties.

Several tools (e.g. Supremica based on finite automata models - http://www.supremica.org/, Graphviz, Desco http://www.research.att.com/sw/tools/graphviz/ used in a control and communication course in Goteborg, Spectool a set of software tools for automatic synthesis of DES controllers of the University of Kentucky - http://www.engr.uky.edu/~holloway) have been developed and several automata-based supervisory techniques for a systematic PLC program design procedure in the IEC 61131-3 compliant environment have been proposed.

4. METHODS

Three methods of supervisory control have been implemented in the tool. The first method synthesizes the supervisor that consists only of places and arcs (Yamalidou et al. 1996). The main synthesis technique is based on the idea that specifications representing desired plant behavior could be enforced by making them invariants of the closed-loop system. The controller is described by an auxiliary Petri net connected to the plant's transitions. Supervisors designed in this way are computed very efficiently by a single matrix multiplication. The resulting controllers are identical to the monitors of Guia et al. (1994), which were derived independently using a different methodology.

The system to be controlled is modeled by a plant Petri net with the incidence matrix D_p: The controller net is the Petri net with incidence matrix D_c made up of the process net's transitions and a separate set of places. The controlled systems (controlled Petri net) are the Petri nets with incidence matrix D made up of both the original process net and the controller. The control goal is to force the process to obey constraints of the form

$$\sum_{i=1}^{n} l_i m_i \leq \beta \qquad (4)$$

where m_i represents the marking of the place p_i, l_i and β are integer constants. The inequality constraints can be transformed into equality by introducing nonnegative slack variables m_{cj} into them.

131

$$\sum_{i=1}^{n} l_i m_i + m_{cij} = \beta \qquad (5)$$

Each constraint will have a slack variable associated with it, and each slack variable will be represented in the controlled net by a control place. The controller D_c is defined by

$$D_c = -LD_p., \qquad (6)$$

Where L consists of the elements l_i. The initial markings of the controller can be written

$$m_{c0} = \beta - \sum_{i=1}^{n} l_i m_{i0}. \qquad (7)$$

The method may be applied to the systems whose constraints are expressed as inequalities or logic expressions involving elements of the marking and/or the firing vectors. It solves directly neither the problem of the uncontrollable transitions nor the problem of required accessibility of home places. These drawbacks can be mitigated partially by transforming the systems specifications to include all uncontrollably reachable states. However it is difficult to obtain maximally permissive controllers via this method.

The second method is based on the reachability tree analysis algorithm (Hrúz 1996b, Flochová, Hrúz 1996, Flochová et al. 1997). The novel aspect of the reachability tree analysis approach, that makes it potentially useful for control synthesis for large plants, is the fact that the reachability tree algorithm can be applied modularly. Under certain conditions it will be possible to generate a maximally permissive supervisory controller using this modular approach.

The supervisory control method based on reachability tree analysis. The basic algorithm consists of the following steps:
Without imposing the return to home states or nodes.
  Forbidden nodes in the reachability tree are labeled as *IA_nodes*.
  Deadlocks in the reachability tree are labeled as *IA_nodes*.
  Repeat until new forbidden nodes or not allowed edges were found
  - If the edge of the reachability three is uncontrollable and its successor is *IA_node*, the preceding node is labeled as *IA_node*.
  - If the edge of the RT is controllable and its successor is *IA_node* the edge is not allowed and labeled as *NA_edge*.
  - If all the edges going out of a node aren't allowed the node is labeled as *IA_node*.
Imposing the guaranteed return to home states or nodes.
  Begin in a home state. Analyze from which admissible nodes it is possible to reach home states. The admissible nodes from which one can't reach home nodes using allowed edges are converted into forbidden nodes.

The reachability tree can be designed modularly (Hudák 1994) and the set of weakly forbidden states (forbidden states + states that can uncontrollably lead to forbidden states) can be constructed without enumerating all nodes of the reachability tree.

The third method (Boel, Stremersch 1998-2000) is being added to the program algorithms. The authors constructed a controlled PN model of the system to be controlled, treated the forbidden state problem for the class of DEDS and general forbidden sets, allowing the specification of a lower and/or upper bounds on the number of tokens in some places. They characterized forbidden markings through general constraint sets obtained from unions and/or intersections of simpler constraint sets expressing that some places of the net cannot contain more than a certain number of tokens. The control laws require that one can observe the marking in certain influencing nets, containing all the places from where a token can uncontrollably reach a place involved in the specification of the forbidden markings. The method uses integer linear algebraic algorithms for describing the set of states reachable from the present state under certain control settings. If the Petri net satisfies certain structural conditions then a simple linear algebraic algorithm exists for enumerating all the markings reachable from a given state, when the control law blocks certain transitions. The method provides a necessary condition for maximal permissivity and allows approximating the worst-case uncontrollable behavior of the original plant Petri net without doing any reachability analysis.

5. ON LINE COURSE

The supervisory control teaching has been included in the graduate (master's) course of FEI STU "Theory of DES" and an on line multimedia course devoted to the Petri net supervisors synthesis has been prepared.

Concepts of DES control taught through classical lectures are completed by laboratory exercises and projects. The exercises have been built on present laboratory equipments and on hardware and software models (simulations of several simplified real plants) and at present they are divided into three basic modules. At the beginning of each module period a new case topic is assigned to student groups. The last, four weeks deal with the supervisory control theory and Petri net supervisors design. Students are expected to become familiar with basic problems of supervisory control; they have to formulate the control problem and the constraints imposed on the operation of their plants. They choose methods; provide a conceptual solution in a simulated environment (Matlab tool SUPCON see bellow), quantitatively test the proposed solutions and write a report – a case study on the assigned problem. Finally the case is presented to all students. The teacher's main tasks are to select good case topics, to present the relevant material, to explain the methods, to help with the supervisors design and with the simulation of the real time control.

Some features of the methodology "Motivation-by-challenge" have been used in teaching and an interactive on line course has been designed to help the students (and the teacher too) to deal with the theoretical problems and with the evaluation. The course includes the basic Petri nets theory, description of relevant Petri net classes, supervisory control principles and a MATLAB-based tool for student training sessions. The content of the on line course is divided into the following chapters:

- Petriu nets - basic definitions, state transition function, firing rules.
- Behavioral and structural properties of nets.
- Analysis techniques (the reachability tree, state equations).
- Reduction techniques.
- Petri nets classes and subclasses.
- Timed and hybrid Petri net models
- Supervisory control of systems modeled by PN.
- Tests.

Target groups for the described on line material are university students, university, high school and industrial schoolteachers.

6. MATLAB-BASED TOOL SUPCON

The supervisory control design can be solved by the program modules SUPCON written in Matlab 5.3/6.1. The programs need as processor at least PC Pentium. The user interface is written in English/Slovak and accepts inputs from the graphical simulation tool PESIM (Češka 1994, Urbášek, Češka 1998) and HPSIM (http://home.t-online.de/home/henryk.a/petrinet /e/hpsim_e.htm). The inputs from several free PN tools are being added. The software consists of several modules, which allow solving the following problems:

- PN analysis (boundedness, L1-liveness, deadlocks, P/T-invariants), reachability tree construction, incidence matrix design and correction.
- The supervisory control problem formulation and supervisory control analysis methods.
- The design of a supervisor, real-time control simulation and testing.

A part of the program solutions is an on line help file. The main menu and windows are fully keyboard and/or mouse driven and are shown in fig.1a-d.

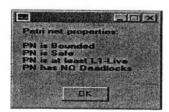

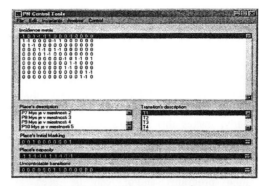

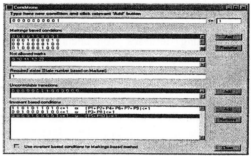

Fig. 1a-d The windows of the program SUPCON.

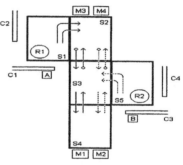

Fig. 2. A small robotic cell successfully solved by all programmed methods included in the tool SUPCON.

7. CONCLUSIONS

It is often necessary to regulate or supervise the behavior of discrete event systems to meet safety or performance criteria. DES supervisors are used to ensure that the behavior of the plant does not violate a set of constraints under a variety of operating conditions. The regulatory actions of the supervisor are based on observations of the plant resulting in feedback control. This paper has presented the description of an on line course being prepared for teaching Petri nets and Petri net based supervisory control of DES, the description of its supplementary training tool SUPCON and of program oriented methods for solving supervisory control problems. Methods described in the paper don't need more than matrix multiplications and basic background in the graph theory; they can be easily programmed in Matlab, C++, and/or IEC6311-3 Structured text. The first and the third method are based on matrix equations, are computationally very efficient, and can be used for a very large Petri net. The second method, based on the reachability tree analysis is enumerative by nature, but the fact, that the reachability tree algorithm can be implemented modularly, makes this method potentially useful for control synthesis for large plants.

The on-line course may be included in distance education programs and/or can be used as a supplementary master study material at the university. The approaches provide particularly simple methods for constructing feedback controllers for untimed discrete event systems modeled by Petri nets. Program modules written in Matlab 5.3/6.1 have been used to simulate the real time supervisory control of a discrete event dynamic system, to analyze and check the proposed controller, and to solve the OFFLINE supervisory control synthesis for discrete event systems. The implementation of control algorithms is performed in the IEC 61131-3 compliant programming environment with help of a quick PLC controller, or a quick PN-dedicated controller.

The proposed approaches to the PLC programs design include verification techniques, they prevent reaching of forbidden states, and after a systematic design procedure ensure the correctness of the underlying logic in the sense of disabling the inputs leading to forbidden possibly dangerous situations, economical damages and dropouts, or avoid some human mistakes. Applying and teaching this theory should be one of the key achievements of the controller programs in the future.

Acknowledgments
This work was supported in part by Slovak grant 7630 – Intelligent method for modelling and control and by the Research fellowship for Central and Eastern Europe of the Belgian federal office OSTC.
The first author thank professor René K. Boel for his help and suggestions on drafts of the paper.

REFERENCES

Boel, R.K., Ben-Naoum, L., Van Breusegem, V. (1995): On Forbidden State Problem for a Class of Controlled Petri Nets. *IEEE Trans. on Automatic Control*, **40**, no. 10, pp. 1717-1731.

Boel, R.K. (2002): Adaptive supervisory control, in: *Synthesis and Control of Discrete Event Systems* (Caillaud et al, Ed.), Kluwer Academic Publisers, 2002, Boston. pp.115-124. ISBN 0-7923-7639-0

Cassandras, C.G., Lafortune, S. (1999): *Introduction to Discrete Event Systems*. Kluwer Academic Publishers, 848 pages, ISBN 0-7923-8609-4

Češka, M. (1994): *Petriho sítě*. Brno: Akademické nakladatelství CERM, 94 s. ISBN 80-85867-35-4.

David, R., Alla, H. (1992): *Petri nets and Grafcet*. Cambridge: Prentice Hall international (UK) Ltd, ISBN 0-13-327537-X.

Flochová, J., Hrúz, B. (1996): Supervisory control for discrete event dynamic systems based on Petri nets. *Proceeding of International conference on Process control*, Horní Bečva, **2**, 80-83.

Guia, A., DiCesare, F., Silva, M.: Generalized mutual exclusion constraints on nets with uncontrollable transitions. In: *Proceeding of Int. conference Systems, Man, Cybernetics*, Chicago, Il, 1992, pp. 974-979.

Ho, Y. (ed.) (1982): Discrete Event Dynamic Systems: Analyzing complexity and performance in the Modern World. A Selected Preprint Volume, The Institute of Electrical and Electronics Engineers, Inc. New York.

Hrúz, B., Niemi, A.J., Virtanen, T.: (1996): Composition of conflict-free Petri net models for control of flexible manufacturing systems. *Proceeding of the 13th World Congress*, San Francisco, USA.

Hudák, Š. (1994): DE-compositional reachability Analysis. *Elektrotechnický časopis*, **45**, 11, pp. 424-431.

Lafortune, S.: http://www.eecs.umich.edu/~stephane/. Publications, 2000.

Li, Y., Wonham, W.M.: Control of Discrete-Event Systems I, II-The based Model. In: *IEEE Transaction on Automatic Control*, **38**, pp. 1214-1227, **39**, pp. 512-531.

Lin, F., Wonham, W.M. (1998): Decentralized supervisory control of DES. In: *Information Science*, **44**, pp.199-224.

Moody J.O., Antsaklis, P.J.: Petri net supervisors for DES with Uncontrollable and Unobservable Transitions. In: *IEEE Trans. on Automatic Control*, **45**, 2000, pp.462-476

Murata, T. (1998): Petri Nets: Properties, Analysis and Applications, In: *Proceedings of the IEEE*, **77**, no.4, pp. 541-580

Ramadge, P., Wonham, W.M. (1997): Supervisory control of a class of discrete event processes, *SIAM J. Control and optimization*. **25**, pp. 206-230.

Ramadge, P.J., Wonham, W.M.(1998): The Control of Discrete Event Systems. In: *Proceeding of the IEEE*, **77**, no.1, pp. 81-98.

Rudie, K., Wonham W.M. (1992): Think globally, act locally: Decentralized Supervisory Control. In: *IEEE Transaction on Automatic Control*, **37**, pp. 1692-1708.

Sreenivas, R.S.: On the existence of supervisory policies that enforce liveness in discrete-event dynamic systems modeled by Petri nets. In: IEEE Transaction on Automatic Control, **42**, pp.928-945. July 1997.

Stremersch, G., Boel, R.K. (1999): Enforcing k-safeness in controlled state machines, In: *Proceeding of 38th IEEE Conf. on Decision and Control*, 1999, pp. 1737-1742

Stremersch, G., Boel, R.K. (2001): Decomposition of supervisory control problem for Petri nets, in: *IEEE Transaction on Automatic Control*, **46**, pp.1490-1496.

Stremersch, G. (2001): *Supervision of Petri nets*. Kluwer Academic Publishers, 200 pages ISBN 0-7923-7486-X.

Urbášek, M., Češka, M.(1998): Extension of the Pesim simulation tool. In: *Proceeding of the XXth International Workshop Advanced Simulation of Systems*, Krnov, Czech Republic, pp. 81-86.

Wong. K.C., Van Schuppen, J. (1996): Decentralized Supervisory Control of DES with Communication, In: Preprints of International Workshop on Discrete Event Systems WODES'1996, London, pp.284-289.

Yamalidou, K., Moody, J. , Lemmon, M., Antsaklis, P. (1996): Feedback Control of Petri Nets Based on Place Invariants. *Automatica*. **32**, No. 1, pp.15-28.

Zhou, M.CH., DiCesare, F., Rudolph. D.L. (1992): Design and Implementation of a Petri Net Based Supervisor for a Flexible Manufacturing System. *Automatica*, vol. 28, **6**, pp. 1199-1208.

COURSE ON DYNAMICS OF MULTIDISCIPLINARY CONTROLLED SYSTEMS IN A VIRTUAL LAB*

Heřman Mann, Michal Ševčenko

Computing and Information Centre
Czech Technical University in Prague
Zikova 4, CZ-166 35 Prague 6, Czech Republic
{mann,sevcenko}@vc.cvut.cz

Abstract: The DynLAB project currently developed by an experienced international consortium aims at motivating young people to engineering study, and at improving engineering training using innovative didactic and technological approaches. The resulting web-based training modules are supported across the Internet by tools like a robust simulation engine DYNAST, publishing and monitoring system, and environment for virtual experiments. DYNAST can be used across the Internet as a modelling toolbox for the MATLAB control design toolset installed on client computers. *Copyright © 2003 IFAC*

Keywords: engineering, education, control, design, Internet

1. INTRODUCTION

The subject of dynamic and control underlies all aspects of modern technology and plays the determining role in the World-market competition of engineering products. Its importance increases with the ever-growing demands on operational speed, efficiency, safety, reliability, or environmental protection. National authorities and entrepreneurs in many countries, however, report lack of professionals well qualified in this field as well as a critical overall decline of interest in engineering study among young people.

Professional associations call for radical changes in the engineering curriculum and for innovative approaches to vocational training (e.g., [1]). The existing courses are criticised namely for discouraging young people from engineering study by overemphasis on theory and mathematics at the expense of practical engineering issues. Dynamics is covered in several courses separated along the borders between the traditional engineering disciplines despite the fact that most of the contemporary engineering products are of multidisciplinary nature. Computers are often used to carry out old exercises without radical modification of the curriculum to exploit capabilities of current software.

Automatic control education is criticised for a very narrow approach [2]. Courses on control are presenting 'textbook' problems engineered to fit the 'underlying' theory without undertaking realistic modelling of control systems. Professors tend to teach more and more sophisticated control algorithms as applied to oversimplified models of controlled plants. On the other hand, the industry mostly resorts to rather simple control, but uses very realistic models to verify the design sufficiently.

To reverse this gap widening between academia and industry, it is necessary to attach greater importance to all phases of the control-design process – namely to modelling and identification – key factors for achieving a good design. Nevertheless, only few engineering schools have introduced realistic modelling as a distinct topic and have given their students the opportunity to deal with real-life problems and practical tasks.

Professors often assume that modelling and simulation is just a matter of routine utilising a ready-made software package. As they consider such activity uninteresting academically, many of them still have had no 'hands-on' personal experience in modelling that they could share with their students.

On the other hand, many engineers working in the industry are very competent in this field, but they very rarely publish. Newcomers in large organisations can fill the gaps in their education and training

---

* This is an outcome of the DynLAB Pilot Project partly supported by the Leonardo da Vinci grant No. CZ/02/B/F/PP/134001.

by learning from old-timers, but those starting in small enterprises must struggle on their own.

2 PROJECT DynLAB

2.1 *Project consortium and background*

The above mentioned analysis gave rise to the Pilot Project Project DynLAB within the Leonardo da Vinci Vocational and Training Programme. The aim is to develop and disseminate a Web-based course on dynamics and control of multidisciplinary engineering systems. The project consortium consists of the following academic and industrial institutions:

- Computing and Information Centre, Czech Technical University in Prague (Co-ordinator)
- Automatisierung und Prozessinformatik, Ruhr-Universität, Bochum
- Institute of Technology Tallaght, Dublin
- Fraunhofer Institut Integrierte Schaltungen, Dresden
- ABB Automation Control, Västerås
- University of Sussex, Brighton

The project builds on the partners' experience gained in the previous projects, namely RichODL and DynaMit. Outcomes of these two projects initiated establishing two Virtual Action Groups – one of them is focused on Multidisciplinary System Simulation, the other one on Teachware. The Groups are parts of the IEEE Control Systems Society Technical Committee on CACSD [3].

1.2 *Project application areas and target groups*

The application areas of the course include dynamics of:

- electrical, electronic and magnetic circuits
- mechanical and automotive systems
- electromechanical devices
- fluid power and acoustic systems
- heat-transfer systems
- energy transducers and sensors
- vibration and damping systems
- robots and manipulators
- mechatronic systems
- manufacturing machinery
- vehicles and transportation systems
- power electronics

The main target groups of the DynLAB course are

- regular students wishing to complement the traditional face-to-face courses
- distance-education students at different levels of vocational study and training
- practising engineers in the context of their continuing education or lifelong learning
- disadvantaged people who want to study from their home
- teachers intending to innovate the courses on dynamics and control they teach
- industrial enterprises interested in enhancing the qualification and efficiency of their staff
- providers of continuing and life-long-learning engineering courses

2. PROJECT OUTLINE

2.1 *Innovations in the project*

The emphasis and style of the proposed course differs from most of the existing courses by

- exposing learners to a novel systematic and efficient methodology for realistic modelling of multidisciplinary system dynamics applicable to electrical, magnetic, thermal, fluid, acoustic and mechanical dynamic effects in a unified way
- introducing learners to the methodology through simple, yet practical, examples to stimulate their interest in engineering before exposing them to rigor math
- giving learners a better 'feel' for the topic by problem graphical visualisation and interactive virtual experiments
- allowing different target groups to select individual paths through the course tailor-made to their actual needs and respecting their background
- allowing both for self-study and remote tutoring combined with investigative and collaborative modes of learning
- integrating computers into the course curriculum consistently and giving learners a hands-on opportunity to acquire the necessary skills
- exploiting the computers not only for equation solving, but also for their formulation minimising thus learners' distraction of from study objectives
- giving learners the opportunity to benefit from 'organisational learning', i.e. from utilising knowledge recorded during previous problem solving both in academia and industry

136

2.2. Presentation of system dynamics

Figure 1 shows examples of three different graphical presentations of system dynamics exploited in DynLAB. Movable 3D virtual-reality geometrical models allow learners to investigate dynamic behaviour of the systems under study qualitatively. Plots of system responses allow them to evaluate the system behaviour quantitatively.

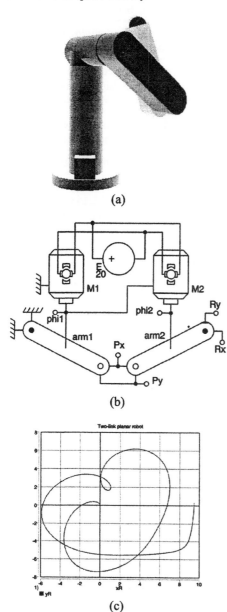

(a)

(b)

(c)

Fig.1: Robot: (*a*) 3D geometric model, (*b*) multipole diagram, (*c*) plot of the robot-arm trajectory.

The plotted responses result from simulation of the system dynamics. In DynLAB, such a simulation exploits *multipole models* of system dynamics represented graphically by *multipole diagrams*. As these diagrams portray directly the configuration of real systems, their set up is easy and straightforward.

Multipole diagrams consist from graphical symbols of multipole models of individual components of the modelled real system. The symbols are interconnected by line segments representing energy interactions between the real components. Each of the line segments is associated with a pair of conjugate variables the product of which expresses the power transferred in the interaction. Interconnections of the line segments respect physical laws governing the energy interactions.

The multipole diagrams can consist from symbols of twopoles like 'pure' resistors, capacitors, dampers as well as from symbols of sophisticated multipole models of complex real components like motors, valves, amplifiers, etc. Learners have at their disposal a large collection of multipole models of different level of abstraction and idealisation for the typical system components.

Topology of multipole diagrams is isomorphic with the geometric configuration of the modelled real systems. Thanks to this, the diagrams can be set up in a kit-like way based on mere inspection of the real systems in the same way in which the systems have been assembled from their real components. There is no need for forming any equation, a block diagram or a bond graph. If necessary, however, the multipole diagrams can be freely combined with equations or block diagrams.

Using the multipole approach is also of several other important advantages:

- multipole models can be developed, debugged, tuned up and validated once for ever for the individual subsystems independently of the rest of the system, and once they are formed they can be stored in submodel libraries to be used any time later
- this job can be done for different types of subsystems (e.g., fluid power devices, electronic elements, electrical machines, mechanisms, etc.) by specialists in the field
- behaviour of the individual submodels can be represented by different descriptions each of them suiting best to the related engineering discipline or application (lagrangian equations in mechanics, circuit diagrams in fluid power or electronics, block diagrams in control, etc.)
- the submodel refinement or subsystem replacement (e.g., replacement of an electrical motor by a hydraulic one) can be taken into account without interfering with the rest of the system model

2.3 Learning modes

Table 1 shows examples of learning modes used in DynLAB.

Table 1: <u>Learning modes in DynLAB</u>

Learning objective	Prerequisites	Course assignment	
		Given	Task
stirring up interest in dynamics	high-school math and physics	3D virtual model of a real system	to modify system parameters and excitation to observe changes in its dynamic behaviour
introduction to dynamic modelling	high-school math and physics	configuration of a real system	to set up the corresponding multipole diagram and to simulate its behaviour
more advanced dynamic modelling	fundamentals of system dynamics	configuration of a real system	to set up the multipole diagram from custom-made submodels and to simulate its behaviour
formulation of system equations	introduction to dynamic modelling	configuration of a real system	to form corresponding equations and to solve them, to set up the multipole diagram, and to compare the solution with simulation results
introduction to control design	formulation of system equations	model of a plant & control objectives	to reduce the model, to design control, and to verify it using the plant unreduced model
introduction to system design	introduction to control design	system specification	to design system configuration and to optimise its parameters
design of virtual experiments	advanced dynamic modelling	experiment specification	to design 3D virtual model, to set up the dynamic model, and to write the simulation script

2.2 *Learning environment*

The DynLAB course is delivered within a Web-based *learning environment* supporting learners' mutual collaboration and communication with their tutor. The investigative way of learning is encouraged by open problems and virtual experiments. The course flexibility is achieved by its modular arrangement with a number of different entry points. In each module, the prerequisite knowledge required for its study is clearly specified.

The course is accompanied by a large collection of *examples* of various problems solved both in academia and industry to imitate knowledge sharing and informal learning typical for large organisations. The examples can be resolved and modified in an interactive way across the Internet. This gives the learners a hands-on opportunity to acquire the necessary skills in solving real-life problems.

Organisational learning imitates knowledge sharing and informal learning typical for large organisations. It is supported in DynLAB by a computer-assisted ontology-based process in which knowledge gained during solution of problems is captured, recorded and later made available to learners 'just in time' when it is relevant to the problems they are supposed to solve.

Self-study is supported by interlacing the course texts with self-assessed quizzes, tests and other motivation elements. For the remote tutoring mode, there is a collection of tutor-marked assignments in each module.

3. INTERNET-BASED TOOLS

3.1 *Modelling, simulation and visualisation*

DynLAB partners have developed a number of innovative tools applicable to the project. one of them is DYNAST – a software package for efficient modelling, simulation and analysis of multidisciplinary systems. It consists of several software components that can be installed either on a single computer, or they can form a distributed system interconnected by the Internet.

The kernel of the package – DYNAST Solver – is a tool for

- solving implicit sets of nonlinear algebro-differential *equations* submitted in a natural textual form
- analysing nonlinear *multipole diagrams* that may be combined with block diagrams or equations and submitted in a graphical form
- linearising the diagrams and providing their semisymbolic analysis in the time- and frequency-domains

In the case of multipole diagrams the underlying equations are formulated automatically and then solved by the DYNAST Solver. The Solver can be accessed across the Internet in a Web-based, on-line and e-mail modes as it is illustrated by figure 2.

Setting up the multipole and block diagrams in a graphical form directly on the Web is enabled by the schematic editor DYNCAD, a Java applet.

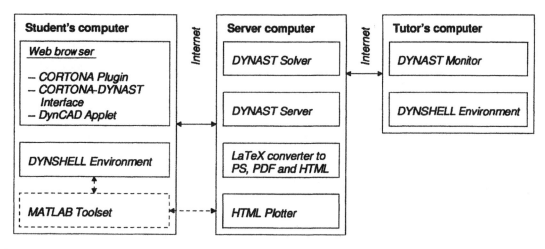

Fig.2: Environment for modelling, simulation, virtual experiments, control design and publishing across the Internet.

DYNCAD converts diagrams into textual files and sends them across the Internet to DYNAST Solver installed on a server. After the computational results are sent back, they are plotted on the client-computer screen. Users can open their free private accounts in DYNCAD and store their simulation problems on the server. DYNCAD is also able to convert the set-up diagrams into PostScript and send them to users by e-mail.

DYNAST can be accessed in an even more comfortable and user-friendly way via the on-line mode. This mode requires, however, downloading and installing working environment called DYNSHELL on client computers with MS Windows. This software has been designed to suit to users of different levels of qualification and experience. Dialog windows (wizards) allow for submitting problems in an intuitive way without learning any simulation language. All operations are supported by a context sensitive help system, and they are continuously checked by a built-in syntax analyser. Dynamic diagrams can be submitted in a graphical form using a built-in schematic editor.

DYNAST Shell can also communicate across the Internet with the LaTeX-based software package for automated publishing reports on simulation experiments. The documents can be published in PostScript, PDF and HTML. The simulation results can be also used for animation of 3D geometric models of the simulated objects using VRML. The learners need to download and install on their PCs only the free CORTONA software.

Another very useful tool is the DYNAST Monitor. Installed on computers of DynLAB tutors, it allows them to observe the data files and diagrams submitted by learners to DYNAST Solver. The tutors can then help the learners to overcome their eventual difficulties and discuss their problems.

3.2. *Control design with MATLAB*

MATLAB – the most popular control-design toolset admits model descriptions in the form of block diagrams or equations. These descriptions suit well to the abstract and idealised models used in control synthesis. Using them, however, for 'virtual prototyping', i.e. for thorough control design verifications and for realistic dynamic studies, is too laborious, cumbersome and error prone. Equations describing the system model as well as a block diagram representing the equations must be formed manually before the block diagram can be submitted to a computer. In addition, the block-diagram-oriented simulators usually encounter numerical problems with causality, algebraic loops, changes of the equation order, etc.

In DynLAB, learners use MATLAB neither for simulation, nor for virtual experiments. They are exploiting its advantages for control design, however. The server-based DYNAST can communicate with MATLAB control-design toolsets installed on learners' computers across the Internet.

Using either DYNCAD or DYNSHELL, the dynamic-diagram model of a plant to be controlled can be easily set up in a graphical form. DYNAST can be then used to simulate the plant and to validate its open-loop model. If the model is nonlinear, DYNAST is capable of linearising it. Then it can compute the required plant transfer-function poles and zeros, and export them to MATLAB in an M-file. After designing an analogue control within the MATLAB control-design environment, the DYNAST model of the plant can be augmented by the designed control structure and thoroughly verified by DYNAST. As an example, figure 3 shows closed-loop model of the inverted pendulum problem specified in [4]. The procedure is described in more detail in [5].

In the case of digital control, there is another option for verification of the designed controlled system. After designing the digital control using the MATLAB control-design toolset, the resulting control structure is implemented in Simulink installed on the client computer while the controlled-plant model remains in the remote DYNAST as shown in figure 4. During the verification, Simulink communicates via its S-function with DYNAST across the Internet at each time step.

CONCLUSIONS

The automated access analysis to the DynLAB project website [5] clearly indicates that the tools available on the server are utilised across the Internet by visitors from all over the world. Their number grows rapidly despite the fact that the project outcomes are still in the development phase.

REFERENCES

[1] *Future Directions in Control Education,* special section in the *IEEE Control Systems,* Vol. 19, No. 5, Oct. 1999).

[2] S. Dormido Bencomo: *Control Learning: Present and Future.* b'02 *IFAC Congress* plenary paper, Barcelona 2002

[3] *IEEE Control Systems Society Technical Committee on Computer Aided Control System Design* - http://www-er.robotic.dlr.de/cacsd/

[4] Messner, B. and D. Tilbury. 2000. *Control Tutorials for MATLAB* at http://www.engin.umich.edu/group/ctm/

[5] Website of the DynLAB project at http://icosym.cvut.cz/dynlab/

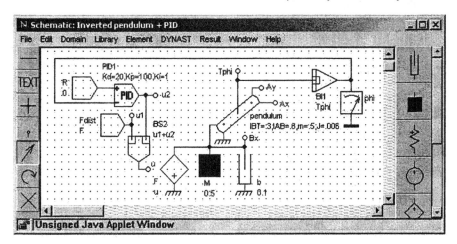

Fig. 3: Analogue PID pendulum control.

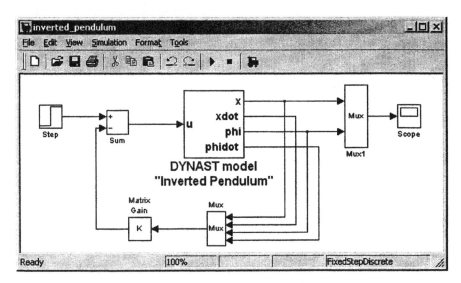

Fig. 4: Digital control of pendulum.

www.elsevier.com/locate/ifac

VIRTUAL-ELECTRO-LAB, A MULTINATIONAL LEONARDO DA VINCI PROJECT

Gheorghe Scutaru [X]), Paul Borza [X]), Vasile Comnac [X]),
Ingmar Tollet [Y]), Seppo Lahti [Y])

[X]) *Transilvania* University, BRASOV, ROMANIA, tel & fax 00 40 268 474718, e-mail. scutaru@unitbv.ro
[Y]) EVITech, ESPOO, FINLAND, tel +358 40 569 2838, fax +358 9 511 9988, e-mail. Ingmart@evitech.fi

Abstract: A 5 - country multinational Leonardo da Vinci project with 4 EU partners and 4 Romanian partners with the acronym *VIRTUAL - ELECTRO - LAB* has been approved for 2002 - 2004. The goal of our project consists of developing a complex training system that includes the correlation of the courses, seminars / workshops and testing systems with the virtual & remote experiment elements. Some aspects of this project are presented. *Copyright © 2003 IFAC*

Keywords: Virtual - Electro - Lab, Virtual and Remote Experiment Elements, Software Tools, Laboratory Education.

1. INTRODUCTION

In January 2002 a Leonardo da Vinci Project with partners from Romania and 4 EU countries with the acronym VIRTUAL - ELECTRO - LAB were initiated for a period of three years. The 8 partners are Hogeschool, Gent, Belgium, EVITech, Espoo - Vantaa, Finland, Laboratorio delle Idee s.a.s, Italy, UNINOVA, Lisbon, Portugal, Transilvania University of Brasov (promoter), County School Inspectorate Brasov, County Agency for Employment and Vocational Training, Brasov and PSE Siemens, Romania.

The goal of our project consists of developing a complex training system that includes the correlation of the courses, seminars / workshops and testing systems with the virtual & remote experiment elements. This is pointed out through the following:

It proposes an innovative approach of the teaching method using a virtual & remote laboratory and it supposes the setting up of new ICT software tools for training in order to implement this innovative approach.

In this presentation the focus of our attention is just on a few items, i. e. the architecture and the operating mode of the Virtual - Laboratory, to give the general description of the Software

Tools which will be generated and present a concrete example. However, the Software Tools are still under development in early 2003.

2. THE ARCHITECTURE AND THE OPERATING MODE OF THE VIRTUAL - LABORATORY

The developing of the remote and virtual laboratory represents a new stage on the way to the informational society. The education and especially the open and distance learning system assure the continuous updating of knowledge for numerous persons along their life. The critical aspect of this form of educational process is the capacity of the system to transmit and build solid practical skills for students.

Our solution represents a first approach in this direction. In addition, our work represents a new implementation for some remote measurement systems in the electrical domain.

The main ideas are:
- To build a system organized on three tiers: Clients, Web Server & SQL Server and Work Bench Server;
- To accomplish a system that has some very useful features: standardization of elements,

functions and languages, scalability and interoperability;
- To state the data (the protocols) in order to assure the features mentioned above;
- To develop a data-base application for the management of the users and requests.
Between the important features of implementation we can mention the following:
• Dynamic re configurability of laboratories in order to implement some optimization functions like as: on line adaptation of work shops at client requests, optimization of global hardware resources of laboratory in

order to realize quickly the measurement requests.
• Allow for the clients of laboratory to change remotely and customize the work shops in order to implement the educational principle what you see, can modify and follow the consequences of own actions.

In Fig. 1 is exhibited the hardware development and Fig. 2 is shown the proposed software architectural solution.

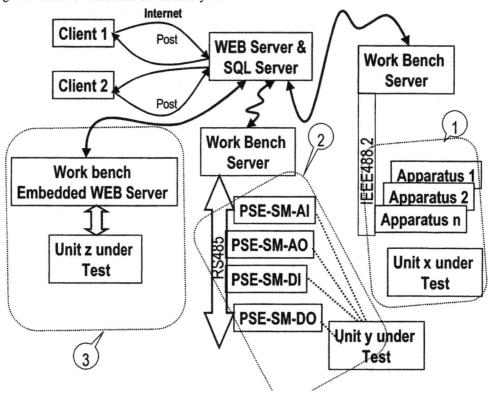

Fig. 1. The structure of remote & virtual laboratory

The structure of the system will be presented below. PSE – SM xx represent dedicated micro-controller systems specialized in processing of the analog input signals, the analog output signals, the digital input and output signals, too. A new micro-system GeMASYS, based on AduC812 processor that allows one to acquire in DMA mode the analogue signals with a sampling rate that attains 200KS/s, has been designed and used.

A different kind of specific workbench servers has been used:
- Sets of measurement instruments linked using the IEEE488.2 bus (1);
- PCs used to manage a LAN developed around the RS485 bus (2);
- Embedded Web Servers, which control directly the measurement process (3).

Both the workbench servers and the http clients use a polling mechanism for posting data and retrieving the results. This will provide a virtual communication channel among http client and workbench servers.

The way to accomplish a remote experiment supposes the following steps, in this stage:
1. Client authentication
   - If he/she is a new client, he/she needs to register; if not, the client is logged in;
2. Chooses the page corresponding to his/her desired workshop;
3. Chooses the parameters and the desired devices;
4. Push the START button for the desired experiment ;
5. The consequence of the command issued by the client is the transmission of a request to the Web Server that stores and parses the content in the data base;
6. The appropriate workbench server is periodically polling the data base for requests addressed to it. In a particular situation the clients can remotely reconfigure the structure of laboratory on the principle of basis cubes that

form primary entities of work shop. (e.g. colonies of apparatus – signal sources, array of multiplexers, colonies, classes of measurement instruments that can be manual or automatically re-configured. In this sense the SQL server include a engine that analyze the current status of all elements of laboratory and that assure the automatic reconfiguration of labs.
7. It finds the request and processes it and then returns the response into the data base through the
   WEB Server;
8. Also, in this time the client periodically polls the data base for response;
9. It finds the response and displays the appropriate data.

In Fig. 2 is shown the architectural image of our system. In order to provide a distributed workflow and syndication capabilities a xml / web service interface has been proposed. This last feature assures the auto-documentation of laboratory (the MS-SQLXML3.0 has been used).

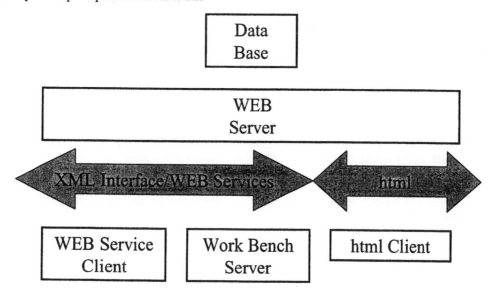

Fig. 2. The architectural view of R&V Lab system

The interface will describe the access methods / parameters and each device under test and meta data.

At the LAN level of each workbench server (variant 1 and 2) is used the structure of the message that is presented in Fig. 3.

Communication Protocol

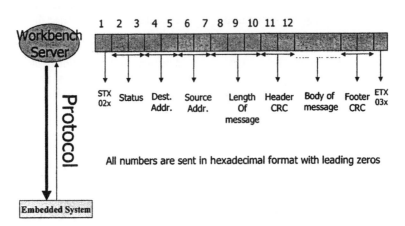

Fig. 3. The structure of protocol

The body of the message could accept all the formats of the command. Two parsers placed on the second and third layer assure the decoding of messages for each element. At the level of SQL Server, the message status that comes from workbench server is analyzed and memorized and on this base, a command message is transmitted to the workbench server. The time-out systems assure the rejection of requests transmitted to the element that is not in use or does not respond.

3. THE GENERAL DESCRIPTION OF THE SOFTWARE TOOLS

The conventional didactical modalities do not ensure a convenient intuitive support in order to reach a deep understanding of the phenomena in electrical domain. Our software-tool set is dedicated to the teachers that perform didactical activities in electrical domain and meets a fundamental didactical need: the lack of an intuitive support from practical life that guides the deep understanding.

The software-tool set consists of the following subjects:

Package 1, Electrical
1. Properties and characteristics of the electrotechnical materials
2. Simulation and computing of electrical circuits
3. The electrical transformer and the induction machine

Package 2, Electronic
4. Measurements of electronic devices and circuits
5. Simulation of the electric drives
6. Home appliance systems (DOMOTICS) and peripheral components

Package 3, Automation
7. Web-oriented applications on databases used in electrical processes
8. Measurement & automated test systems
9. Simulation of the control systems used in electrical processes

Each software tool is enhanced by applications performed in the virtual laboratory described above.

Pedagogical approach

Multiple complexity levels of contents will be provided, for each software tool, in order to allow a flexible and open access. Each software tool is organized on three levels: basic, intermediate and advanced.

The contents will be conceived according to the dynamic requirements of the professions on the labor market and with development of trends in technological fields.

All the software products developed in the frame of the project will meet the following requirements:

1. Start from real life technical problems in defining pedagogical objectives of the training.
2. Use individual's effort and personal experience as a base in constructing new knowledge.
3. Prove support in experiment and visualization of very abstract electrical phenomena in order to help a
    deeper understanding.
4. Use specific interactivity of educational software to obtain a customization of the teaching / learning
    process.

The following procedure will be used to produce a software tool:

- Accomplishing of a detailed need analysis for each training type in order to identify the study disciplines
    for which these software products will determine a maximal efficiency.
- Designing and making up of the software products.
- Testing of the software tool produced in order to improve it.

After project fulfillment, the primary target group's person that wants to use the results of the project will have to follow the procedure presented in Fig. 4.

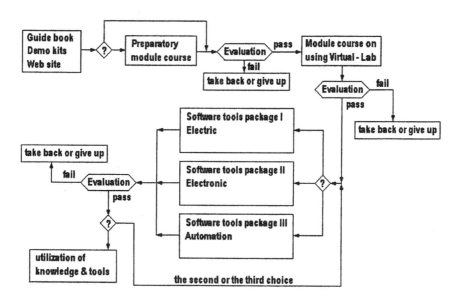

Fig. 4. The procedure for using the developed tools

4. A CONCRETE EXAMPLE - DC MOTOR SPEED MODELLING IN SIMULINK

A common actuator in control systems is the DC motor. It directly provides rotary motion and, coupled with wheels or drums and cables, can provide transitional motion. The electric circuit of the armature and the free body diagram of the rotor are shown in Fig. 5:

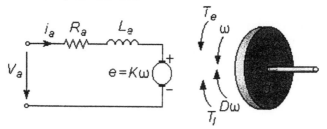

Fig. 5. The diagram of the DC electric motor

For this example, the following values for the physical parameters are assumed:

The moment of inertia of the rotor $J = 0.01 \, \mathrm{Nms}^2$, damping ratio of the mechanical system $D = 0.1 \, \mathrm{Nms}$, motor constant (= electromotive force constant = torque constant) $K = K_e = K_t = 0.01$ Vs or Nm/Amp, resistance $R_a = 1$ ohm, inductance $L_a = 0.5$ H, input V_a = Source Voltage, output (omega): angular velocity of shaft.

The rotor and shaft are assumed to be rigid. The motor torque, T_e, is related to the armature current, i_a, by a constant factor K_t : $T_e = K_t i_a$. The back emf, e, is related to the rotational velocity: $e = K_e \, \omega$. In SI units (which will be used), K_t (torque or armature constant) is equal to K_e (emf constant).

Kirchhoff's voltage law describes the behavior of the electric circuit armature:

$$L_a \frac{di_a}{dt} + R_a i_a + K_e \omega = v_a \, . \qquad (1)$$

The free-body diagram for the rotor, shown in Fig. 5, defines the positive direction and shows the three applied torques, T_e (motor torque), T_l (load torque) and $D\omega$ (damping torque). Application of Newton's law yields the model of the motor mechanical part described by the differential equation:

$$J \frac{d\omega}{dt} + D\omega + T_l = K_t i_a \, . \qquad (2)$$

Using Laplace transforms for the equations (1) and (2) one gets the relations:

$$I_a(s) = \frac{1}{sL_a + R_a} [V_a(s) - K_e \Omega(s)] \, , \qquad (3)$$

$$\Omega(s) = \frac{1}{sJ + D} [K_t I_a(s) - T_l(s)] \, . \qquad (4)$$

The equations (3) and (4) are implemented as a SIMULINK Diagram in Fig. 6.

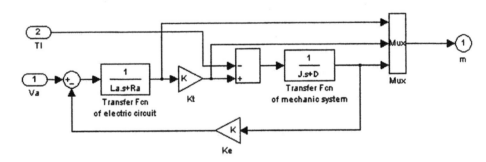

Fig. 6. The SIMULINK Diagram of the DC electric motor

5. CONCLUSIONS

An innovative approach of the vocational teaching method using a virtual & remote laboratory, which consists of developing a complex training system that includes the correlation of the courses, seminars / workshops and testing systems with the virtual & remote experiment elements has been presented. Trainees and teachers that perform didactical activity in electrical domain are the main target group. During the first project year, i.e. 2002, the Software Tools have been developed and the first ones are scheduled to be ready by March 2003, hence, there are not yet experiences about their use in teaching environments.

ELSEVIER
IFAC
PUBLICATIONS
www.elsevier.com/locate/ifac

THE EXPERIENCE OF CREATING OF THE ELECTRONIC TEXTBOOK ON THE HISTORY OF AUTOMATIC CONTROL THEORY

N.A. Pakshina *

Nizhny Novgorod State Technical University at Arzamas, 19, Kalinina Str., Arzamas, 607220, Russia. E-mail: pakshina@afngtu.nnov.ru

Abstract: The experience of creating of the electronic textbook on "The History of Automatic Control Theory" is discussed. First part of the textbook devoted to the founders of automatic control theory in Russia, such as I. A. Vyshnegradskii, A. M. Lyapunov, I. N. Voznesenskii, A. A. Andronov and others. Second part tells about the life and activity of brilliant scientists from other European countries: J. C. Maxwell, A. Stodola, E. Routh, A. Hurwitz . At the end of each part there is a test with questions on the given material. Much attention is paid to historical connections among scientists. *Copyright ©2003 IFAC*

Keywords: history, education, textbook, Web pages, brilliant scientists, control theory.

1. INTRODUCTION

The humanity has entered the XXI-st century, the century of further development in the sphere of science and technology. The names of the great scientists, who presented people with the results of their brilliant researches and inventions and made possible new professions to appear and new scientific perspectives to be realized, are becoming the history now. Who are they, "the movers of progress"? What were their fate and the circumstances in which they had to realize their ideas?

Among the senior students of our university, having specializations in applied mathematics and electrical engineering was held an anonymous questioning, which can be called "Who, What, When? " for short.

The results was surprising for us. It turns out that many important historical perspectives in the field of computer science and control theory are unknown for our students. Taking into account this fact we try to create the electronic textbook devoted to history of automatic control theory. From our point of view the electronic form is the best for this goal.

The paper is organized as follows. In Section 2 we give the the results of the questioning among the

senior students. In section 3 we analyze this results. In section 4 we describe our electronic textbook. The organization of this textbook is based on hyper textual technology. The section 6 explains the structure of the textbook by the example of the Web pages devoted to I. A. Vyshnegradskii. Some short concluding remarks ends the paper.

2. THE RESULTS OF THE QUESTIONING

The students were given a list of famous scientists, specialists in mathematics, computer science, and control theory. The students were to answer the following questions on each scientist:

1) In what spheres were gained the main results?

2) In what century and what country?

The attention was paid not to the biographical or creative activity details. In other words, the question was "Who is this, in your opinion? " For the information processing was made a special programm, which gives the opportunity of multilateral analysis of the information. See (Pakshina, 1998) for more detail information about this programm and this questioning. The results of the questioning showed that the majority of the students have rather vague ideas of the valuable contribution to the development of automatic control theory that made such outstanding scientists as Maxwell and Stodola. The scientists worked in our Nizhny Novgorod region were even less lucky. It turns out that the founder of the famous Nizhny Novgorod scientific school in the field of nonlinear oscillation and control theory, academician A.A.Andronov, is very bad known for most of the students. The former Andronov's student, professor Yu. I Neimark, actively working in University of Nizhny Novgorod, as a lot of students think, lives in Israel. The names of other distinguished scientists from Andronov school, such as Ya. Z. Tsypkin, M.A. Aizerman, were practically unknown for testing students too.

3. THE GAPS OF KNOWLEDGE

In our opinion, these gaps of knowledge are the results of dealing our students in the course of studies with lectures, not with textbooks. There are plenty of books on automatic control theory but there are very few ones on the history of this science. There is historical perspectives in IEEE Control Systems Magazine, nice book by Andronov and Voznesenskii (1949), others journal and conference papers (Dorato, 2002), but these publications are not available for all the students. For educational process one needs well structured and organized textbook.

The thing is quite different with computer science. Though there are not so many books on the history of computer science and computer technology, there exist plenty of virtual museums of computer science and computer technology where one can find biographies of prominent scientists and programmers, see (Davydova, 2000). The Internet can be of great help here.

As far as automatic control theory is concerned, there is lack such information on Russian Web sites. The material about the scientists' achievements is very voluminous and cannot be delivered by lectures. Taking into consideration that the efficient ability of the human's visual analyzer is 100 times larger than of the acoustic one, see (Krechetnikov, 2002), it is better to give preference to individual work with an electronic textbook. Thus was decided to create such a textbook.

4. THE TEXTBOOK

The organization of this textbook is based on hyper textual technology. The pages are in HTML format with embedded hiperlinks. The use of hyper references makes it possible to cut down the time for searching of the necessary information. The information in such a textbook has a more visual look, see (Poindexter and Heck, 1999). Initially, some Web pages used background pictures and animation effects but later they were rejected as they increase fatigue and make the apprehension of the material less effective. The range of colors was chosen according to physiological peculiarities of color perception. The preference was given to the yellow color as it belongs to the long-wavelength part of the spectrum and facilitates power activation and mobilization and creates optimistic mood as well. To produce the impression of coherence and continuity, the range of colors is the same on all the Web pages. Interactive Web pages with grafical displays for inputting data are written using Java applets.

At the moment the textbook consists of two parts: "The Founders of Automatic Control Theory in Russia", devoted to such prominent scientists as I. A. Vyshnegradskii, A. M. Lyapunov, I. N. Voznesenskii, A. A. Andronov, and "The Founders of Automatic Control Theory in Europe" which tells about the life and activity of foreign brilliant scientists J. C. Maxwell, A. Stodola, A. Hurwitz, E. Routh. Let us pay attention to the first part.

5. THE FIRST PART

On the first page there is a scientist's portrait gallery.

The structure of the textbook can be understood by the example of the Web pages devoted to the founder of automatic control theory, creator of the famous stability diagram and distinguished Russian statesman (the minister of finance) Ivan Alekseevich Vyshnegradski. Thus, each scientist has his own Web page with his portrait and the list of the main achievements in the sphere of automatic control theory (see fig. 1,2,3,4). Besides, skipping from link to link one can come to the Web pages where there is information about Vyshnegradski's achievements in other spheres and interesting facts of his biography. Viewing of these additional documents helps to relax a little. Such pages are of another color. Each Web page has a full list of literary and Internet source books which were used for its creation. That is why there is no links in this article. At the end of each part there is a test with questions on the given material. Much attention is paid to historical connections among scientists. We develop here some ideas by Bissell (1998, 1999). Tough there is a lot of interesting facts concerning scientists' life and activity; the tests are based on their contribution to automatic control theory. If a person gives a wrong answer, the hyper reference will lead him/she to the Web page with the right answer. At present this part contains information about scientists of the early period of the control theory and consists of HTML documents and accompanying files (pictures, elements of design). The volume of two parts is a bit larger than 2 Mb. Thus the information can be easily transferred from one computer to another.

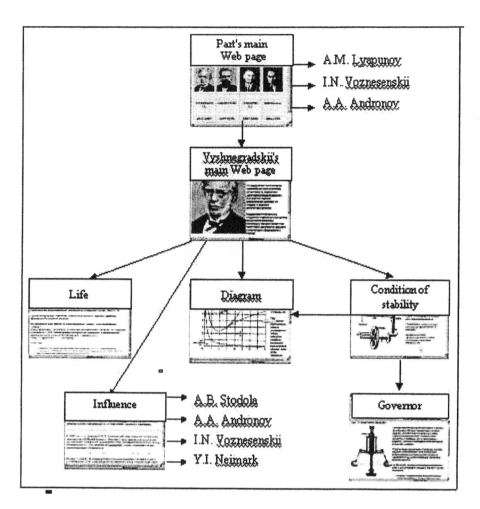

Fig. 1. The structure of the part.

Fig. 2. The scientist's portrait gallery.

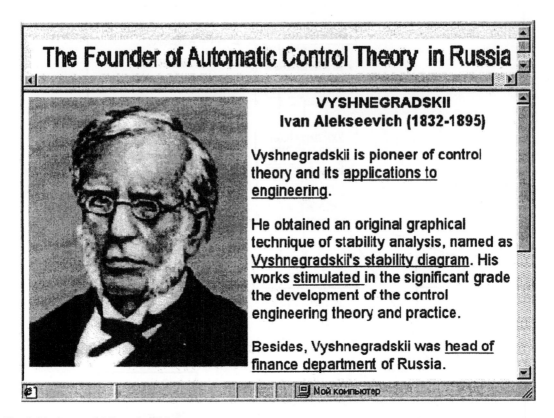

Fig. 3. Vyshnegradski's main Web-page.

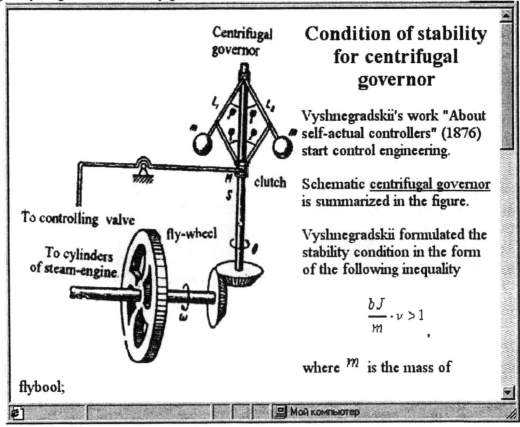

Fig. 4. Condition of stability.

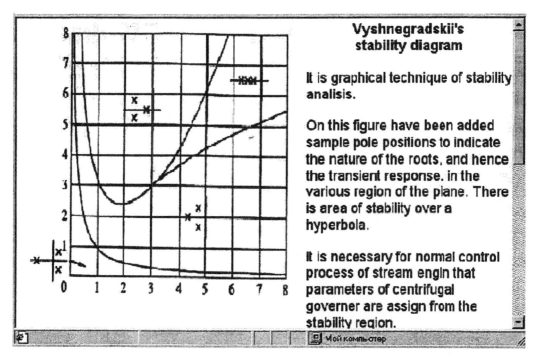

Fig. 5. The stability region.

6. CONCLUSION

In the nearest future we have a plan to expand this textbook with 1 more part "The History of Optimal Control Theory". The Web pages of this part will be devoted to such scientists as John and Jacob Bernoulli, L. S. Pontryagin and other. Though the primary idea was to make an electronic textbook for the students of our University in the Russian language, we are planning to create a Web site both in Russian and English. We hope that in time our electronic textbook will become a peculiar virtual museum.

We would like to finish this article with the words of the academician Dmitri Sergeevich Likhachev:

"Every scientist must posses gratefulness to forerunners, respect for contemporaries and responsibility towards future scientists".

7. REFERENCES

Andronov, A.A. and I.N. Voznesenskii (1949). *J.C. Maxwell, I.A. Vyshnegradskii, A. Stodola. Teoriya Avtomaticheskogo Regulirovaniya,* AN USSR, Moscow.

Bissell, C.C. (1998). A.A. Andronov and the Development of Soviet Control Engineering. *IEEE Control Systems Magazine,* **1**, 56-62.

Bissell, C.C. (1999). Control engineearing in former USSR: some ideological aspects of the early years. *IEEE Control Systems Magazine,* **1**, 111-117.

Davydova, E.V. (2000). Virtual museum of computer science and computer technology. *Informatics and Education,* **2**, 78-81 (In Russian).

Dorato, P. (2002). A History of Analytic Feedback Design. In: *Preprints of the 15-th World Congress of IFAC,* (E.F Camacho, L Basanez, J.A. de la Puente, Ed.), Barcelona.

Krechetnikov, K.G. (2002). Specific of Interface Design in Educatinal Tools. *Informatics and education,* **4**, 65-74 (In Russian).

Pakshina, N.A. (1998) About role of the test-questioning in the learning process. In: *Proc. of the Conference on Science to Industry.,* Arzamas, Russia, pp. 316-318 (In Russian).

Poindexter, S.E. and B.S. Heck (1999). Using the Web in Your Courses: What Can You Do? What Should You Do? *IEEE Control Systems Magazine,* **1**, 83-92.

ELSEVIER

IFAC
PUBLICATIONS
www.elsevier.com/locate/ifac

EXTENSIONS OF LINEAR ALGEBRAIC METHODS TO NONLINEAR SYSTEMS: AN EDUCATIONAL PERSPECTIVE

C. H. Moog,* Ü. Kotta,**,[1] S. Nõmm**,*
and M. Tõnso**

* Institut de Recherche en Communications et Cybernètique de
Nantes, Division Image, Signal et Automatique, 1 rue de la Noë,
BP 92101, 44321 Nantes Cedex 3, France
** Institute of Cybernetics, Tallinn Technical University,
Akadeemia tee 21, Tallinn, 12618, Estonia, email:
kotta@cc.ioc.ee

Abstract: Linear algebraic approach provides crucial conceptual tools and a simple theoretical framework for several, typical problems of nonlinear control systems theory which makes it useful for educational purposes. Additional assistance is provided by *Mathematica* functions developed by us that accommodate a set of symbolic computation tools for modelling, analysis and control system design for nonlinear systems. *Copyright © 2003 IFAC*

Keywords: education, nonlinear control, linear algebraic method

1. INTRODUCTION

Certain recent trends in control education can be identified (Dorato, 1999; Heck, 1999). First, movement of modern control design techniques into the undergraduate curriculum. Second is making the computer an integral part of the educational process. Many students today come from nontraditional backgrounds, and they often are less well prepared in mathematics while being better prepared to work with modern computing techniques.

Any development in control theory which can be clearly linked to classical control without the need to introduce many new theoretical concepts and mathematical machinery are good candidates for consideration to be included into control curricula, especially if the software is available to support the teaching. One such topic is the linear algebraic approach in nonlinear control systems

(Conte *et al.*, 1999; Aranda-Bricaire *et al.*, 1996). The purpose of this paper is to show how these methods can be easily introduced into the control course and how the teaching can be supported by symbolic software.

One of the distinctive characteristics which makes the linear algebraic approach useful is its inherent simplicity. In comparison with the mathematical background needed for employing profitably differential geometric or differential algebraic methods in nonlinear control, the knowledge required for using the linear algebraic tools is very limited.

A significant example of this is offered in this paper by the way in which the notion of accessibility (controllability), model irreducibility and reduction, the problem of feedback linearization and the problem of realization the higher order input-output (i/o) equation in the classical state space form are dealt with. In all cases, a single tool, based either on elementary shifting a function (in the discrete-time case) or on differentiating a function (in the continuous-time case), namely a

---
[1] The work of Ü. Kotta, S. Nõmm and M. Tõnso was partially supported by the Estonian Science Foundation, through grant No 5404

notion of relative degree, gives the key for carrying on a deep analysis and for characterizing relevant dynamical properties.

In addition, simplicity facilitates the development of efficient algorithmic procedures, and implementing them via the use of the computer algebra systems like Maple and Mathematica. The Mathematica functions have been developed to assist the course.

Another positive quality of the algebraic approach is its wide applicability. The tools and methods apply successfully both to continuous-time, and discrete-time systems.

The paper is organized as follows. The mathematical preliminaries including notions from exterior differentiation are described and illustrated in Section 2. Section 3 illustrates briefly the necessity of nonlinear control. As an example of system analysis a fundamental property of the system – accessibility – is treated in Section 4. In Section 5 the algebraic approach is employed for solving an important control problem – feedback linearization. In Section 6, two fundamental modelling problems for single-input single-output nonlinear control systems, described by i/o differential or difference equations are studied. These problems are the equivalence of i/o models and the problem of realization the i/o model in the state space form.

2. MATHEMATICAL PRELIMINARIES

The mathematical background assumed for taking the course is the basic knowledge of linear algebra which is taught in most technical universities during the first year. Though the methods in this course calculate the vector spaces not over the field of real numbers but over the field of (meromorphic) functions, and the elements of the vector spaces are differential one-forms, the concepts and ideas are still the same as in the standard linear algebra. The main difference is in calculations – instead of numerical routines symbolic (computer algebra) software has to be used – but once this software is available, not much new understanding is required from the students.

Besides the basic knowledge of linear algebra, elementary knowledge of differential equations and functions of several variables are required. Results from exterior differentiation are necessary like the notion of one-forms and their extension to k-forms plus Poincare Lemma and Frobenius Theorem. The latter is related to the problem of finding the integrable basis for the vector space of one-forms, whenever possible, i. e. presenting the basis vectors as the closed one-forms.

The course tackles both the cases of continuous-time $\dot{x} = f(x, u)$ and discrete-time $x^+ = f(x, u)$ systems. Note that in the above equation $^+$ denotes the forward shift. We assume that the function $f(\cdot, \cdot)$ defining the model is analytic, since these functions and their derivatives can only vanish at isolated points. Because we are often interested in various system-theoretic properties that can be characterized by the non-vanishing of specific functions defined by the system equations, this restriction allows us to characterize *generic* system properties that hold on an open and dense subset of some suitable domain of definition. The distinction between such generic characterizations and *global* characterizations is that the latter are required to hold everywhere, without exception. In connection with the problem of integrating one-forms, by focusing on generic properties, we require that the one-forms be integrable everywhere except possibly at a set of isolated singular points.

Our vector spaces are built over the field $\mathcal{K}$ of meromorphic functions in a finite number of variables $\{x, u^{(k)}, k \geq 0\}$ or $\{x(0), u(k), k \geq 0\}$ in the continuous-time and in the discrete-time case, respectively. The reason to work with the field of meromorphic functions is that analytic functions do not have, in general, analytic inverses, and also the quotient of two analytic functions f/g is not necessarily analytic anymore, but meromorphic.

The linear algebraic approach is built up by introducing the notion of differential form in an abstract and formal way. Over the field $\mathcal{K}$ one can define a vector space $\mathcal{E} := \mathrm{span}_{\mathcal{K}}\{\mathrm{d}\varphi \mid \varphi \in \mathcal{K}\}$.

Any element in $\mathcal{E}$ is a vector of the form $\omega = \sum_{i=1}^{n} F_i \mathrm{d}x_i + \sum_{j=1}^{m} \sum_{k \geq 0} F_{jk} \mathrm{d}u_j^{(k)}$ or $\omega = \sum_{i=1}^{n} F_i \mathrm{d}x_i(0) + \sum_{j=1}^{m} \sum_{k \geq 0} F_{jk} \mathrm{d}u_j(k)$ where only a finite number of coefficients F are nonzero elements in $\mathcal{K}$. The elements of $\mathcal{E}$ are called one-forms and we say that ω is exact if $\omega = \mathrm{d}\varphi$ for some $\varphi \in \mathcal{K}$.

The subspaces $\mathcal{H}_k$ of one-forms

The relative degree r of one-form ω in $\mathcal{X} = \mathrm{span}_{\mathcal{K}}\{\mathrm{d}x\}$ is given by $r = \min\{k \in \mathbb{N} \mid \mathrm{span}_{\mathcal{K}}\{\omega, \ldots, \omega^{(k)}\} \not\subset \mathcal{X}\}$ or $r = \min\{k \in \mathbb{N} \mid \mathrm{span}_{\mathcal{K}}\{\omega(0), \ldots, \omega(k)\} \not\subset \mathcal{X}\}$.

Let us define a decreasing sequence of subspaces $\mathcal{H}_0 \supset \mathcal{H}_1 \supset \mathcal{H}_2 \supset \ldots$ such that each $\mathcal{H}_k$, for $k > 0$, is the set of all one-forms with relative degree at least k:

$$\mathcal{H}_0 = \mathrm{span}_{\mathcal{K}}\{\mathrm{d}x, \mathrm{d}u\}$$
$$\mathcal{H}_k = \{\omega \in \mathcal{H}_{k-1} \mid \dot{\omega} \in \mathcal{H}_{k-1}\}$$

or

$$\mathcal{H}_k = \{\omega \in \mathcal{H}_{k-1} \mid \omega^+ \in \mathcal{H}_{k-1}\}.$$

There exists an integer $k^* > 0$ such that $\mathcal{H}_k \supset \mathcal{H}_{k+1}$, for $k \leq k^*$, and $\mathcal{H}_{k^*+1} = \mathcal{H}_{k^*+2} = \ldots \mathcal{H}_\infty$, $\mathcal{H}_{k^*} \not\supseteq \mathcal{H}_\infty$. The existence of k^* comes from the fact that each $\mathcal{H}_k$ is a finite dimensional vector space so that, at each step either the dimension decreases by at least one or $\mathcal{H}_{k+1} = \mathcal{H}_k$.

$\mathcal{H}_\infty$ contains one-forms with infinite relative degree so that these one-forms will never be influenced by the control.

The Mathematica function HSpaces finds the sequence of subspaces $\mathcal{H}_k$.

Integrability of a subspace

An one-form $\omega \in \mathcal{E}$ is closed if $d\omega = 0$. We say that the subspace is completely integrable if it admits the basis which consists only of closed forms.

In a nonlinear case, integrability property plays a prominent role. Many constructions which are always possible in the linear context, depend in the nonlinear case on integrability of certain subspaces of one-forms. Only in case the subspace is completely integrable, the required state transformations can be actually found. The examples in our paper of that kind are transformation the system equations into the controller canonical form and finding the state equations for higher order i/o differential (difference) equation.

Any exact one-form is closed but the converse is true only locally.

Poincaré Lemma. If ω is a closed one-form then there exists locally $\varphi \in \mathcal{K}$ such that $\omega = d\varphi$.

A requirement weaker that exactness for an one-form ω is than of being collinear to an exact form, i. e. that there exist λ and φ in $\mathcal{K}$ such that $\lambda\omega = d\varphi$ or, equivalently, such that $\mathrm{span}_\mathcal{K}\{\omega\} = \mathrm{span}_\mathcal{K}\{d\varphi\}$. A function λ is called an integrating factor. The above holds iff $d\omega \wedge \omega = 0$ where by $\wedge$ we denote the wedge product. This property is a special case of the Frobenius Theorem.

Frobenius Theorem. Let $V = \mathrm{span}_\mathcal{K}\{\omega_1, \ldots, \omega_r\}$ be a subspace of $\mathcal{E}$. V is closed iff for any $i = 1, \ldots, r, d\omega_i \wedge \omega_1 \wedge \ldots \wedge \omega_r = 0$.

The Mathematica function Integrability checks if the set of one-forms is completely integrable or not.

3. NECESSITY OF NONLINEAR CONTROL

Nonlinear control systems have been traditionally approached via their linear approximations about a working point. Though this approach works in many situations there are cases in which linear approximations may cause serious drawbacks due to neglected nonlinear terms. Consider the simplified cart described in Fig. 1 below, where only the

rear wheels are represented. The controlled inputs are the longitudinal velocity u_1 and the angular velocity u_2. Let x_1 and x_2 denote the cartesian coordinates of the middle of the rear axle, and x_3 denote the angular position of the cart. Its state

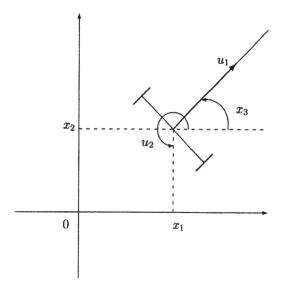

Fig. 1. The unicycle

representation is

$$\begin{aligned} \dot{x}_1 &= (\cos x_3)u_1 \\ \dot{x}_2 &= (\sin x_3)u_1 \\ \dot{x}_3 &= u_2. \end{aligned} \qquad (1)$$

The linear approximation of (1) about the origin is

$$\begin{aligned} \dot{z}_1 &= v_1 \\ \dot{z}_2 &= 0 \\ \dot{z}_3 &= v_2 \end{aligned}$$

and the variable z_2 is not controllable, the cart can not move with a motion perpendicular to the axle of the cart. Nevertheless, the cart, and its nonlinear model (1), are controllable. In this example, higher order nonlinear terms can not be considered as a "slight" disturbance but thanks to nonlinearities, the system enjoys basic important properties such as controllability. Therefore, for this case, and other systems, linear control designed on the basis of linear approximations is disqualified and nonlinear design is mandatory.

4. ANALYSIS PROBLEM: ACCESSIBILITY

Controllable linear systems have the property that any final state can be reached from any initial state, following any trajectory in some arbitrary finite time. For nonlinear systems this general property is not universal. For instance, the cart

155

(2) can move to any final point, but the trajectory that has to be followed is not arbitrary. This is due to the fact that the equations (2) yield the non-holonomic constraint $\dot{x}_1^2 + \dot{x}_2^2 = u_1^2$: the middle point of the axle has necessarily a trajectory which is tangent to the perpendicular axis. The terminology on controllability is denoted to this property of reaching any arbitrary final point. This property did not get a general characterization in the nonlinear setting. A weaker property is the so-called accessibility which got a complete algebraic characterization analogous to the celebrated Kalman's criterion for controllability of linear time invariant systems. Accessibility represents the property the set of states which are reachable in finite time, is dense. This is much weaker than controllability and is able to cope with a number of constraints on the state trajectories. Such obvious constraints may even be due to the physics of the system. Consider for instance a three tank system whose state variables are the three levels in the tanks. These variables have to be positive and there is no question to reach a negative value!

Theorem 1. (Accessibility) The following statements are equivalent

(i) The nonlinear system is strongly accessible
(ii) $\mathcal{H}_\infty = \{0\}$

Note that the above theorem holds both in continuous and discrete-time case (Conte *et al.*, 1999; Aranda-Bricaire *et al.*, 1996) though the rules to calculate the subspaces $\mathcal{H}_k$ are different.

To check this property, a Mathematica function `Accessibility` can be applied.

For linear approximation of (1) we get $\mathcal{H}_\infty = \mathrm{span}_K\{dz_2\}$ which means that the approximation is not accessible, though system (1) itself is accessible as $\mathcal{H}_\infty = \{0\}$.

Note that in the linear case the subspace of non-autonomous states $\mathcal{H}_\infty \perp [B \vdots AB \vdots \ldots \vdots A^{n-1}B]$ which agrees with the well-known Kalman criterion.

The subspace $\mathcal{H}_\infty$ is a nonaccessible subspace and the factor space $\mathcal{X}_a := \mathcal{X}/\mathcal{H}_\infty$ such that $\mathcal{X}_a \oplus \mathcal{H}_\infty = \mathcal{X}$ precisely describes the accessible part of the system. Although $\mathcal{H}_k$ are, in general, not completely integrable, the limit $\mathcal{H}_\infty$ is and so there exist locally r functions, say $\zeta_1, \ldots, \zeta_r$ so that $\mathcal{H}_\infty = \mathrm{span}_K\{d\zeta_1, \ldots, d\zeta_r\} := \mathrm{span}_K\{d\zeta^1\}$. Since $\mathcal{H}_\infty$ is invariant under applying differentiation (in continuous-time case) or forward shift operator (in discrete-time case) one has in particular

$$\dot{\zeta}^1 = f_1(\zeta^1)$$
$$\dot{\zeta}^2 = f_2(\zeta, u)$$

or

$$\zeta^{1+} = f_1(\zeta^1)$$
$$\zeta^{2+} = f_2(\zeta, u).$$

A Mathematica function `AccessibilityDecomposition` decomposes the state equations (1) into an accessible and nonaccessible subsystem.

5. CONTROLLER DESIGN PROBLEM: INPUT-TO-STATE LINEARIZATION

Some nonlinear systems are transformable by state space change of coordinates $\zeta = T(x)$ and nonlinear state feedback $u = \alpha(x, v)$ into a linear controllable system $\dot{\zeta} = Ay + Bv$. This transformation is not far from the spirit of the pole placement idea which states that any linear controllable system is transformable into a system with desired eigenvalues by means of state feedback. The proof of the above result relies on transforming the linear system into the controller canonical form

$$\dot{\zeta} = \begin{bmatrix} 0 & I_{n-1} \\ -a_1 & -a_2 \ldots -a_n \end{bmatrix} \zeta + \begin{bmatrix} 0 \\ \vdots \\ 0 \\ 1 \end{bmatrix} v$$

which is always possible.

Since not every nonlinear system enjoys the advantageous property to be state feedback linearizable, it is important to characterize those systems which are state feedback linearizable.

The nonlinear controller canonical form

$$\begin{aligned} \dot{z}_1 &= z_2 \\ &\vdots \\ \dot{z}_{n-1} &= z_n \\ \dot{z}_n &= \tilde{f}(z, u) \end{aligned} \qquad (2)$$

can be considered in analogy to the corresponding linear form. Equation (2) can be linearized by a state feedback into the linear controllable system. The applicability of this design method is the characteristic property of the nonlinear controller canonical form. In the nonlinear case, unlike the linear case, not every accessible system can be transformed into the form (2). Calculation of the $\mathcal{H}_n$ subspace yields an one-form which defines the infinitesimal controller form

$$d\dot{\zeta} = \begin{bmatrix} 0 & I_{n-1} \\ -a_1(\zeta) \ldots -a_n(\zeta) \end{bmatrix} d\zeta + \begin{bmatrix} 0 \\ \vdots \\ 0 \\ b(\zeta) \end{bmatrix} dv(3)$$

In order to be able to transform (3) into (2) the one-form in $\mathcal{H}_n$ must be integrated. The latter is possible only when $\mathcal{H}_n$ (or equivalently $\mathcal{H}_1, \ldots, \mathcal{H}_n$) is integrable, or

$$\mathcal{H}_n = \operatorname{span}_{\mathcal{K}}\{\omega\} = \operatorname{span}_{\mathcal{K}}\{\mathrm{d}\varphi(x)\}.$$

In that case one can define the state transformation as

$$z = \begin{bmatrix} \varphi(x) \\ \dot{\varphi}(x) \\ \vdots \\ \varphi^{(n-1)}(x) \end{bmatrix}.$$

The approach can be extended to the multi-input case.

Theorem 2. (Feedback linearization) The nonlinear system is locally feedback linearizable iff

(i) $\mathcal{H}_\infty = \{0\}$
(ii) $\mathcal{H}_k$ for $k = 1, \ldots, k^*$ is completely integrable

Again, this theorem is valid both for continuous-time and discrete-time case (Conte *et al.*, 1999; Aranda-Bricaire *et al.*, 1996).

Example 3. (Synchronous motor) (Marino and Tomei, 1995)

$$\dot{x}_1 = x_2$$
$$\dot{x}_2 = c_1 x_3 \sin(c_2 x_1) + c_1 x_4 \cos(c_2 x_1) + c_3 x_2 + c_4$$
$$\dot{x}_3 = c_5 x_3 + c_6 x_2 \sin(c_2 x_1) + c_7 u_1$$
$$\dot{x}_4 = c_5 x_4 - c_6 x_2 \cos(c_2 x_1) + c_7 u_2$$

in which x_1 and x_2 are rotor position and speed, (x_3, x_4) and (u_1, u_2) are stator currents and stator voltages expressed in a fixed stator frame. This system is fully linearizable since $\mathcal{H}_2 = span\{dx_1, dx_2\}$ and $\mathcal{H}_{k^*} = \mathcal{H}_3 = span\{dx_1\}$ are integrable.

The Mathematica functions `Linearizability` and `Linearization` check if the control system is feedback linearizable and find the linearized system together with the state transformation and feedback.

For those systems which are not feedback linearizable, a natural problem is the characterization of the part which can be made linear by feedback. This task can be approached by *Mathematica* function `PartialLinearization`.

6. MODELLING PROBLEMS: EQUIVALENCE AND REALIZATION

Consider a higher i/o differential or difference equation

$$y^{(n)} = \Phi(y, \ldots, y^{(n-1)}, u, \ldots, u^{(s)}) \quad (4)$$

or

$$y(t+n) = \Phi(y(t), \ldots, y(t+n-1), \\ u(t), \ldots, u(t+s)) \quad (5)$$

where n and s are nonnegative integers, $s < n$ and Φ is a real analytic function. We will associate with the system (4) ((5)) an extended state-space system with input $v = u^{(s+1)}$ ($v(t) = u(t + s + 1)$) and state $z = [y, \ldots, y^{(n-1)}, u, \ldots, u^{(s)}]^T$ ($z(t) = [y(t), \ldots, y(t+n-1), u(t), \ldots, u(t+s)]^T$) defined as $\dot{z} = f_e(z, v)$ ($z^+ = f_e(z(t), v(t))$) where the state transition map $f_e(\cdot) = [z_2, \ldots, z_n, \Phi(z), z_{n+2}, \ldots, z_{n+s+1}, v]^T$. The extended system will play a key role in the definition of equivalence and in the realization procedure.

A definition of equivalence is given which generates the notion of transfer equivalence, well-known for the linear case.

Definition 4. A function $\varphi_r \in \mathcal{K}$ is said to be an autonomous element for system (4) ((5)) if there exist an integer μ and a non-zero meromorphic function F so that $F(\varphi_r, \dot{\varphi}_2, \ldots, \varphi_2^{(\mu)}) = 0$. ($F(\varphi_r, \ldots, \delta^\mu \varphi_r) = 0$)

Definition 5. The i/o system is said to be irreducible if there does not exist any non-zero autonomous element for it in $\mathcal{K}$.

Theorem 6. The i/o system is irreducible iff, for the extended system $\mathcal{H}_\infty = \{0\}$.

In case the i/o system does not satisfy the irreducibility condition, we may pick any closed form $\mathrm{d}\varphi_r \not\equiv 0$ from $\mathcal{H}_\infty$ which exists because $\mathcal{H}_\infty$ admits locally an exact basis. Finding $\mathrm{d}\varphi_r$ may sometimes require finding the integrating factor(s). We call $\mathrm{d}\varphi_r$ a reduced differential form and $\varphi_r(\cdot) = 0$ a reduced equation of i/o system. We may repeat the reduction procedure for the system $\varphi_r(\cdot) = 0$ provided that this new equation can be solved explicitly for $y^{(n)}$ in the continuous-time case or for $y(t+n)$ in the discrete-time case. The system $\varphi_r(\cdot) = 0$ can be either irreducible or not. At each step the order n of the i/o equation decreases and finally the reduction procedure converges to an irreducible i/o equation $\varphi_{ir}(\cdot) = 0$. The form $\mathrm{d}\varphi_{ir}$ is said to be an irreducible differential form of i/o systems.

Note that, for the special case of linear time-invariant systems, the reduction procedure corresponds to pole/zero cancellation in the transfer function.

Definition 7. (Transfer equivalence). Two systems which are assumed to admit an irreducible i/o equation are said to be transfer equivalent if they have the same irreducible i/o equation.

The Mathematica functions Irreducibility and Reduction determine whether the i/o equation is irreducible or not and find the reduced i/o equation respectively.

The realization problem is to construct the state equations

$$\dot{x} = f(x,u) \qquad x^+ = f(x,u)$$
$$y = h(x) \qquad\quad y = h(x)$$

for the i/o equation (4) or (5), respectively

The i/o difference equation is assumed to be in the irreducible form so that one can obtain a realization, which is both accessible and observable. If the system is not in the irreducible form, one has first to apply the reduction procedure to transform the system into the irreducible form, otherwise the state equations will be nonaccessible.

Theorem 8. The irreducible i/o equation has an observable and accessible state-space realization iff for $1 \leq k \leq s + 2$ the subspaces $\mathcal{H}_k$ for the extended system are completely integrable. The state coordinates can be found by integrating the basis functions of $\mathcal{H}_{s+2}$.

The functions Realizability and Realization check whether the i/o equation can realized in the classical state space form, and find the state equations, respectively.

7. CONCLUSION

The objective of this paper has been to draw attention to the new developments in the field of nonlinear control systems and to show how these ideas can be introduced into an undergraduate control course, related closely to the course on classical control. These new theoretical developments can be understood as the extensions of well-known classical methods like the Kalman controllability criterion, decomposition the control system into controllable and non-controllable subsystems, pole placement and realization of higher order i/o differential equation in the classical state space form by removing the input derivatives of the state equations (Kailath, 1980) (pp. 39–42). By the above reason, the new concepts are easy to understand. It has also been shown that with suitable symbolic software, the students need not have to be experienced in checking the integrability of one-forms and in integrating them or alternatively, solving the partial differential equations which can be the most difficult task in applying the new methods for nonlinear control systems. More detailed information about the Mathematica functions can be found in (Kotta and Tonso,

1999; Kotta and Tonso, 2002), see also (de Jager, 1995; Rothfußand Zeitz, 1996; Glumineau and Graciani, 1996; Aranda-Bricaire, 1996; Kotta *et al.*, 1999).

8. REFERENCES

Aranda-Bricaire, E. (1996). Computer algebra analysis of discrete-time nonlinear systems. In: *Proc. of IFAC 13th Triennal World Congress*. pp. 305–309. San Francisco, USA.

Aranda-Bricaire, E., Ü. Kotta and C. H. Moog (1996). Linearization of discrete-time systems. *SIAM J. Control and Optimization* **34**(6), 1999–2023.

Conte, G., C. H. Moog and A. M. Perdon (1999). Nonlinear control systems. In: *Lecture Notes in Control and Inf. Sci.* Springer. London.

de Jager, Bram (1995). The use of symbolic computation in nonlinear control: is it viable?. *IEEE Transactions on Automatic Control* **40**(1), 84–89.

Dorato, P. (1999). Undergraduate control education in the U. S.. *IEEE Control Systems Magazine* pp. 38–39.

Glumineau, A. and L. Graciani (1996). Symbolic nonlinear analysis and control package. In: *Proc. of IFAC 13th Triennal World Congress*. pp. 295–298. San Francisco, USA.

Heck, B. S. (1999). Future directions in control education. *IEEE Control Systems Magazine* pp. 36–37.

Kailath, T. (1980). *Linear systems*. Prentice Hall. London.

Kotta, Ü. and M. Tonso (1999). Transfer equivalence and realization of nonlinear higher order input/output difference equations using mathematica. *Journal of Circuits, Systems and Computers* **9**(1-2), 23–35.

Kotta, Ü. and M. Tonso (2002). Linear algebraic tools for discrete-time nonlinear control systems with mathematica. In: *Nonlinear and Adaptive Control: NCN4*. pp. 195–206. Springer.

Kotta, Ü., P. Liu and A. S. I. Zinober (1999). Transfer equivalence and realization of nonlinear higher order i/o difference equations using maple. In: *Proc. of the 14th IFAC World Congress*. Vol. E. pp. 249–254. Beijing.

Marino, R. and P. Tomei (1995). *Nonlinear control design*. Prentice Hall. London.

Rothfuß, R. and M. Zeitz (1996). A toolbox for symbolic nonlinear feedback design. In: *Proc. of IFAC 13th Triennal World Congress*. pp. 283–288. San Francisco, USA.

AN ELECTRONIC LEARNING
ENVIRONMENT FOR CONTROL THEORY

Martin Horn, Josef Zehetner

Department of Automatic Control[1]
Christian Doppler Laboratory
Graz University of Technology
Inffeldgasse 16c/II, A-8010 Graz, Austria
horn@irt.tu-graz.ac.at

Abstract: E-learning has become a very important part of modern education in the last few years. But there is still a lot of development work to do to create systems which can be used in a broader context. The current applications do not meet the special requirements of universities. This paper describes a possible way to close the gap between traditional learning aids and future e-learning environments. It focuses on connecting various standard software applications to a modern e-learning system. *Copyright © 2003 IFAC*

Keywords: Control Education, Documents, Educational Aids, Learning Systems

1. INTRODUCTION

Due to the permanently increasing availability of computers with steadily improving software, it appears obvious to use electronic learning aids for training purposes. Basically, new media and formats offer lecturers the possibility to provide students with modern training documents in a substantially simpler way.

A growing number of tools, viewers and standards is available, concerning themselves with the topic of electronic learning. But, however there exist hardly any concepts, which support documents especially for students. For large companies it is rather easy to establish learning environments e.g. via their intranets. These systems can be maintained centrally. All potential users work with the same system so that the learning content is highly re-usable.

For universities this approach is not generally suitable. In order to justify considerably high

development costs (2000 to 20000 euro per hour of content (P. Baumgartner and Maier-Haefele, 2002)) of such a system it should be used by all lecturers and of course by all students. This scenario seems to be rather unrealistic.

According to our point of view it is presently not a reasonable approach to develop a completely new, sophisticated system to fulfill the very special requirements of the target user group, but to use existing, wide spread tools instead.

Our system currently offers the possibility to extend existing conventional paper-based manuscripts (which are still preferred by students, see below) by a large number of additional features for the use on a PC. This combination of traditional and modern media helps students to deeply understand the learned material.

In this paper the structure of a typical document is outlined with the help of an example from control engineering. To demonstrate the versatility of the system the interaction of different applications is shown.

---

[1] http://www.cis.tugraz.at/irt/

2. E-LEARNING CONCEPTS

In recent years a large number of different e-learning systems and standards have been developed. These systems often use proprietary formats and/or viewers, many are based on the XML standard (*eXtensible Markup Language*)[2]. They can be used for example with conventional internet browsers (and appropriate style sheets).

The systems are generally designed for use on a computer screen. This (still) does not correspond to current learning habits. Most users prefer printed versions of the presented material e.g. due to tiring screen work.

A problem is however, (especially in HTML (*HyperText Markup Language*)[3] based systems) the unsatisfactory ability to print documents adequately. Printing from a web-browser often yields poor results (e.g. no line wrapping). The generation of special printable documents (e.g. printer-ready formatted web-pages, PDF (*Portable Document Format*)[4] documents) is a time consuming task and requires additional administrative expenditure. Unfortunately, the printed document often differs dramatically from the layout on the screen. This fact makes it unnecessarily difficult to toggle between both media, e.g. if extended functionalities of the electronic version are used.

At the moment no standard has been generally accepted. A lot of development work has to be done on the technical realization of the planned systems.

This is the reason why we tried to find a new concept for simple electronic learning systems, which are based on available standard software. This symbiosis of different software tools permits us to create high quality multimedia documents which are ready to use, independent of future systems.

It is planned to link the best features of existing systems and combine them in a meaningful way. Ideally, the entire software needed for the representation of our concept is already installed on the user's computer.

3. PROPOSED CONCEPT

3.1 Overview

The electronic learning system can be started directly from the internet via a hyperlink. It can also be obtained as a package from the internet, and/or

on CD–ROM. There are also arbitrary combinations possible, for example the document can be stored on a local data medium, the extensions on the internet.

3.2 Structure of a document

Our system is currently built by the following applications: Acrobat Reader of Adobe[5], a Java[6]-prepared internet browser (e.g. Microsoft Internet Explorer[7], Netscape Navigator[8] ...) as well as Matlab[9].

Acrobat Reader is used as a document viewer. This software is available for almost all operating systems for free and has attained an enormous high spreading. The supported format is PDF, an open standard for electronic document distribution developed by Adobe. It is very similar to PostScript[10]. It permits high quality printing. PDF documents can be created e.g. directly from TeX[11]/LaTeX[12] and/or PostScript/DVI (*DeVice Independent format*)[13] documents and/or by means of special printer drivers from arbitrary software applications, e.g. from Microsoft Word. As a lot of manuscripts and especially academic texts are written in LaTeX, the production of PDF documents is rather straight forward.

Moreover the format offers the possibility to merge hyperlinks into a document, which can be processed by Acrobat Reader. These hyperlinks can refer, e.g. to other PDF documents and additionally to all kinds of documents, which can be handled by an internet browser. By following a hyperlink a browser application will be started. In the context of the browser then e.g. videos can be played or Java applets can be executed. Local applications, like Matlab, can be remote controlled by the help of so-called *signed applets*[14].

The advantage of this structure is that documents can be printed as independent manuscripts on the one hand, for example to be used during a lecture or in order to help preparing for an examination. On the other hand they are available on every PC, where they have the same visual appearance like the printed version. So the user has fast access to the above mentioned features.

[2] http://www.w3.org/XML/

[3] http://www.w3.org/MarkUp/

[4] http://www.adobe.com/products/acrobat/adobepdf.html

[5] http://www.adobe.com/products/acrobat/readermain.html

[6] http://java.sun.com

[7] http://www.microsoft.com/ie/

[8] http://www.netscape.com/browsers/

[9] http://www.mathworks.com

[10] http://www.adobe.com/products/postscript/

[11] http://www.tug.org

[12] http://www.latex-project.org

[13] http://www.cs.berkeley.edu/~phelps/Multivalent/doc/dvi/DVI.html

[14] http://developer.java.sun.com/developer/technicalArticles/Security/Signed/

In order to examine the validity of our concept, we made a study of the students' behaviour in that direction. About one hundred students were interviewed. We used a questionnaire which was arranged in two major parts:

- questions on experience and visions on electronic learn-management systems
- questions on the available technical equipment and internet connection

It turned out that most of the students prefer manuscripts in traditional hard copy (83%), however, a substantial percentage of the students (71%) would also use electronic extensions of manuscripts.

Text and pictures (in each case 80%) are the preferred way to absorb information as well as interactive simulations and animations (66%).

In detail students expect from (electronic) manuscripts:

- available in printed form
- additional features on a computer
- the entire material of the lectures is included
- they are up to date
- commented examples with solutions
- simulations and animations
- extended search functions
- cross–linking to further documents
- FAQ's (Frequently Asked Questions)
- notes
- also available off-line

The answers on technical equipment have shown that 97% of the students asked use Microsoft Windows, almost 100% have a connection to the internet, 88% use an internet browser and 93% use Acrobat Reader.

The results of the questionnaire show that the students are very interested in modern training documents. The computer is seen as a supporting medium. In the foreseeable future traditional (printed) documents will still play an important role in e-learning.

4. THE ROTATIONAL FLEXIBLE JOINT MODEL

4.1 Introduction

As an example a laboratory plant for the investigation of parameter varying systems was chosen. The purpose of the experiment is to control the angle of an arm which is attached via a flexible joint to a rotating body. The load inertia can be changed by moving a small cart on the arm. The

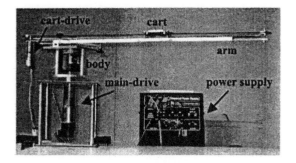

Fig. 1. laboratory experiment

cart is driven by a DC-motor so that the moment of inertia can be varied during operation (fig. 1).

4.2 Modeling

Considering the simplified model in figure 2 the equations describing the motion of the system can easily be derived. Let φ be the angle of the body and α the angle between body and the arm. The body can be rotated with the help of a DC-drive on which axis it is attached (Doczy, 2000).

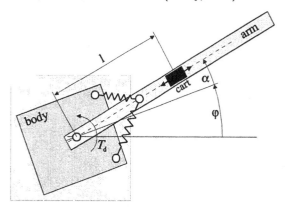

Fig. 2. rotational flexible joint

Neglecting the reaction moment of the cart drive, the equations of motion are:

$$J_l\left(\ddot{\alpha} + \ddot{\varphi}\right) + \dot{J}_l\left(\dot{\alpha} + \dot{\varphi}\right) = -k_s\alpha$$
$$J_l\left(\ddot{\alpha} + \ddot{\varphi}\right) + \dot{J}_l\left(\dot{\alpha} + \dot{\varphi}\right) + J_b\ddot{\varphi} = T_d \quad (1)$$

where J_b denotes inertia of the body, J_l is the load inertia and T_d is the torque of the electrical drive. The right side of the first differential equation represents the linearized restoring moment of the flexible joint, i.e. k_s denotes the joint stiffness. The time dependent load inertia J_l is composed of the constant arm inertia J_a and the inertia J_c of the movable cart with mass m_c, i.e.

$$J_l = J_a + J_c = J_a + m_c l^2 \Rightarrow \dot{J}_l = 2m_c l\dot{l}. \quad (2)$$

The distance between the cart and the axis of rotation is denoted by l. The torque T_d of the electrical drive is given by

$$T_d = \frac{k_d}{R_m}u - \frac{k_d^2}{R_m}\dot{\varphi}, \quad (3)$$

where k_d and R_m are the motor constant and the armature resistance respectively. The drive input voltage is denoted by u, the motor inductance was neglected in (3). The dynamics of the electrical drive which is used to move the cart can be neglected as well. The differential equations (1) together with (2) and (3) define a fourth order system. Introducing the state vector

$$\mathbf{x} := \begin{bmatrix} \varphi & \alpha & \dot{\varphi} & \dot{\alpha} \end{bmatrix}^T \qquad (4)$$

the system can be represented in the general form

$$\dot{\mathbf{x}} = \mathbf{A}(\delta)\mathbf{x} + \mathbf{b}\,u. \qquad (5)$$

The matrix $\mathbf{A}$ is a function of the real valued parameter vector $\delta = \begin{bmatrix} \delta_1 & \ldots & \delta_k \end{bmatrix}^T$ which represents the perturbations in the system parameters. In the present case the vector δ has the dimension 2 and is made up of the time dependent inertia J_l and its time derivative $\dot{J}_l$. Table 1 shows the numerical values for the above introduced constants, $l_{\max}$ is the length of the arm while $\dot{l}_{\max}$ is the maximum speed of the cart.

$k_d = 0.9\,\mathrm{Nm/A}$	$J_b = 85799 \cdot 10^{-6}\,\mathrm{kg\,m^2}$
$k_s = 2.0\,\mathrm{Nm/rad}$	$J_a = 13618 \cdot 10^{-6}\,\mathrm{kg\,m^2}$
$l_{\max} = 0.5\,\mathrm{m}$	$m_c = 0.15\,\mathrm{kg}$
$\dot{l}_{\max} = 0.2\,\mathrm{m/s}$	$R_m = 1.34\,\Omega$

Table 1: numerical data for flexible joint

As the cart moves along the arm, the moment of inertia J_l varies within the bounds given by

$$13618 \cdot 10^{-6} \le J_l \le 51118 \cdot 10^{-6}\,\mathrm{kg\,m^2}, \qquad (6)$$

the upper and lower bounds for $\dot{J}_l$ can be computed as

$$-30000 \cdot 10^{-6} \le \dot{J}_l \le 30000 \cdot 10^{-6}\,\mathrm{kg\,m^2/s}. \qquad (7)$$

The presented laboratory experiment is a valuable tool for implementing and testing robust control system design techniques.

A suitable approach to the stabilization of the plant is to design a state space controller. In the simplest case the controller is found by pole placement methods, more sophisticated procedures are based e.g. on linear matrix inequalities (LMIs (Hofer, 1998), (S. Boyd and Balakrishnan, 1994)).

Fig. 3. Acrobat Reader with document rotflex.pdf

5. PRESENTATION

Students are able to download a multimedia document which they can use to prepare for a practical experiment.

As a typical task, a robust state space controller based on LMIs has to be designed for the model (5) ((P. Gahinet and Chilali, 1995),(C. Scherer, 1997)). For example, such a controller which guarantees the robust stability of the controlled system is given by (M. Horn, 2001):

$$\mathbf{k}^T = \begin{bmatrix} -12.2925 & 162.1516 & -15.1540 & -7.6227 \end{bmatrix} \qquad (8)$$

The numerical simulation can be carried out with Matlab and Simulink (fig. 6). To get a real "feeling" of the dynamic behaviour of the system, a Java applet for the visualization of the system was developed (fig. 5).

The original PDF document is written in LATEX and created with the use of *pdflatex*.

At first sight, the document is an ordinary PDF document. It can be viewed on a PC-screen (fig. 3) and printed as well.

Additional hyperlinks are merged into the document, which are colored and highlighted in Acrobat Reader. Fig. 4 shows a hyperlink to an internet address with additional information on the rotational flexible joint model. By following this hyperlink an internet browser application is launched and the corresponding web-page is displayed.

Of course, a hyperlink can also be linked directly to an internet address showing an Java applet. Fig. 5 shows a Java applet simulating the dynamics of the rotational flexible joint model with the designed state controller (8).

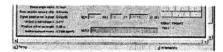

Additional information to RotFlex model on the Internet

http://www.cis.tugraz.at/regis/applets/RotFlex/index.html

A word or two to conclude, and this even includes

Fig. 4. hyperlink to an internet address with additional information on the rotational flexible joint model

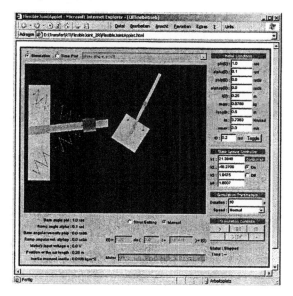

Fig. 5. Java applet for rotational flexible joint model

With a hyperlink to special kinds of Java applets (so-called *signed applets*), it is possible to control applications, which are installed on the local host of the user. In our example we use Matlab to simulate the dynamic behaviour of the rotational flexible joint model with the help of a Simulink model. Fig. 6 shows a picture of the block diagram and an opened scope window. The user is able to start the simulation and to change parameters in the block diagram.

6. CONCLUSION

Our simple but efficient system does not introduce any new technology, it just employs existing well-known techniques.

And this is the strength of our system. The "old" techniques are interconnected in a new context. It is fully operational at the current time. The different tools have already been in use for a long time and are well established. So users have already gained experience with them. The tools exist on almost all PCs of the target group members. A lot of existing manuscripts

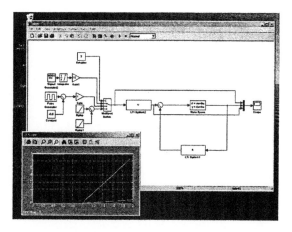

Fig. 6. Matlab Simulink model started from a Java applet

can easily be upgraded to our system. Lecturers are usually familiar with the used formats. A challenge for them is the interconnection between the individual components.

The system presented is certainly not the ultimate solution to all problems arising in electronic learning (besides, an over-all solution was never our intention) but it closes the gap between traditional learning aids and future e-learning environments which are developed presently.

REFERENCES

C. Scherer, P. Gahinet, M. Chilali (1997). Multi-objective output-feedback control via lmi optimization. In: *IEEE Transactions on Automatic Control.* Vol. 42. pp. 896–911.

Doczy, S. (2000). *Control of parameter varying systems, PhD-thesis.* Graz University of Technology.

Hofer, A. (1998). Multivariable control system design with linear matrix inequalities. *Proceedings of the 7th Electrotechnical and Computer Science Conference ERK'98* pp. 267–270.

M. Horn, S. Doczy, N. Dourdoumas (2001). *A laboratory prototype for robust control experiments, Internal Report.* TU Graz.

P. Baumgartner, H. Haefele and K. Maier-Haefele (2002). *E-Learning Standards aus didaktischer Perspektive.* Campus.

P. Gahinet, A. Nemirovski, A.J. Laub and M. Chilali (1995). *LMI Control Toolbox.* The MathWorks Inc.. Natick MA.

S. Boyd, L. El Ghaoui, E. Feron and V. Balakrishnan (1994). Linear matrix inequalities in system and control theory. In: *SIAM Studies in Applied Mathematics.* Vol. 15.

ELSEVIER

IFAC
PUBLICATIONS
www.elsevier.com/locate/ifac

CONTROL THEORY EDUCATION IN THE DISTANCE EDUCATION B.ENG. STUDIES AT WARSAW UNIVERSITY OF TECHNOLOGY

Andrzej Dzieliński* Włodzimierz Dąbrowski*
Rafał Łopatka*

* *Warsaw University of Technology*
Institute of Control and Industrial Electronics,
ul. Koszykowa 75, 00-662 Warszawa, Poland

Abstract: In recent years there is a growing interest of both academic and corporate education centers in the use of telemedia technologies in teaching. The greatest potential of successful application lies in asynchronous education via INTERNET. *Copyright © 2003 IFAC*

Keywords: Distant learning, INTERNET, distant control education.

1. INTRODUCTION

Distance Learning is not a new phenomenon. Its roots date back to 1700, when in the USA the first advertisement on correspondence learning was published in the press. This kind of knowledge acquiring is still quite popular and successfully used to these days. In the last decade a rapid development of information technology, telecommunications and Internet has opened new horizons for Distance Learning providing new magnificent tools for this type of teaching. One can notice a significant and growing interest of not only academic but also corporate centers in the use of telemedia technologies in teaching. The most commonly used is the asynchronous education using the Internet. Basing on other universities experience (e.g. Fern Universitaete Hagen - one of the largest European universities applying the distant learning scheme) Warsaw University of Technology, as a first among the public higher education establishments in Poland has decided to start, in September 2001, full 4-year B.Eng studies covering four courses: computer engineering, industrial computer engineering, mechatronics and multimedia technology. Currently this is the biggest achievement of this kind in all the Polish universities in the area of Technology. Con-

trol theory and engineering is the part of all the courses and belongs to the core subjects (so called "big modules").

2. SPRINT MODEL IN DISTANT LEARNING

Part-time B.Eng. Distant Studies based on SPrInt (Studies via Internet) model proposed at Warsaw University of Technology is a special form of part-time studies, where all the knowledge, skills and information is being transmitted to the student at a distance using the computer, Internet and the lecturers advises (see www.okno.pw.edu.pl). It takes four years to complete the studies. After this period of time one can obtain a B.Eng. degree in one of the specified areas and specializations. The time it takes the student to complete the studies may be adjusted individually by the student. Basic tool necessary to study in this form is the computer that allows:

- Internet connection
- E-mail sending and receiving
- Reading teaching materials from CD-ROM
- Solving problems, preparing reports, projects etc.
- Internet meetings

- Discussions with lecturers and other students

Main teaching material for each module is prepared by the Warsaw University of Technology professors and experienced lecturers and is distributed to the students on CD-ROM, as a multimedia textbook. It is also published on a web-site. The students may pass their exams during special exam sessions at the University. There are also some additional meetings to complete laboratory experiments and some of the projects.

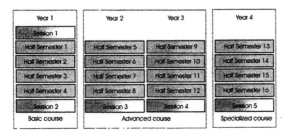

Fig. 1. Distant learning model of SPrINT

The academic year is divided into 4 half-semesters: autumn, winter, spring and summer. Each of the half-semesters lasts 8 weeks and at the end of it there is a one-week examination session. The subdivision of the academic year into four and not parts as it is normally in the case of full-time studies has been motivated by the assumption that the student should study as little modules as possible at one time. In this case the student takes two modules at a time. The studies are divided into three stages:

- Basic studies, this is the first year of studies devoted to the basic engineering subjects and mathematics and physics. It is normally assumed that it takes a year to complete this stage.
- Advanced studies, this stage usually takes two years to complete this stage.
- Specialized studies, it is assumed that this stage should take a year, one half-semester should be devoted to the preparation of the diploma thesis.

The structure of studies is shown in Figure 2. "Big modules" are assigned 8 points, "small modules" are given 5 points. These points are independent of the marks and they allow to verify the progress of studies. The student during his/her studies can have three types of modules according its difficulty and complexity:

- "Big module", with examination.

- "Small module", with examination.
- Laboratory/project session (one-week)

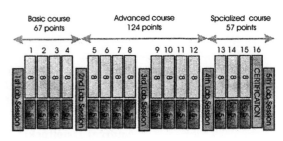

Fig. 2. Structure of the SPrINT model

Part-time B.Eng. Distant Studies via Internet (SPrINT model) are offered by Warsaw University of Technology since 2001 and are completely new both in terms of its contents and in terms of students' recruitment. The most unusual feature of this form of studies is its distant character. There is a physical distance between the lecturer and the student and there are different from traditional communications channels. This way of communication requires from the students certain skills. These are:

- Basic computer skills,
- Internet literacy,
- Self-discipline and good organization of work.

Taking into account the specific model of studies (SPrINT) the special program of studies has been prepared for teaching automatic control and control theory. For these modules it is vital that the new teaching model incorporates the traditional one-week laboratory meeting. This allows the longer direct contact with the lecturer and also lets the lecturer to verify the experimental skills of the students. The students can also run some of the experiments that they would not be able to perform virtually (remotely).

3. TECHNICAL INFRASTRUCTURE OF THE SPRINT MODEL

The following software modules have been incorporated into SPrINT model in order to support teaching:

- software module supporting didactic process management based on Lotus LearningSpace 5.0 platform. The module provides materials necessary for students in the form of textbooks in HTML format with other additional supplements (updated during the course) and tools allowing to trace students progress and activity.

- software module for creation of teaching materials on CD. This module is mostly designated for authors of student textbooks and allows keeping them up to date.
- IBM DB2 data base. Data base stores information for LearningSpace module necessary for the proper usage of it, information about students and lecturers.
- module serving students and lecturers. This module contains functionality for serving students from perspective of Dean's Office, i.e. student registration, management of student grades card, tracing of payments etc.

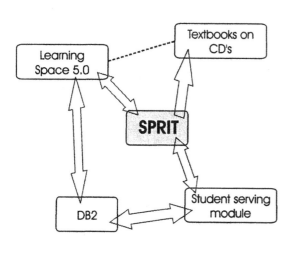

Fig. 3. SPrINT model

General structure of the above mentioned model is presented in figure 4. Management of design teamwork was a separate issue during preparation for launching part-time studies via Internet. Members of the SPrINT team involved in this undertaking were divided into sub-teams dealing with different aspects of the problem. There were few groups created: technical support team, programmers team, management team and finally many teams concerning specific learning tasks. All teams were supervised by SPrINT project manager. Origins of most difficulties and problems are implications of lack of experience in this area and lack of reliable tools.

An important goal was to create high quality final product. Quality control methods were split into two parts according to groups of created products. For part of didactic products the well proved traditional methods using feedback and three month updating cycle have been applied.

4. CONTROL COURSE

Control courses are essential part of distant learning. Students of this course are expected to have good mathematical and computer science background and therefore Control Course is scheduled on third year of studies. The main goal of the control course is to teach students basic techniques used in control system design. This course covers all main topics in control theory and engineering. This includes:

- definition of basic notions, i.e. signal, regulator, plant, feedback etc.
- models of linear systems (state-space, transfer function, frequency characteristics)
- models of nonlinear systems (linearization, static characteristics)
- basic dynamical elements
- stability problems (linear and nonlinear systems)
- regulation criteria
- control system design
- discrete signals and discrete systems
- robotics (Lagrange-Euler equation, control and design)
- economy issues of robotics

Control course has been classified as a big module because of its complexity and good mathematical background expectations. All students having accomplished distant course are expected to attend one week laboratory to get an in-depth feeling of theoretical knowledge gained throughout the year and get some practical experience. Currently in the control course framework lab only simulations are performed remotely so that timing and availability problems can be tackled effectively. Obviously remote access to real laboratory experiments would require more sophisticated timing and avaialabilty control approaches.

5. CONCLUSIONS

There are about 200 students involved in distant learning program at first year. Many of them are active employees of various companies aged between 20 and 35. Some of the students are Polish citizens residing in other countries (i.e. Germany, Sweden or even Australia) who want to graduate from Warsaw University of Technology. Relatively large percent (about 5%) of student community is composed of disabled people. The Distance Education is a great chance for them due to difficulties of visiting the university on a day-to-day basis.

The SPrINT initiative has been met with a very good response of the students and high school graduates. Many young people in Poland want to study at Warsaw University of Technology, but not all of them can afford to do this full-time. On the other hand classical part-time studies are not flexible enough for many people in the starting phase of their professional career. Therefore, new form of studies proposed within the SPrINT scheme is an interesting opportunity for this people. The experience of Warsaw University of Technology is being used by some other Polish universities as a basis of their own distance learning trials.

REFERENCES

Minoli D., Distance Learning, Technology and Applications, Artech House, Boston - London 1996.

B. Galwas: "The Development of New Forms of Continuing Professional Education for Telecommunication Engineers in Poland", Proceedings of 7th World Conference on Continuing Engineering Education, Torino-Italy, May 10-13, 1998, p. 411.

Dąbrowski W., Galwas B., Nowak S., Morawski M. Analiza stanu i moliwoci zastosowania ODL w edukacji, WPW 2001 (in Polish)

IEEE P1484.1/D8, 2001-04-06 Draft Standard for Learning Technology - Learning Technology Systems Architecture (LTSA), February 2002

Dąbrowski W., Wykorzystanie technik telemedialnych w procesie dydaktycznym w rodowiskach akademickich, w materiaach seminariw z zakresu baz danych (in Polish)

MCPC – AN ANALYSIS AND CONTROL TOOL FOR EDUCATION AND SMALL ENTERPRISES

Brian Roffel, Ben H.L. Betlem and Frederik van der Wolk

University of Twente, Faculty of Chemical engineering
P.O. Box 217, 7500 AE Enschede, The Netherlands
e-mail: dcp@ct.utwente.nl

Abstract: Model-based predictive control is a well-established technique for multivariable control of industrial processes. Various packages are available for academic and commercial use. Prices are dependent on the possibilities and features of the software. Some packages are difficult to use, we developed therefore MCPC – Multivariable Constrained Predictive Control – a versatile and easy to use package to develop models from process data and study multivariable control. The package is attractive since a quick assessment of the feasibility of the control application can be made before purchasing a commercial software package. *Copyright © 2003 IFAC*

Keywords: predictive control, multivariable systems, constrained, software, object-oriented programming

1. INTRODUCTION

Model predictive control (MPC) is a control technique for systems with multiple inputs and outputs, based on a predictive model (Qin and Badgwell, 1996). The current generation of MPC technology has evolved to a powerful predictive control technique originating from software such as Identification and Command (IDCOM) and Dynamic Matrix Control (DMC). A model representation often used in predictive control is a non-parametric step-weights representation. By using a least squares minimization approach, step weight models can easily be developed. The control algorithm in a MPC application is based on the process model, on possible constraints and on other settings such as weighting factors, and is based on a least squares minimization of the prediction error and changes in process inputs.

A software package (MCPC – Multivariable Constrained Predictive Control) has been developed that is easy to use and provides a graphical user interface using modern standards and that can be executed in a Windows (95 or higher) environment.

-   The functionality is based on the three key aspects of MPC: identification of stepweight models, controller design and simulation of the controller performance

The user interface will give a comprehensive overview of the data and a clear presentation of the results in charts with zooming and scaling functionality.

2. THE FUNCTIONALITY STRUCTURE

The allowable sequences of events for multivariable control functionality are specified by the availability of the right data at the right time. When analysing the stages in the development of a multivariable controller system the following four phases can be identified: ***Identification:*** identification of the relations between the inputs, disturbances and outputs resulting in a process model; ***Controller design:***

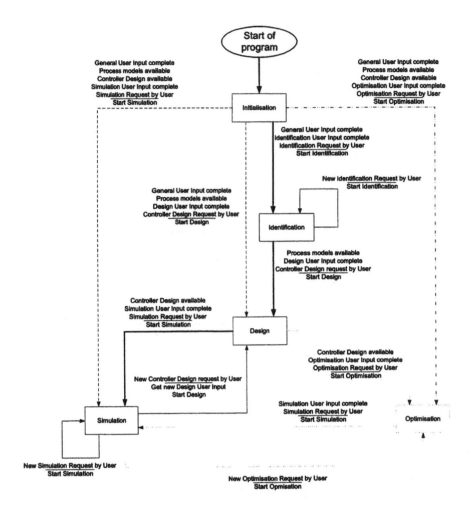

Figure 1. State transition diagram[1] for desired MCPC functionality

identification of the controller algorithm making use of the process model and information about the process constraints; **_Simulation_**: simulation of the effect of the controller algorithm by making predictions with use of the process model and the controller algorithm and **_Optimisation_**: calculation of the optimal process setpoints making use of the process model and the controller algorithm

As can be seen from the description of the phases, Controller Design is only possible after identification of the process model, but Simulation and Optimisation can be executed both directly after the controller algorithm has been identified. This sequence of phases or events can be represented schematically in the state transition diagram (Yourdon, 1989) in Figure 1.

The extra initialisation phase is added since it is a description of the phases of the software and not merely the controller system.

The thick arrows indicate the normal sequence of phases in the development of a multivariable controller system. The loops on Identification, Controller Design and Simulations are an indication

for the iterative way of working for getting specific results. The dashed lines from the Initialisation phase indicate the loading of a saved state, because the user should have the opportunity to stop the program at any moment, save the current state and continue some other time.

3. MULTIVARIABLE CONTROL IN OBJECTS

A multivariable control system has as main constituents the multivariable process, a process model and a controller algorithm. Each of these constituents can be described with a set of objects. These objects are each a representation of an entity or concept from the system. The object structure is designed such that it is in agreement with the actual relationships between these entities or concepts. This results in a structure that can easily be adjusted or maintained. The following objects describe the multivariable process: **_Process_**: describes the structure of the process with the number of inputs, outputs and disturbances together forming the process; **_Input_**: describes the characteristics (such as

---

[1] In this type of diagram the phase is represented by a rectangle and a transition between phases by an arrow. The text with the arrow gives the conditions for a phase transition above the line and the corresponding event below the line.

the name and measurement data) of each process variable that can be adjusted to influence the process outputs; **_Disturbance:_** describes the characteristics of each process variable that influences the process outputs but can not be adjusted. **_Output_**: describes the characteristics of each process variable of which the value is defined by the combination of values of inputs and disturbances. These variables usually need to be controlled at a certain value.

In the case of multivariable predictive control a simple process model in the form of stepweights or mathematical transfer functions is sufficient to predict process values adequately. Because the program will be based on process model identification from measurement data, the process model can be described as a set of StepweightModel objects: giving the relation between an input and output or between a measurable disturbance and an output in the form of stepweights.

The characteristics of the multivariable controller algorithm will be stored in an object called MultivariableController. Therefore the multivariable control system can be described with the object structure as shown in the lower left part of Figure 2.

4. THE IDENTIFIED DATA

The data required for the multivariable control functionality has been divided over the objects as discussed in the previous section. The data to keep

track of the system state and the specific data for each phase has been divided over a another set of objects (see lower right part of Figure 2).

Another addition to the data structure is the generalisation class ProcessVariable for the classes Input, Output and Disturbance. During the data identification it became clear that these classes have some member variables (and functionality) in common. This common data and functionality is stored in the ProcessVariable class to prevent redundancy in the code.

5. DESIGN AND DEVELOPMENT

The previous section described *what* the system needs to consist of and what the functionality sequence and the data backbone have to be, this section will describe *how* this system can be designed and developed. The structures for the functionality and the data from the previous sections can be hosted in a variety of different software architectures. The basic criterion for choosing a type of software architecture is that it can host the functionality the software architecture as object oriented with the capability to focus on contents with little windows programming overhead. Software architecture that can comply with the criteria is the Single Document Interface (SDI) This architecture is provided by a. structure and the data backbone and can be the basis for a program that complies with the general system

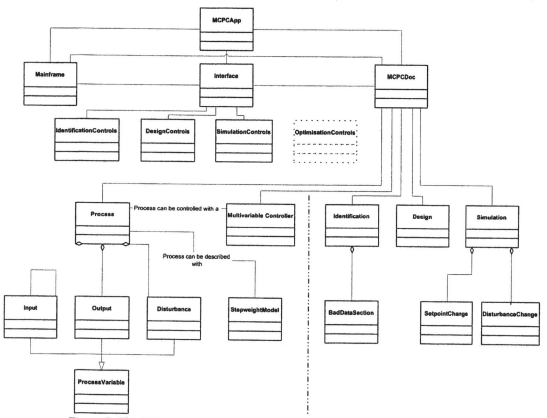

Figure 2. The SDI structure for MCPC showing the overall object structure for data.

171

specification. This leads to the general description of C++ class library called the Microsoft Foundation Classes (MFC), (Prosise, 1999). This library was designed as an object-oriented interface to the Windows operating system. It can be used as an application framework in which the structure can be easily defined and routine chores such as opening and closing windows and creating message loops is taken care of. This application framework is based on a document/view architecture, which defines a program structure that relies on document objects to hold application's data and on view objects to render graphical representations of that data. In conjunction with objects for the application and the main window frame this architecture can be represented graphically as in Figure 3.

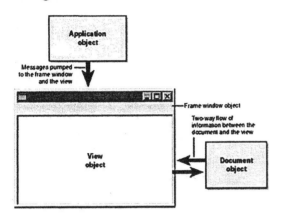

Figure 3. Document/View architecture

The SDI-architecture is also based on the structure displayed in Figure 3, because SDI is a special case of Document/View architecture. The four main objects of this structure all have their specific role in the program functionality:

1. The application object is responsible for maintaining the structure between the other three objects and for the flow of messages between these objects.
2. The document object has as role to archive the data of the program and to communicate these data to and from the view object. This object is also the basis for initialising, storing and

reloading the current state and data of the program.
3. The view object is responsible for the correct representation of the data in the document on the screen and for the translation of user input into commands that operate on the document's data.
4. The frame window object defines the application's physical workspace on the screen and serves as a container for the view. It has functionality as resized.

6. SDI-ARCHITECTURE FOR MCPC

Implementing MCPC based on the SDI architecture has as main aspect that the data structure as proposed previously will be hosted by the document object. The functionality structure can be guaranteed by storing the system state in this document object and by creating a view that is dependent on the phase of the system. This can be implemented through creating child views of the main view object for each phase. Each child view is responsible for representation of the data of the active phase it belongs to and the main view object will be responsible for translation of the user input to commands working on the document data. The overall SDI-structure for MCPC is represented graphically in Figure 2. A short overview of the main data objects of this structure is given in Figure 4.

The selected platform for the program is Microsoft Visual C++ 6.0 (Kruglinski, 1998; Petzold, 1998). The desired charting functionality (such as zooming and scaling) is achieved with the TeeChart Pro v5 ActiveX control.

7. THE GRAPHICAL USER-INTERFACE

The system specification describes the user interface as giving an overview of the used data and a clear presentation of the results in charts with zooming and scaling functionality. How these specifications are secured by the currently developed system, will be made clear with a description of several key aspects of the user interface.

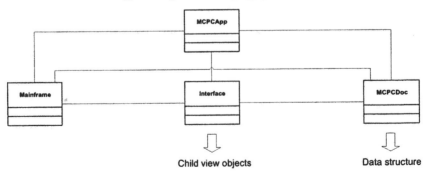

Figure 4. Overview of main data objects[2] of SDI for MCPC

---

[2] MCPCApp = application object, Mainframe = frame window object, Interface = view object and MCPCDoc = document object

172

8. DATA OVERVIEW

According to the design considerations, the overview of the data is to be guaranteed through creation of a view that is dependent of the system phase. However, a view that changes completely after a certain phase has been finished is not desirable because the user must have the opportunity to look back at the data of previous phases. Therefore the view of each phase must be accessible to the user at all times. This can be guaranteed with a so-called tabbed view in which the user can select a tab for each system phase (see Appendix A).

As can be seen, all the data for the active phase are represented together on the screen. As can be seen, all the data for the active phase are represented together on the screen. Because the user interface would become too complex if all data editing were to be happening in the view for the phase itself, it has been decided to do some of the data editing through dialogs. This dialog displays the active settings for the selected item and is also dependent of these settings for the (de)-activation of certain data fields.

9. CHARTING FUNCTIONALITY

The representation of data in charts is given through the small previews and the large size plots. Both types have the same functionality in zooming and scrolling and are automatically scaled. The plots can be resized and positioned on the screen by the user so that a comparison of different plots is possible. On the identification tab plots are available for all the measured inputs, outputs and disturbances, for the identified stepweight relations and for the predictions of the outputs. The simulation tab contains the possibility for plots of the simulated inputs, controlled and uncontrolled outputs and measurable and unmeasurable disturbances. The plots for the controlled outputs also contain the belonging setpoint values.

10. USER GUIDELINES

General: The user interface has been designed in such a way that the user can see easily which path to follow to get results. It is possible to exit the program at any moment and save the current state and data. This gives the opportunity to load a saved state and continue with the saved data at another moment.

Zooming and scrolling plots: All the plots (preview and large size) can be zoomed or scrolled and are automatically scaled for the currently displayed values. Scrolling the data in the chart is also possible.

11. APPLICATION

The software has been applied in control of two heat-integrated cryogenic distillation columns, producing oxygen from air for a coal gasification plant, as described by Roffel et al (2000). The control application includes five outputs, six inputs and one measurable disturbance. The process models are shown in Figure 5 (process gains and constant increments are also shown). The application controls the levels in the bottoms of the two towers and three concentrations.

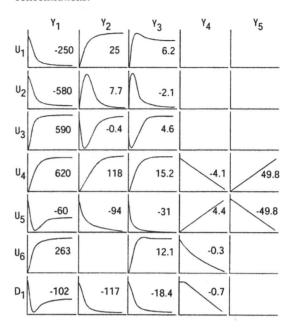

Figure 5. Models for heat-integrated distillation towers

All process inputs are subject to low and high constraints. The nitrogen concentration in the produced oxygen flow has a maximum value, the levels in both distillation columns have to remain between minimum and maximum values.

The control variables are all flows to and from the towers. The disturbance is the oxygen flow that is produced by the towers and is a variable that is set by the operator. The control application performs well and has now been online for over three years.

12. CONCUSIONS

A new software package MCPC has been created. Data interaction is based on an internal object structure. The most suitable software architecture for the desired structure is the SDI architecture because it supplies a framework for data structure, graphical representation and basic Windows functionality. The multivariable control algorithms could be implemented in this structure by compiling them to a dynamic link library. For further structural improvements to the program this could be replaced by functions within the data structure that directly access the data objects. The software package has a modern user interface. This could be realised through following the complete path of structured analysis, design and development. The resulting object-oriented code is therefore according to a clearly defined structure, which can be modified easily for possible future additions or changes to the functionality.

The most important future additions are help- and printing-functionality. Another feature that could be added is the possibility to import and export data for the system. This will give the user the chance to use data from other software.

REFERENCES

Kruglinski, D.J., S. Wingo and G. Shepherd, Programming Microsoft Visual C++, *5th edition,* 1998

Petzold, C., Programming Windows, *5th edition,* 1998

Prosise, J., Programming Windows with MFC, *2nd edition,* 1999

Qin, S.J., T.A. Badgwell, An Overview of Industrial Model Predictive Control Technology, *Chemical Process Control-V, Jan 7-12, Tahoe, California,* 1996

Roffel, B., B.H.L. Betlem, and J.A.F.de Ruijter, First principles modeling and multivariable control of a cryogenic distillation tower, *Comp. & Chem. Eng.,* 24 (2000), 111

Yourdon, E., Modern Structured Analysis, *Prentice Hall,* 1989

Appendix A: View for identification and simulation phase.

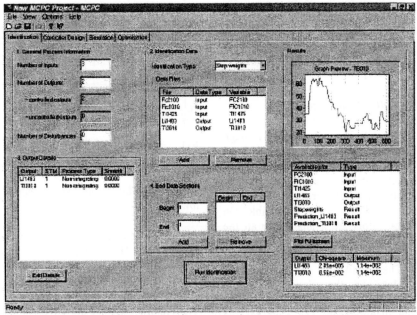

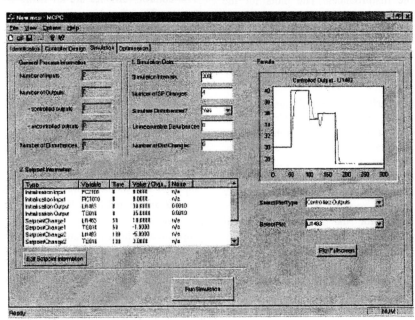

www.elsevier.com/locate/ifac

INTERACTIVE LEARNING OF CONSTRAINED GENERALIZED PREDICTIVE CONTROL

S. Dormido[*], M. Berenguel[#], S. Dormido-Canto[*], F. Rodríguez[#]

[*]Facultad de Ciencias, Universidad Nacional de Educación a
Distancia. Avda. Senda del Rey, 7. Madrid-28040. Spain
E-mail: [sdormido,sebas]@dia.uned.es
[#]Dpto. Lenguajes y Computación,, Universidad de Almería.
Carretera Sacramento s/n. Almería-04120. Spain
E-mail: [beren,frrodrig]@ual.es

Abstract: This paper presents an interactive control education tool developed in SysQuake (Piguet, 2000) aimed at introducing basic concepts of constrained generalized predictive control. The tool helps the students to understand the theory and elements involved in this control technique and to interactively analyze how changes in different design parameters (weighting factors, control and prediction horizons, sampling time, etc.) and constraints (input, output, performance, etc.) affect the closed loop system response using a receding horizon control strategy. Some application examples have been selected and included to show the main features of the tool. *Copyright © 2003 IFAC*

Keywords: Control education, predictive control, linear systems, quadratic programming.

1. INTRODUCTION

New information technologies can be a useful tool in control education as a way to appropriately design tools in order to help the students to understand basic concepts and practical aspects under a "learning by doing" basis. As pointed out by (Copinga *et al.*, 2000; Poindexter and Hech, 1999), information technologies cannot be a solution to educational needs unless the creative component is included. These control education tools incorporate interesting features such as better man-machine interaction, natural and intuitive graphical user interfaces and high degree of interactivity. Computers are quite useful in control engineering to calculate and represent different magnitudes which are closely related and constitute different visions of a single reality. The understanding of these relationships is one of the keys to achieve a good learning of the basic concepts and allows the student to be in disposition of carrying out control-systems design accurately (Dormido *et al.*, 2002).

The design of control systems has been traditionally carried out by an iterative process in which each iteration is divided into two phases (synthesis and analysis) until the specifications are fulfilled. This iterative procedure can be confusing and laborious for the student. It is possible, however, to merge both phases into one in such a way that the modification of one parameter in the system produces an immediate effect and thus the design procedure becomes really dynamic and the student perceives the gradient of the change of the performance criteria with regard to the elements he/she manipulates. This interactive capacity allows us to identify much more easily the compromises that can be achieved. As pointed out in (Dormido et al., 2002), many tools for control education have been developed over the years incorporating the concepts of dynamic pictures and virtual interactive systems (Wittenmark *et al.*, 1998). The main objective of these tools, which are accessible to the students at any time through the Internet, is to make students more active and concerned in control courses (stimulating the student's engineering intuition). These interactive tools attempt to "demystify" abstract mathematical concepts through visualization in specifically chosen examples. At the present time a new generation of software packages have generated an interesting alternative for the interactive learning of automatic control (García and Heck, 1999).

In this paper, these characteristics of new interactive control education tools are exploited in the field of constrained generalized predictive control as a very useful "user-friendly" way to learn the main concepts. It is essential to have access to suitable software in order to solve any non-trivial predictive control problems and to gain experience of how it actually works. As pointed out by (Maciejowski, 2002), predictive control was developed and used in industry for nearly 20 years before attracting much serious attention from the academic control community. This community tended to ignore its potential for dealing with constraints, thus missing its main advantage. In addition, it tended to point to the fact that, when constraints are ignored, predictive control is equivalent to conventional, though generally 'advanced' linear control, and hence apparently nothing new. Fortunately, the academic community has for some years now appreciated that predictive control really does offer something new for control in the presence of constraints, and has provided much analysis, and new ideas, to such an extent that it has gone beyond current industrial practice and is preparing the ground for much wider application of predictive control – potentially to almost all control engineering problems. The constant increase in computing speed and power certainly makes that a real prospect, as shown in this paper with the interactive calculation of QP problems in order to solve constrained predictive control.

The paper is organised as follows. Section 2 is devoted to introduce basic concepts of constrained generalized predictive control. The tool is explained in section 3 and some illustrative examples are described in section 4 Finally, section 5 presents some conclusions.

2. BASIC CONCEPTS OF CONSTRAINED PREDICTIVE CONTROL

Model Predictive Control techniques are powerful tools to include constraints in the control design stage, accepting in general any kind of models, objective functions or constraints, taking into account the different operating criteria present in the process industry. Predictive control integrates disciplines such as optimal and stochastic control, control of processes with dead-times, multivariable control, constrained control, etc. constituting the most general way to represent the control problem in the time domain. The main advantages of this methodology are (Camacho and Bordóns, 1999; Maciejowski, 2002):

1. it is attractive to staff with only a limited knowledge of control because the concepts are very intuitive and the tuning is relatively easy.
2. it can be used to control a great variety of processes, from those with relatively simple dynamics to other more complex ones (with long delay times, non-minimum phase, unstable, etc.)
3. its basic formulation extends to multivariable plants with almost no modification.
4. it intrinsically has compensation for dead times.
5. it introduces feedforward control in a natural way to compensate for measurable disturbances.
6. the resulting controller is a linear control law without taking constraints into account.

7. its extension to constraints handling is conceptually simple and these can be included during the design process.
8. it is useful when future references are known.
9. it is more powerful than PID control, even for single loops without constraints, without being much more difficult to tune, even on 'difficult' loops such as those containing long time delays.

In Europe, one of the most famous techniques included at University Education levels is Generalized Predictive Control, GPC (Clarke *et al.*, 1987; Clarke and Mohtadi, 1989). It is characterised by the strategy presented in Fig. 1, where at each sampling instant and using a model of the process, the future outputs are predicted for a determined horizon. With these predicted outputs and using an objective function, the future control increments are calculated taking into account the constraints acting on the process. Finally, the first control signal calculated is implemented, and the rest are rejected; then the horizon is displaced and the procedure is repeated the next sampling instant as the new output is known (all the sequences are brought to date). This is the known receding horizon approach. In order to implement this strategy, the control structure shown in Fig. 2 is implemented, using a model to predict the future outputs of the system ($\hat{y}(t+k|t)$, $k=1...N$), based on past inputs and outputs and future control signals ($u(t+k|t)$, $k=0...N-1$), these last obtained from the optimisation stage of the algorithm (taking into account the cost function and constraints).

Model: the model used by the GPC methodology in this work is the CARIMA one to achieve offset-free closed loop control (Clarke *et al.*, 1987), given by:

$$A(z^{-1})y(t) = z^{-d}B(z^{-1})u(t-1) + C(z^{-1})\frac{e(t)}{\Delta}, \Delta = 1 - z^{-1} \quad (1)$$

where $u(t)$ and $y(t)$ are the input and output sequences, d is de dead time and $e(t)$ is a zero mean white noise. A, B and C are polynomials in the backward shift operator z^{-1}. In the usual formulation of GPC, the polynomial $C(z^{-1})$ usually equals 1. When robustness is to be enhanced, this polynomial is used as a design one ("*T*-polynomial") that can be treated as a filter to attenuate the components of the prediction errors due to modelling uncertainties, as far as unmeasurable load disturbances.

GPC Cost function:

$$J(N_1, N_2, N_u) = E\{\sum_{j=N_1}^{N_2} \delta(j)[\hat{y}(t+j|t) - w(t+j)]^2 + \sum_{j=1}^{N_u} \lambda(j)[\Delta u(t+j-1)]^2\}$$

$$(2)$$

where $E\{.\}$ is the mathematical expectation, $\hat{y}(t+j|t)$: sequence of j optimal predictions of the output of the system performed at instant t, $\Delta u(t+j-1)$: sequence of future control increments, obtained from the minimisation of the cost function. Parameters: N_1 and N_2 are the minimum and maximum prediction horizons and N_u is the control horizon. $\delta(j)$ and $\lambda(j)$ are weighting sequences that penalize the future tracking and control effort respectively.

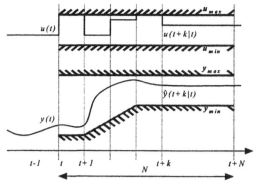

Fig. 1. GPC strategy

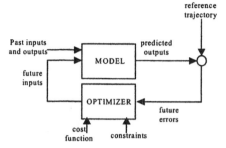

Fig. 2. Basic structure of GPC

Variable	Linear constraint
Control signal amplitude $u_{min} \leq u(t) \leq u_{max}, \ \forall t.$	$\mathbf{1} u_{min} \leq \mathbf{T} \Delta u + u(t-1)\mathbf{1} \leq \mathbf{1} u_{max}$
Control signal increment $\Delta u_{min} \leq u(t)-u(t-1) \leq \Delta u_{max} \ \forall t$	$\mathbf{1} \Delta u_{min} \leq \Delta u \leq \mathbf{1} \Delta u_{max}$
Output signal amplitude $y_{min} \leq y(t) \leq y_{max} \ \forall t$	$\mathbf{1} y_{min} \leq \mathbf{G} \Delta u + \mathbf{f} \leq \mathbf{1} y_{max}$
Output band constraints $y_{min}(t) \leq y(t) \leq y_{max}(t) \ \forall t$	$\mathbf{G} \ \Delta \mathbf{u} \leq \mathbf{y_{max}}\text{-}\mathbf{f}; \quad \mathbf{y_{max}}=[y_{max}(t+1)$ $y_{max}(t+2) \ ... \ y_{max}(t+N)]$ $\mathbf{G} \ \Delta \mathbf{u} \geq \mathbf{y_{min}}\text{-}\mathbf{f}; \quad \mathbf{y_{min}}=[y_{min}(t+1)$ $y_{min}(t+2) \ ... \ y_{min}(t+N)]$
Output overshoot $y(t+j) \leq \gamma w(t); \ j=N_{o1}...N_{o2},$	$\mathbf{G} \ \Delta \mathbf{u} \leq \mathbf{1} \gamma w(t)\text{-}\mathbf{f}$
Output monotonous behaviour $y(t+j) \leq y(t+j+1) \ if \ y(t)<w(t)$ $y(t+j) \geq y(t+j+1) \ if \ y(t)>w(t)$	$\mathbf{G}\Delta\mathbf{u}+\mathbf{f} \leq \begin{bmatrix} \mathbf{0}^T \\ \mathbf{G'} \end{bmatrix} \Delta\mathbf{u} + \begin{bmatrix} y(t) \\ \mathbf{f'} \end{bmatrix}$ $\mathbf{G'}, \mathbf{f'}$ are the result of eliminating the last row of $\mathbf{G}$ and $\mathbf{f}$.
Avoid NMP behaviour $y(t+j) \leq y(t) \ if \ y(t)>w(t)$ $y(t+j) \geq y(t) \ if \ y(t)<w(t)$	$\mathbf{G} \ \Delta \mathbf{u} \geq \mathbf{1} y(t)\text{-}\mathbf{f}$
Final state $y(t+N+1) \ ... \ y(t+N+m) = w$	$\mathbf{y_m} = [y(t+N+1) \ ... \ y(t+N+m)]^T:$ $\mathbf{y_m} = \mathbf{G_m}\Delta\mathbf{u}+\mathbf{f_m}, \ \mathbf{G_m}\Delta\mathbf{u}=\mathbf{w_m}-\mathbf{f_m}$
Output integral $T_m \sum_{t+1}^{t+N_i} y(t) = I$	$\mathbf{1}^T \begin{bmatrix} y(t+1) \\ \vdots \\ y(t+N_i) \end{bmatrix} = \mathbf{G_i}\Delta\mathbf{u}+\mathbf{f_i} \geq I$

All these parameters are design ones known as tuning knobs. The reference trajectory $w(t+k)$ can be the set point itself or a smooth approximation from the actual value of the output $y(t)$ to the set point.

System and performance constraints: MPC techniques allow the inclusion of constraints in the design stage of the algorithm. It can be easily observed (Camacho and Bordóns, 1999; Berenguel, 1997) how different constraints can be represented. As has been shown, the control increments calculated by the GPC strategy are obtained by the minimisation of a quadratic objective function: $J = (\mathbf{y}-\mathbf{w})^T(\mathbf{y}-\mathbf{w})+\lambda \Delta\mathbf{u}^T\Delta\mathbf{u}$, where $\mathbf{w}$ is the sequence of future references, $\mathbf{y}$ the sequence of future predictions constituted by the forced and free response ($\mathbf{y}=\mathbf{G}\Delta\mathbf{u}+\mathbf{f}$) and λ is a control effort weighting factor. $\Delta\mathbf{u}$ are future control increments, matrix $\mathbf{G}$ contains the coefficients of the open loop step response of the system and $\mathbf{f}$ contains terms depending on present and past outputs and past inputs of the plant (output of the system if future control increments are zero). By substituting in the cost function the sequence of future outputs: $J = \frac{1}{2}\Delta\mathbf{u}^T\mathbf{H}\Delta\mathbf{u}+\mathbf{b}^T\Delta\mathbf{u}+f_0$, with $\mathbf{H}=2(\mathbf{G}^T\mathbf{G}+\lambda\mathbf{I})$, $\mathbf{b}^T=2(\mathbf{f}-\mathbf{w})^T\mathbf{G}$ and $f_0=(\mathbf{f}-\mathbf{w})^T(\mathbf{f}-\mathbf{w})$, the optimal solution without constraints is a linear one: $\mathbf{H}\Delta\mathbf{u}=-\mathbf{b}$. When constraints are taken into account, there is no explicit solution and optimisation algorithms have to be used. In this case, the problem can be posed as a quadratic cost function with linear inequality and equality constraints in the control increment $\Delta\mathbf{u}$ ($\mathbf{R} \ \Delta\mathbf{u} \leq \mathbf{c}$ and $\mathbf{A} \ \Delta\mathbf{u} = \mathbf{a}$), that is, a quadratic programming problem. The different constraints that can be taken into account are the following, where $\mathbf{1}$ is a N-dimensional vector which elements are all one and $\mathbf{T}$ is a NxN lower triangular matrix with all elements equal 1.

3. BASIC ELEMENTS OF THE TOOL

This section briefly describes the main functions of the developed tool programmed with SysQuake (Piguet, 2000). When developing a tool of this kind, the programming of the algorithms is a time consuming aspect, but one of the most important things that the teacher has to have in mind is the organization of the main windows and menus of the tool to facilitate the student the understanding of the control technique (Dormido, 2002). In this case, the main window is divided into several sections represented in Fig. 3 (basic screen of the developed interactive tool). The sections are the following:

Process and control parameters: In this window the configuration parameters can easily be defined:

System parameters: number of poles and zeros of the CARIMA linear plant ZerosGp and PolesGp, dead-time (Delay), sampling time (Tm) for continuous/discrete time translation. Some of these parameters can also be modified by using the Settings menu.

Unconstrained control parameters: δ-Delta, λ-Lambda, number of zeros of the T-polynomial (ZerosT), also modifiable using the Settings menu, which also allows the introduction of a set-point filter. N_1 and N_2 can be interactively modified by using vertical lines that appear in the Y/Output amplitude constraints plot. N_u can be modified in the same way by accessing the vertical line shown in the U/Input amplitude constraints plot. When the mouse pointer is placed on these lines, the actual value of the selected horizon is shown in the left –bottom corner of the main screen.

Test parameters: final simulation time Nend, number of GPC pre-programmed set-point steps (SP), which amplitude and shape can be changed by accessing the active points of the reference signal, represented by small circles in the Y/Output amplitude constraints plot.

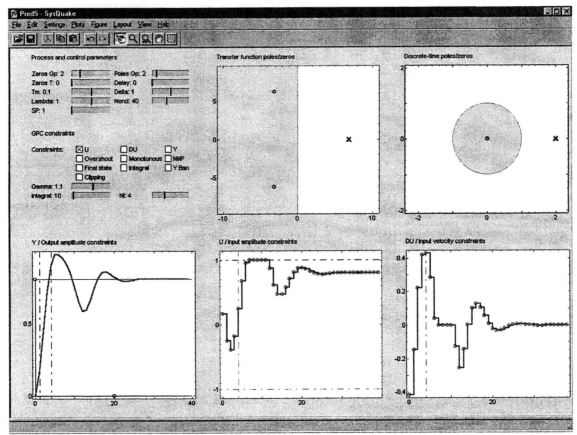

Fig. 3. Main window of one example

GPC Constraints: Different active boxes have been included to allow the user the selection of the different type of constraints briefly explained in the previous section: input amplitude constraints (U), input rate constrains (DU), output amplitude constraints (Y), Overshoot constraints (degree of overshoot determined by parameter γ-Gamma included below in this section and modifiable using a slider), Monotonous behaviour constraints, non-minimum phase constraints (NMP), Final state constraints, output Integral constraints (which values can be selected using a slider), output YBand constraints, etc. Also, the possibility of obtaining the unconstrained solution with clipping (saturating) the control signal has been included.

Continuous transfer function poles/zeros: This window allows to graphically define the poles (x) and zeros (o) of the linear plant in the *s*-plane.

Discrete time transfer function poles/zeros: allows defining the location of poles/zeros in the *z*-plane.

Closed loop system step response and output amplitude constraints: This plot shows the evolution of the closed loop system step response. The number of step changes in the reference can be modified by using the SP slider in the process and control parameters section. The values of the N_1 and N_2, output amplitude constraints (represented as horizontal discontinuous lines) can be interactively modified. Another interesting possibility is to include an output unmeasured disturbance to see the effect of the T-polynomial in the design.

Closed loop control signal and input amplitude constraints: The evolution of the closed loop input signal is shown in this plot, as far as the corresponding amplitude constraints.

Closed loop control signal increments and input rate constraints: Represents the evolution of the closed loop control increments and related constraints.

4. ILLUSTRATIVE EXAMPLES

In this section, some examples are shown in order to demonstrate some of the capabilities of the application, although its main characteristic – the interactivity – is difficult to be reflected in a written text. Other examples can be found in (Dormido and Berenguel, 2002).

4.1. Unconstrained case

Effect of tuning knobs: The first simple example is extracted from the text (Camacho and Bordóns, 1999, pp.59-61) in which, a complete design of an unconstrained GPC controller is carried out for a plant given by $(1-0.8z^{-1})$ $y(t)=(0.4+0.6z^{-1})u(t-1)+e(t)/\Delta$. In the book, the effect of including a filter in the set-point is commented. With the help of interactivity, other parameters such as prediction and control horizons, weighting factors, etc. can be changed in such a way that the student can easily select the most appropriate ones for his application and control specifications. Fig. 4 shows results with different values of tuning knobs and reference filter.

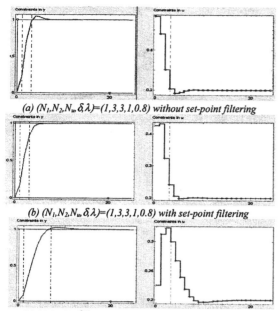

(a) $(N_1,N_2,N_u,\delta,\lambda)=(1,3,3,1,0.8)$ without set-point filtering

(b) $(N_1,N_2,N_u,\delta,\lambda)=(1,3,3,1,0.8)$ with set-point filtering

(c) $(N_1,N_2,N_u,\delta,\lambda)=(1,7,3,1,10)$ without set-point filtering
Fig. 4. Interactively modification of tuning knobs

Effect of T-polynomial: An example has been selected from (Clarke and Mohtadi, 1989). As has been pointed out, the use of a filtering polynomial enhances robustness against unmodelled dynamics and enables independent tailoring of set-point and disturbance responses. To illustrate that consider the first order plant $(1-0.9z^{-1})y(t) = (0.1\ z^{-1}+0.2\ z^{-2})\ u(t)$. Fig. 5 shows the results obtained when applying a GPC controller with $(N_1,N_2,N_u,\delta,\lambda)=(1,10,1,1,0)$ and respectively *(a)* $T(z^{-1})=1$ and *(b)* $T(z^{-1})=(1-0.8\ z^{-1})$. To illustrate the effect of disturbances, the experiment consisted of an initial step in the set-point, followed by an additive step of load-disturbance acting on the input, and finally a period of uncorrelated random noise added to the plant output (e.g. measurement noise). Note that the step responses of the controllers in Fig. 5 are the same, but for $T=1$ the load-disturbance is rapidly eliminated but the output noise induces a control signal with high variance and hence output fluctuations, while for $T=1-0.8\ z^{-1}$ the controller takes longer to eliminate the load-disturbance but now the response to output noise is better in that the control variance is reduced.

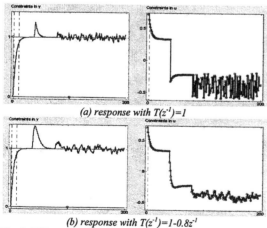

(a) response with $T(z^{-1})=1$

(b) response with $T(z^{-1})=1-0.8z^{-1}$
Fig. 5. Effect of T-polynomial

4.2. Constrained case

As has been commented, the constraints in processes can be included by physical or security limitations in the operation or by looking for a desired performance. The predictive nature of GPC allows anticipating the constraints violation, that can lead to even to instability.

Physical and security constraints: An unstable SISO system is used in this section, given by $G(s)=s^2+6.13s+48.04/s^2-13.86s+48.04$, with discrete time description (sampling time $T_m=0.1$s.) given by $A(z^{-1})=1-4\ z^{-1}\ +4\ z^{-2}$, $B(z^{-1})=1$. The GPC methodology can be used to control such system with tuning knobs $(N_1,N_2,N_u,\delta,\lambda)=(1,4,4,1,1)$. The future reference is a step input of amplitude 0.8. The results can be found in Fig. 6(a).

<u>Output amplitude constraints</u>: the output of the controlled system in the unconstrained case presents overshoot. This can be diminished or avoided by imposing constraints on output amplitude or on overshoot. Fig. 6(b) shows the results when imposing limits in the output signal evolution $y_{max}=0.8$, $y_{min}=0$. Although this is not the case, feasibility problems can appear with this kind of contraints. Performance constraints as NMP constraints and overshoot constraints (not shown here due to space limitations) can be also used (Dormido and Berenguel, 2002).

<u>Input amplitude constraints</u>: in this example, the student can see how constraints are not only important for optimum performance, but also to achieve stable behaviour when saturations occur in the control signal. In this case, the limits in the control signal amplitude are $u_{min}=-1$ and $u_{max}=1$ (the control signal slightly surpasses this value in the unconstrained case). The usual industrial control practice is to saturate that signal and apply the clipped signal to the system. As can be seen in Fig. 6(c), this practice leads to instability, while if the predictive capability of constrained GPC is used to advance future constraints violation, the behaviour of the system can be improved and, what is more important, instability is avoided (Fig. 6(d)).

Constraints and stability: As pointed out by (Maciejowski, 2002) predictive control, using the receding horizon idea, is a feedback control policy. There is therefore a risk that the resulting closed loop might be unstable, mostly in the presence of constraints. One technique relating GPC with stability (to assure stability under different assumptions) is included in this section. GPC$^\infty$ (Scokaert, 1994) is a compromise between the LQ law obtained with infinite horizons and finite horizon strategies where an infinite upper costing horizon is still used but the control horizon is reduced to a finite value. As an example (Scokaert, 1994, pp. 51-53) a second-order NMP underdamped system is considered and described by $(1-1.5z^{-1}+0.7z^{-2})y(t)=(-1+2z^{-1})u(t-d)$ with $d=4$. First, mean level control is illustrated with horizons $N_2=\infty$ (in the tool a large value is selected) and $N_u=1$ (Fig 7(a)).[1]

---

[1] with $N_2=\infty,N_u=1$, the values of N_1 and λ do not affect the control

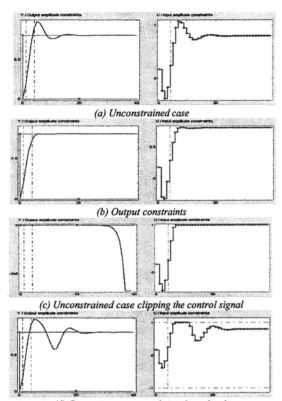

(a) Unconstrained case

(b) Output constraints

(c) Unconstrained case clipping the control signal

(d) Constraints on control signal amplitude

Fig. 6. Example of physical constraints in GPC

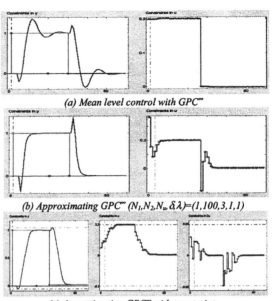

(a) Mean level control with GPC$^\infty$

(b) Approximating GPC$^\infty$ ($N_1, N_2, N_u, \delta, \lambda$)=(1,100,3,1,1)

(c) Approximating GPC$^\infty$ with constraints

Fig. 7. Examples with GPC$^\infty$

ACKNOWLEDGEMENTS

This work has been supported by the Spanish CICYT under grants DPI2001-1012, DPI2001-2380-C02-02 and DPI-2002-04375-C03-03. The authors would also thank Prof. Eduardo F. Camacho for his help and suggestions.

To illustrate the effects of a different choice of tuning parameters $(N_1, N_2, N_u, \delta, \lambda)$= $(1,100,3,1,1)$ accordingly GPC behaves as shown in Fig. 7(*b*). Notice that, in this simulation an exact GPC$^\infty$ cannot be achieved; for the process considered, N_2=100 is however deemed an adequate approximation of N_2=∞. The technique can also be applied by including constrains. Fig. 7(*c*) shows the results obtained when applying constraints on **y** ([-0.05 1.05]), **u** ([-0.2 0.2]) and Δ**u** ([-0.05 0.05]).

5. CONCLUSIONS AND FUTURE WORKS

An educational tool for introductory constrained generalized predictive control interactive teaching has been presented in this paper. The tool developed in SysQuake helps the students to understand the basic concepts and to gain an insight into the selection of design parameters and how the controlled system performance is influenced by this selection. Future works in this tool will include:

❑ Use of different models to represent the real plant and the model used for control purposes to allow the analysis of unmodelled dynamics and the extension to robust predictive control.
❑ Extension to the multivariable case.
❑ Inclusion of feedforward control.
❑ On-line adaptation capabilities.
❑ Testing of different QP algorithms.
❑ Account for feasibility problems: set-point conditioning, controller/plant shutdown, constraints softening or removal, etc.

REFERENCES

Berenguel, M. (1997) Constrained Predictive Control. *Lectures from the PhD Course "Adaptive, predictive and robust control"*, University of Seville.

Camacho E.F. and C. Bordóns (1999). *Model Predictive Control*, Springer.

Clarke, D.W., C. Mohtadi and P.S. Tuffs (1987), *Generalized Predictive Control - Parts I and II*, Automatica, **23**, 2, 137-160.

Clarke D.W., Mohtadi, C. (1989). Properties of generalized predictive control. Automatica, 25(6):859-875.

Copinga, G. J., M. H. Verhaegen and M. J. van de Ven (2000). Toward a web-based study support environment for teaching automatic control. *IEEE Control Systems Magazine* **10**(4), 8–19.

Dormido, S. (2002) Control Learning: Present and Future. Plenary Lecture. *15th Triennial World Congress of IFAC, Barcelona, Spain*

Dormido, S., F. Gordillo, S. Dormido-Cantó, J. Aracil (2002). An interactive tool for introductory nonlinear control systems education. *15th Triennial World Congress of IFAC, Barcelona, Spain.*

Dormido, S. and M. Berenguel (2002). An interactive tool for teaching constrained generalized predictive control. Internal Report UNED/U. Almería.

García, R.C. and B.S. Heck (1999). Enhancing classical control education via interactive GUI design. *IEEE Control Systems Magazine* **19**(3), 77-82.

Maciejowski J. (2002). *Constrained Predictive Control*. Academic Press.

Piguet, Y. (2000). *SysQuake: User Manual*. Calerga.

Poindexter, S.E. and B.S. Heck (1999). Using the web in your courses: What can you do? What should you do?. *IEEE Control Systems Magazine* **19**(1), 83-92.

Scokaert, P.O.M. (1994), *Constrained Predictive Control*. PhD Thesis. U. Oxford.

Wittenmark, B., H. Häglund and M. Johansson (1998). Dynamic pictures and interactive learning. *IEEE Control Systems Magazine* 18(3), 26–32.

ELSEVIER

IFAC

PUBLICATIONS
www.elsevier.com/locate/ifac

A LABORATORY PROJECT FOR ADVANCED CONTROL METHODS: CONTROL OF A NEUTRALIZATION PROCESS

V. Bartolozzi, A. Picciotto, M. Galluzzo*

*Dipartimento di Ingegneria Chimica, Università di Palermo, Viale delle Scienze,
90128 Palermo, Italy.
E-mail:galluzzo@unipa.it

Abstract: A neutralization process is used in a laboratory project for experimenting advanced control techniques. Students may choose among several techniques to experience the behaviour of advanced controller action: adaptive control, GMG control, fuzzy control, self tuning fuzzy control, neural control, neuro-fuzzy control. The experience in advanced self tuning controller based on fuzzy logic is proposed for the control of a neutralization process.

The test of a self tuning controller based on fuzzy logic is a very important case because the resulting structure of the controller is easy to implement and modify: the final results are compared with those deriving from the application of a conventional PI controller for the control of level and pH. The architecture of a self tuning fuzzy controller is actually very flexible: it is possible to modify easily its constituents, set of rules, membership functions and scale factors. The interactive comparison of the various control loops actions on the system has allowed a deeper understanding of process dynamics and control methodology. Moreover the innovative student approach encourages the problem-based learning. *Copyright © 2003 IFAC*

Keywords: fuzzy control, neutralization process, non-linear system, simulation.

1. INTRODUCTION

The first objective for the laboratory course was providing students with a realistic application of control engineering, using computer control systems similar to those which are used in industrial application.

Many real processes are not linear nor stationary. The scheme of conventional control uses linearized models of process for a particular condition of stationary state in order to achieve good performances in that particular operating point in which the parameters of the process do not undergo any modification. The controller needs the tuning if the operating point changes, and it has to be tuned periodically if the process changes in the time.

This led to the planning of the controllers with adaptation mechanism, so that they can automatically be adapted in order to modernize the parameters according to the current characteristics of the process. This goal can be achieved more easily with advanced controller. In particular, controllers based on fuzzy logic with an adaptation mechanism are very suitable. Conventional adaptive control

systems are presented by Åstrom [Åstrom, Wittenmark, 1989]. The adaptation mechanism must modify the parameters of the controller in order to improve the performance of the controller on the output based on the data obtained by monitoring the system. The adaptation mechanisms can be classified according to the fixed parameters.

If the tunable parameters are the scale factors of the input and output, and the membership functions of the variables, these controllers are called self-tuning .

Another type of adaptation method consists in modifying the rules if-then: in this case the controller is named auto-arranged.

Fuzzy logic controllers are more robust than PI controllers because they can cover a wider range of operating conditions: FLC are suitable to control processes that are non linear and ill-defined.

Controlled variables are the pH of outflow and the level in the reactor.

pH control is very complex because of the high non-linearity of the process and because of the buffer effect of the solution. Since the shape of the

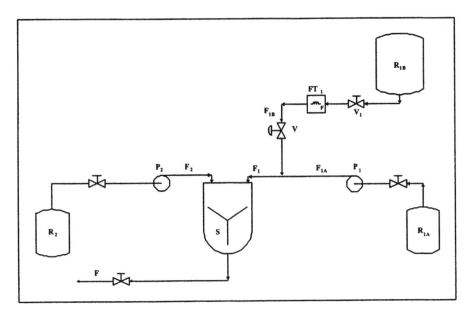

FIG.1 *Experimental system*

neutralization curve considerably changes with time, the control is remarkably difficult and it is opportune to implement a self tuning control to produce on line the controller parameter variations, required by the process changing.

The final architecture of the complete control system consists of a fuzzy PI controller for the level control and a fuzzy PI self tuning controller for the pH control. Moreover, both control loops are interacting. The on-line control has been realized with LabWiew and MatLab toolbox.

2. STRUCTURE OF THE COURSE AND METHODOLOGIES

During the course "Instrumentation and equipment for industrial process", students have been able to choose among several techniques to experience the behaviour of advanced controller action: adaptive control, fuzzy control, self tuning fuzzy control, neural control, neuro-fuzzy control. (Picciotto, 1996;Gulotta, 2000; La Placa, 2001). The study regarding the development of advanced self tuning controller based on fuzzy logic is presented as a clear case.

The course has been divided into various modules. Some of them have been followed by all students, they have regarded:
- study of the system
- assemblage of the instrumentation and the equipment of the prototype
- study of dynamics of the not linear systems
- modellation of the operation
- study of the controllers for the control of the experimental system
- study of FLC methodology
- study of the controller proposed from Mudi and Paul and plan of the architecture of the controller to be used.

From this point, the job has been divided in more groups, that have carried out various tasks:
- definition of the controller fuzzy for the level
- definition of the controller fuzzy for the pH and of the supervisor
- definition of the parameters of the traditional PI controllers to be used for the control of the level and the pH.

The controllers have been therefore implemented to the computer and tested in simulation making the modifications needed in order to improve the performance. Finally they have been tested on the experimental system and final results have been discussed critically with all students. Main experiences and studies are described in more details in the following paragraphs.

3. DESCRIPTION OF THE EXPERIMENTAL SYSTEM AND PROCESS DYNAMICS

3.1 Description of the system

The system studied is specified in fig.1
Inside the S container acid current F1 has been neutralized. The current F1 is sulfuric acid, acetic acid and sodic acetate, and has been neutralized by the F2 current, constituted by sodic hydroxid. The final solution is obtained by F1b, current of water and concentrated acid F1a.

3.2 Dynamics of the system

The study of dynamics of the system has been carried out elaborating the equilibrium relations and the balance equations of the system, and by the experimental verification of some parameters contained in some equations. The model of the neutralization processes presents difficulties in the description of the reactions; the control of pH is very complex because of the high non-linearity of the process and because of the buffer effect of the solution.

Maintaining the concentration of the solutions and considering the fields of variability of the flow capacities, it turns out that the variation of the capacities flow F1a and F2 has not considerable effects on the level, while F1b influences the level and the pH. Moreover, the two control loop are interacting, although their interaction can be considered unidirectional. Initially the loops have been studied singularly.

Relatively to level control, the model has been implemented using RLS method (Recoursive Least Squares) for the parameters estimation, supposing the process to be in a steady state and the constant sampling time. In the steady state it has been supposed that the contributions of the capacities F2 and F1a are negligible and that the fundamental contribution comes from the F1b capacity flow.

The model of the system is based on a simple mass balance in the S container, supposing the solution inside to be uniformly stirred. The characteristic curves of the proportional valve have been found experimentally. The obtained model therefore is represented with not linear first order linearized equation.

As far as the neutralization reaction is concerned, the chemical reactor is modelled considering an invariant part and a part that causes a fast reaction.

balance mass equations and the equilibrium conditions, Gustafsson and Waller Kurt (1983).

The simplifying assumptions for the model are: fast reactions, constant density, perfect mixing, volume control constant for variations F1a and F2.

A variation in the pure water capacity F1B can have a meaningful effect on the level and on the pH of the outflow solution.

The equations have been obtained by S-function of Matlab. Then the answer of the system to some disturbances on the acid capacity flow F1a has been studied using the toolbox of Matlab SIMULINK. Finally two interacting loops have been represented by means of a complex system of equations, corresponding to a model in Matlab.

4. SELF TUNING FUZZY CONTROL

The choice of self a tuning controller based on fuzzy logic is justified by the resulting structure that is easy to implement and modify, it is possible to modify easily its constituents, set of rules, membership functions and scale factors.

The FLC controller have a prefixed set of rules for the control, generally derived from the heuristic knowledge of the experts and codified by linguistics terms

The proposed controller has been constructed according to the method suggested by Mudi and Pal

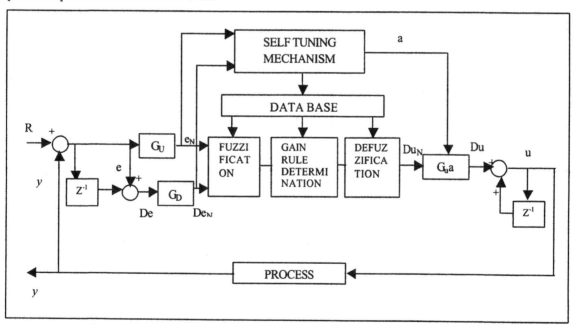

FIG.2 The self tuning fuzzy PI controller

(1999), that considers a self tuning fuzzy PI controller (STFPIC) based on the tuning of the scale factors, optimizing its influence on the performance and stability of the system.

The final architecture of the complete control system consists of a fuzzy PI controller for the level control and a fuzzy PI self tuning controller for the pH control.

If the reaction is in continuous equilibrium, the invariant part of the reactor can be the only one to be taken into consideration. The model has been obtained using the invariant ones of reaction, the

The proposed controller is tuned dynamically by adjusting its output SF through an updating factor a. The value of a is determined by fuzzy rules defined on the error and on the error variation. These rules set up the self tuning mechanism of a supervisor fuzzy controller, regulating on line the parameters of the slave controller.

The structure of the self tuning fuzzy PI controller is shown in figure 2.

4.1 Fuzzy controller used for the control of the level

Error and its variations have been considered as the input for the definition of the fuzzy controller of the level. Signal variation (dV) from the valve has been chosen as the manipulation variable. Membership functions are triangular and trapezoid. The defuzzification method is the barycentre, the chosen method of inference is the Mamdani (1979) method, the operator of minimum is used for the implication, while the aggregation of the rules is carried out with the operator of maximum. The 49 used rules are reported in table 1.

e/De	nb	nm	ns	ze	ps	pm	pb
nb	pb	pb	pb	pm	ps	ps	ze
nm	pb	pm	pm	pm	ps	ze	ns
ns	pb	pm	ps	ps	ze	ns	ns
ze	pm	pm	ps	ze	ns	nm	nm
ps	ps	ps	ze	ns	ns	nm	nb
pm	ps	ze	ns	nm	nm	nm	nb
pb	ze	ns	ns	nm	nb	nb	nb

Tab.1: rules of the controller PI fuzzy for the level.
The qualitative terms presented in the table are defined as:
nb big negative, **nm** medium negative, **ns** small negative
ze zero
ps small positive, **pm** medium positive, **pb** big positive

4.2 Fuzzy controller used for the pH control

The method proposed from Mudi and Pal has been selected for this controller: error and its variations have been used as an input for the primary controller, the variation of basic flow capacity Fb has been used as a variable of manipulation. Also the rules has been suggested from Mudi: the scale factor must be adapted according to the process progress in order to improve the controller performance in the interval in which the disturbances are active. As an example, if the error is small and positive and the error variation turns out negative medium, the value of the scale factor must be increased in order not to have undesired presence of high over elongation. In the Mudi scheme the output control has been obtained by the continuous and non linear variation of updating factor of the gain.

Simulation results using the STFPIC were not satisfactory: therefore, the controller has been modified by introducing a mechanism that reduces strong changes of the updating factor, varying the membership functions in order to consider the high non linearity of the process, and changing the weight of the rules.

The self tuning mechanism is realized updating of scale factors; the other parameters is made proceeding by trial and error.

Table 2 shows the rules that have been used.

e/De	nb	nm	ns	ze	ps	pm	pb
nb	pb	pb	pb	pm	ps	ps	ze
nm	pb	pm	pm	pm	ps	ze	ns
ns	pb	pm	pm	ps	ze	ns	ns
ze	pb	ps	ps	ze	ns	ns	nb
ps	ps	ps	ze	ns	nm	nm	nb
pm	ps	ze	ns	nm	nm	nm	nb
pb	ze	ns	ns	nm	nb	nb	nb

Tab.2: rules of the PI controller fuzzy for the pH.
The qualitative terms presented in the table are defined as:
nb big negative, **nm** medium negative,
ns small negative
ze zero
ps small positive, pm medium positive, pb big positive

The supervisor has following characteristic: error and its variations as input; the gain of the updating factor as an output; seven triangular membership functions for variables representation, de and dFB; the defuzzification method used is the barycentre method; inference of Mamdani type; minimum operator for the applicability and for implication.

The 49 rules used for the control of the scale factor are reported in the following table.

The numerical value that appears in some cells represents the different weight attributed to some rules.

Finally the control of the pH, as a consequence of the non linear behaviour of the system, needs more severe control actions for positive error values, correspondent to pH values greater than 7.

Moreover in implementing the self tuning fuzzy controller, input and output membership functions have been arranged in asymmetrical way, modifying the shape where a marked action was necessary.

e/De	nb	nm	ns	ze	ps	pm	pb
nb	vb	vb	vb	b	sb	s	ze
nm	vb	vb	b	b	mb	s	vs 0,85
ns	vb	mb	b	vs	vs	s	vs
ze	s	sb	mb	ze	mb	sb	s
ps	vs 0,85	s	vs	vs 0,85	b	mb	vb
pm	vs	s	mb 0,85	b	b	vb	vb
pb	ze	s	sb	b	vb	vb	vb

Tab. 3: rules of the supervisor PI fuzzy controller of pH.

The qualitative terms presented in the table are defined as:

nb big negative, **nm** medium negative, **ns** small negative

ze zero

ps small positive, **pm** medium positive, **pb** big positive

5. EXPERIMENTAL RESULTS - SIMULATIONS

Traditional PI controller and FLC have been applied in simulation and also on the experimental system. In the diagrams reported at the following page are shown the answers of the controllers.

After level control tests the offset is completely eliminated in all cases of disturbances up to 50% and no oscillations are present. Moreover, in the event of disturbances of greater entity, the controller succeeds in eliminating the offset although oscillating phenomena are introduced.

The controller fuzzy turns out more ready than the traditional controller PI. Furthermore, it shows no oscillation.

With regard to the adaptive PI fuzzy controller, it has been sufficient to arrange the membership functions of the input and output in an asymmetrical way, modifying them whenever a stronger action was required. The experimental results obtained with large variations of the considered disturbances (approximately 30% of the range), show that the complete control system is able to obtain an efficient control with a fast off-set elimination, a non oscillating behaviour and a reduced interaction between the control loops. It has been observed, moreover, that for disturbances of greater entity, the controller succeeds in eliminating the offset, but oscillating phenomena are remarkable.

6. CONCLUSION

The results of experimental advanced control techniques obtained in a laboratory project course has been presented. The development of a fuzzy self-tuning controller characterized from a high speed of execution and a fast implementation is the case of study reported. The laboratory course has presented a reduced scale version of some industrially relevant problems, providing realistic engineering control problems to our students.

A self tuning controller based on fuzzy logic is proposed as case of study. The development of the FLC is model-free, derivation of mathematical model is not required.

The proposed project provides an opportunity for the students to experience the behaviour of advanced controller action: the package used has the essential characteristics of being interactive, user-friendly and very flexibly structured.

The package has been evaluated by regular meetings and through observations of the students. Most students are in favour of having more computer teaching aids for different subjects, especially topics involving graphical simulations and complex calculations. Good students' performance in class and in the final text and examinations, further confirms the validity of the experimental course.

REFERENCES

Åström K. J., Wittenmark B., (1989) "Adaptive Control", Addison, Wesley.

Picciotto A. (1996), "Sviluppo di un algoritmo di controllo GMC per un sistema del 1° ordine con rilevante tempo morto", Università degli Studi di Palermo, Dipartimento di Ingegneria Chimica dei Processi e dei Materiali, Palermo.

Gulotta P., (2000) "Controllo self-tuning di un processo di neutralizzazione", Università degli Studi di Palermo, Dipartimento di Ingegneria Chimica dei Processi e dei Materiali, Palermo.

La Placa V., (2001) "Controllo adattivo fuzzy di un processo di neutralizzazione", Università degli Studi di Palermo, Dipartimento di Ingegneria Chimica dei Processi e dei Materiali, Palermo.

Gustafsson Tore K., Waller Kurt V. (1983)"Nonlinear and adaptive Control of pH", Process Engineering and Design, Process Control Laboratory, Department of Chemical Engineering, Abo, Finland.

Mudi, R.K. and N.R. Pal (1999) "A Robust self Tuning Scheme for PI and PID Type Fuzzy Controllers", IEEE Transactions on Fuzzy Systems, 7, 2-16.

Mamdani E. H., Procyk T., Baaklini N., (1979) "Application of Fuzzy Logic to Controller Design Based on Linguistic Protocol", Discrete Systems and Fuzzy Reasoning, Queen Mary College, University of London.

VARIATION TO STEP OF 50% OF THE INPUT CAPACITY FLOW, SIMULATION AND EXPERIMENTAL RESULTS

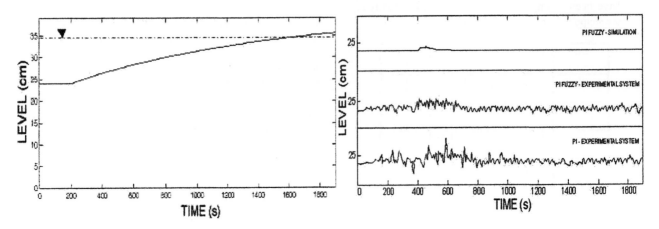

VARIATION TO STEP OF 33% OF THE ACID CAPACITY FLOW. SIMULATION AND EXPERIMENTAL RESULTS.

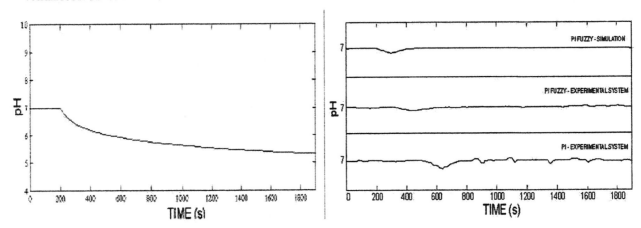

ELSEVIER

IFAC
PUBLICATIONS
www.elsevier.com/locate/ifac

TEACHING ADAPTIVE CONTROL WITH DS1102 DSP CONTROLLER BOARD

Andrzej Dzieliński and Dominik Sierociuk

Institute of Control and Industrial Electronics (ISEP)
Warsaw University of Technology
Koszykowa 75, 00-662 Warszawa, POLAND
email: dsieroci@ee.pw.edu.pl

Abstract: The paper describes the experimental setup for teaching adaptive control to undergraduate students. This has been built using the standard Pentium III – based PC, DS1102 DSP Controller Board and software both standard and purpose written. The experimental results to show the capabilities of the lab are presented. This presents how the students can prepare their own experiments with different adaptive control schema. The laboratory presented has been applied in teaching modern control to the third year students of Electrical Engineering. *Copyright © 2003 IFAC*

Keywords: adaptive control, DSP controller, control education

1. INTRODUCTION

Control theory is a well established subject in modern electrical engineering undergraduate curriculum. However, there is a growing need to teach the undergraduate students more advanced topics in control theory and engineering. This includes robust control, optimal control, nonlinear control and adaptive control. This includes both theoretical considerations and laboratory experiments. In this paper we describe the efforts undertaken at Warsaw University of Technology in order to introduce the basic laboratory setup for undergraduate students to learn fundamentals of adaptive control. The aim of this effort was to prepare a laboratory equipment composed of a Pentium III based PC with a DS1102 DSP controller board, electronic external plant model and software to implement different adaptive control algorithms. The algorithms implemented were uploaded to the DSP board internal memory and this allowed the hardware-in-the-loop configuration. This way a flavour of real-time adaptive control was given to the students.

2. EXPERIMENT DESCRIPTION

In our laboratory setup, a DS1102 Controller Board by dSPACE GmbH and a Pentium III - based PC are used. The board is designed as an ISA card and is based on the Texas Instruments TS320C31 floating-point processor. This card contain also:

- SRAM memory for loaded programs
- 4 A/C converters (two 16-bits and two 12-bits) and 4 12-bits C/A converters, which allow to connect the real object.
- Slave signal processor 320P14, which is responsible for binary input/output, PWM output, encoder input,
- ISA interface,
- serial connector,

The ISA interface allows us to adjust the loaded program variables. It means that we can change the control system parameters on-line without having to stop the execution of the loaded control algorithm. It also allows us to track the system variables and parameters, also those being estimated throughout the adaptive control process. The block diagram of the board is given in Figure 2 (dSPACE GmbH, 1999b) (dSPACE GmbH, 1999a)

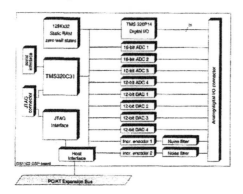

Figure 1. DS1102 Controller Board block diagram

This control system environment enables the demonstration of several basic adaptive control algorithms, (Åström Karl Johan, 1995) like:

- deterministic self-tuning regulator (STR) – direct and indirect
- minimum variance regulator (self-tuning) (MVR) (Tordsen, 1994)
- Model Reference Adaprive Control (MRAC)

Adaptive control algorithms were implemented in C/C++ language (Kernighan and Ritchie, 2000) (dSPACE GmbH, 1999c) .

As a model we used "I or II degree electronic model board" which was based on operational amplifiers.

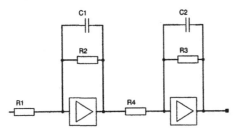

Figure 2. I or II degree electronic model board scheme.

The model's transfer function is given by

$$G(s) = \frac{\frac{R_1}{R_2}}{\frac{C_1}{R_2}s + 1} \frac{\frac{R_3}{R_4}}{\frac{C_2}{R_4}s + 1}. \tag{1}$$

This board allows us to change the model parameters, that makes it possible to better explain and understand how adaptive algorithms cope with the changing plant parameters. In examples which are shown below, we will consider two groups of parameters describing two objects. These parameters are:

I) $R_1 = R_2 = R_3 = 10k\Omega$,
$C_1 = 0.1\mu F$, $C_2 = 0.7\mu F$

II) $R_1 = R_2 = R_3 = 10k\Omega$,
$C_1 = 0.1\mu F$, $C_2 = 0.9\mu F$,

giving two transfer function:

$$T_I(s) = \frac{1}{7 * 10^{-22}s^2 + 8^{-11}s + 1} \tag{I}$$

$$T_{II}(s) = \frac{1}{9 * 10^{-22}s^2 + 10^{-11}s + 1}. \tag{II}$$

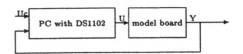

Figure 3. General control scheme.

Estimation model is given by the following discrete transfer function

$$T(z) = \frac{b_0 z + b_1}{z^2 + a_1 z + a_2}, \tag{2}$$

with sampling time Ts=0.001s.

Control law used in all the algorithms is given by the equation (Åström Karl Johan, 1997)

$$Ru(t) = Tu_c(t) - Sy(t) \tag{3}$$

The figure 4 shows object response (upper waveform) and reference (lower waveform). Because reference signal will be the same in all examples in next pictures instead of this signal we will use a control signal. It allows also to show regulator adaptation to the different objects more clearly.

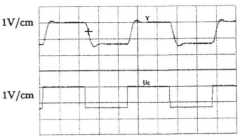

Figure 4. Response (Y) and reference (U_c) from object with indirect STR with zero cancellation

In figure 5 we may see the general idea of adaptive control setup. Parameters of the object are estimated and the estimates are passed to an algorithm which calculates parameters of the controller. These parameters are used in feedback control loop and are adjusted on-line using DS1102 board.

Now we will present experimental results of the selected adaptive control algorithms.

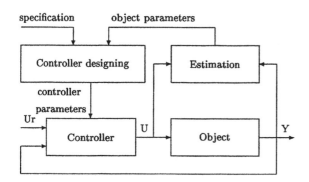

Figure 5. General indirect adaptive control algorithms block diagram.

2.1 Deterministic indirect STR with zero cancel

Estimated parameters for:

object I

$$\hat{b}_0 = 0.06510 \qquad \hat{b}_1 = 0.02614$$
$$\hat{a}_1 = -0.81179 \qquad \hat{a}_2 = -0.04758$$

object II

$$\hat{b}_0 = 0.052412 \qquad \hat{b}_1 = 0.03074$$
$$\hat{a}_1 = -0.611700 \qquad \hat{a}_2 = -0.25741$$

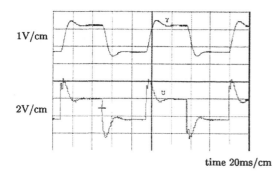

time 20ms/cm

Figure 6. Object I with indirect STR with zero cancellation.

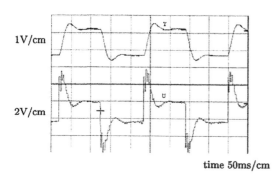

time 50ms/cm

Figure 7. Object II with indirect STR with zero cancellation.

Picture 8 shows output and control signal for object I. Picture 9 shows the same for object II. In this pictures we can see that despite changes of objects parameters output is nearly the same. Different control signal better explains adaptation of regulator parameters.

2.2 Deterministic indirect STR without zero cancel

Object's parameters estimates:

for object I

$$\hat{b}_0 = 0.06490 \qquad \hat{b}_1 = -0.00803$$
$$\hat{a}_1 = -1.16142 \qquad \hat{a}_2 = 0.26039$$

for object II

$$\hat{b}_0 = 0.11730 \qquad \hat{b}_1 = -0.03601$$
$$\hat{a}_1 = -0.57607 \qquad \hat{a}_2 = -0.29338$$

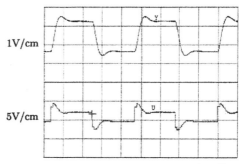

time 50ms/cm

Figure 8. Response (Y) and control signal (U) from object I with direct STR without zeros cancellations.

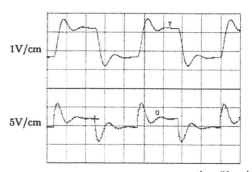

time 50ms/cm

Figure 9. Response (Y) and control signal (U) from object II with direct STR without zeros cancellation.

2.3 Deterministic direct STR

Estimated parameters

for object I

$$\hat{r}_0 = 0.041 \qquad \hat{r}_1 = 0.0229$$
$$\hat{s}_0 = 0.0296 \qquad \hat{s}_1 = -0.124$$
$$\hat{t}_0 = -0.228$$

for object II

$$\hat{r}_0 = 0.0306 \qquad \hat{r}_1 = 0.0156$$
$$\hat{s}_0 = 0.0297 \qquad \hat{s}_1 = -0.105$$
$$\hat{t}_0 = -0.228$$

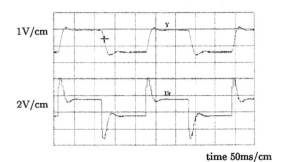

Figure 10. Object I with direct STR.

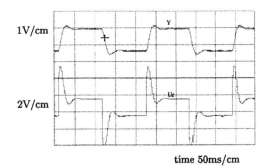

Figure 11. Object II with direct STR.

3. CONCLUSIONS

The results presented show the capabilities for the laboratory setup for teaching the adaptive control for undergraduate students at Warsaw University of Technology. The aim of the laboratory is to show the students the basics of adaptive control using standard PC and DSP controller board attached to the control plant. This setup allows us to introduce the fairly advanced control concepts to the third year Electrical Engineering students without the need of introducing some dedicated control, modeling and simulation tools. The software used is a standard MS Windows, user friendly environment augmented with adaptive control algorithms prepared at our group. The experimental results presented in the paper show the examples of the students laboratory exercise with deterministic STR. These exercises were aimed at demonstrating the capabilities of adaptive control and expose the very ideas of it. For this purpose different control algorithms were presented, both direct and indirect, with and without the zeros cancellations. The plant changes used are just example and the parameters changes of 20% presented can be increased. The laboratory setup described in the paper is easy to use and easy to explain to the undergraduates. It assumes only the basic knowledge of classical control plus fundamental knowledge of linear algebra, calculus and probability. There is also an option of remote access to the experimental PC via local area network (Gigabit Ethernet) and also using Internet. This is still the subject of further research.

REFERENCES

Åström Karl Johan, Wittenmark Björn (1995). "Adaptive Control -II Edition ". Addison-Wesley Publishing Company.

Åström Karl Johan, Wittenmark Björn (1997). "Computer-Controlled Systems: Theory and Design - II Edition". Addison-Wesley Publishing Company.

dSPACE GmbH (1999a). "DS1102 DSP Controller Board. Installation and Configuration Guide". dSPACE digital signal processing and control engineering GmbH.

dSPACE GmbH (1999b). "DS1102 User's Guide ver. 3.0". dSPACE digital signal processing and control engineering GmbH.

dSPACE GmbH (1999c). "Real-Time Interface (RTI and RTI-MP). Implementation Guide". dSPACE digital signal processing and control engineering GmbH.

Kernighan, Brian W. and Dennis M. Ritchie (2000). "The C Programing Language". WNT.

Tordsen, Söderström (1994). "Discrete-time Stochastic Systems. Estimation & Control". Prentice Hall International.

SELF-TUNING CONTROL: LABORATORY REAL - TIME EDUCATION

Vladimír Bobál, Petr Chalupa, Marek Kubalčík, Petr Dostál

*Department of Control Theory, Institute of Information Technologies,
Tomas Bata University in Zlín, Nám. TGM 275, 762 72 Zlín, Czech Republic,
tel.: +420 57 603 3217, E-mail: bobal@ft.utb.cz*

Abstract: The combination of the automatic control theory courses and practical implementation of the designed controller algorithms in real-time conditions is very important for training of the control engineers. This contribution presents a MATLAB-Toolbox for design, simulation verification and especially real-time implementation of single input - single output (SISO) discrete self-tuning controllers. The proposed adaptive controllers what are included into a Toolbox can be divided into three groups. The first group covers PID adaptive algorithms using traditional Ziegler-Nichols method for the setting of the controller parameters, the second group of described controllers is based on the pole placement design and the third group contains the controllers derived on the other approaches (dead-beat, minimum variance etc.). The controllers are implemented as an encapsulated SIMULINK blocks and thus allows users simple integration into existing SIMULINK schemas. The process of developing real-time applications using MATLAB Real-Time Workshop and several control courses is also presented. The MATLAB-Toolbox is very successfully used in Adaptive Control Course in education practice for design and verification of self-tuning control systems in real-time. It is suitable for design and verification of the industrial controllers, too. The MATLAB-Toolbox is available free of charge at Internet site - **http://www.utb.cz/stctool/**. *Copyright © 2003 IFAC.*

Key words: Self-tuning control; ARX model; Recursive least squares; PID control; Pole assignment; Real-time control; MATLAB–Toolbox.

1. INTRODUCTION

Adaptive control methods have been developing significantly over the past four decades. The aim of this research was to solve the problem of designing a controller for systems where the characteristics are not completely known or vary. Different approaches were proposed and utilized. One successful approach is represented by self-tuning controllers (STC). The main idea of an STC is based on the combination of a recursive identification procedure and a selected controller synthesis. Different branches of the STC differ mainly in the method of the controller design while the identification recursive least squares method (RLSM) applied to a regression (ARX) model forms a standard.

Despite intensive research activity, there is a lack of controllers which are able to cope with practical requirements. Industrial applications on a large scale are still missing, however, many realized cases were successful. It is evident that until to now application has required a control engineer skilled in specific technology as well as versed in modern control methods.

Also the aim of this contribution is to inform of potential users about Self-Tuning Controllers Simulink Library - STCSL (see Bobál and Chalupa, 2002) and to help for practical usage in the simulation and real time conditions. The individual self-tuning algorithms are introduced in the brief form in User's Guide that is attached into the STCSL. This library can be also the suitable bridge between theory and practice by presenting some simple controller algorithms in a form acceptable for industrial users. The theoretical background describes not only these simple algorithms, but also algorithms based on polynomial solutions of controller synthesis and controllers that are derived from the use of minimization of linear quadratic criterion, are given in Bobál, et al. (1999a, 1999b).

All the explicit self-tuning controllers that are included into STCSL have been algorithmically modified in the form of mathematical relations or as flow diagrams so as to make them easy to program and apply. Some are original algorithms based on a modified Ziegler-Nichols criterion, others have been culled from publications and adapted to make them more accessible to the user.

2. THEORETICAL BACKGROUND

2.1 Recursive identification

The regression (ARX) model of the following form

$$y(k) = \Theta^T(k)\phi(k-1) + n(k) \qquad (1)$$

is used in the identification part of the designed controller algorithms, where

$$\Theta^T(k) = [a_1 \ a_2 \ ... \ a_{na} \ b_1 \ b_2 \ ... \ b_{nb}] \qquad (2)$$

is the vector of the parameter estimates and

$$\phi^T(k-1) = [-y(k-1) -y(k-2) ... -y(k-na) \\ u(k-1) \ u(k-2) ... \ u(k-nb)] \qquad (3)$$

is the regression vector ($y(k)$ is the process output variable, $u(k)$ is the controller output variable). The non-measurable random component $n(k)$ is assumed to have zero mean value $E[n(k)] = 0$ and constant covariance (dispersion) $R = E[n^2(k)]$. The recursive least squares method for calculating of the parameter estimates (2) is utilized, where adaptation is supported by directional forgetting (Kulhavý, 1987).

2.2 Algorithms of digital PID Ziegler-Nichols controllers

The PID controllers are still widely used in industry. These types of controllers are more convenient for users owing to their simplicity of implementation, which is generally well known. Provided the controller parameters are well chosen they can control a considerable part of continuous technological processes. To get a digital version of the PID controller, it is necessary to discretize the integral and derivative component of the continuous-time controller. For discretizing the

integral component we usually employ the forward rectangular method (FRM), backward rectangular method (BRM) or trapezoidal method (TRM). The derivative component is mostly replaced by the 1st order difference (two-point difference). The recurrent control algorithms which compute the actual value of the controller output $u(k)$ from the previous value $u(k-1)$ and from compensation increment seem to be suitable for practical use

$$u(k) = q_0 e(k) + q_1 e(k-1) + q_2 e(k-2) + u(k-1) \quad (4)$$

where q_0, q_1, q_2 are the controller parameters. The advantage of algorithm (4) is matter of fact that it is not necessary to storage last input and output data in the computer storage.

It is subsequently possible to derive further variants of digital PID controllers. First group of proposed self-tuning PID controllers is based on the classical Ziegler and Nichols (1942) method. In this well-known approach the parameters of the controller are calculated from the ultimate (critical) gain K_{pu} and the ultimate period of oscillations T_u of the closed loop system. The analytical expressions for computing of these critical parameters are derived in Bobál, et al. (1999a, 1999b).

2.3 Algorithms of pole placement controllers

A controller based on the pole placement method in a closed feedback control loop is designed to stabilise the closed control loop whilst the characteristic polynomial should have a previously determined pole. Digital controllers is possible can be expressed in the form of a discrete transfer function

$$G_R(z) = \frac{U(z)}{E(z)} = \frac{Q(z^{-1})}{P(z^{-1})} \qquad (5)$$

Let the controlled process be given by the transfer function

$$G_P(z) = \frac{Y(z)}{U(z)} = \frac{B(z^{-1})}{A(z^{-1})} \qquad (6)$$

with the polynomials $A(z^{-1})$, $B(z^{-1})$ for degree $n = 2$.

PID - A controller structure. This controller structure is shown in Fig. 2a). The controllers polynomial in equation have the form

$$Q(z^{-1}) = q_0 + q_1 z^{-1} + q_2 z^{-2}$$
$$P(z^{-1}) = (1 - z^{-1})(1 + \gamma z^{-1}) \qquad (7)$$

and then the characteristic polynomial of the closed-loop system with a PID-A controller (see Fig. 1a)) is in the form

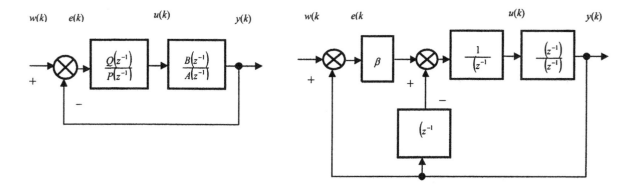

a) PID - A controller   b) PID - B controller

Fig. 1 Block diagrams of control loops with pole placement PID controllers

$$A(z^{-1})P(z^{-1}) + B(z^{-1})Q(z^{-1}) = D(z^{-1}) \qquad (8)$$

The pole placement of characteristic polynomial (8) determines the dynamic behaviour of the closed - loop system. The characteristic polynomial $D(z^{-1})$ can be specified by different methods. It is offen described by the dominant poles for a second order continuous model (controller PID A-1)

$$D(s) = s^2 + 2\xi\omega_n s + \omega_n^2 \qquad (9)$$

The dominant poles are given by the desired damping factor ξ and the natural frequency ω_n of the closed-loop.

A PID-A controller with the characteristic polynomial (controller PID A-2)

$$D(z) = (z - \alpha)^2 [z - (\alpha + j\omega)][z - (\alpha - j\omega)] \qquad (10)$$

is proposed in Bobál, et al. (1999a). Characteristic polynomial (10) has a double real pole $z_{1,2} = \alpha$ within interval $0 \leq \alpha < 1$ and a pair of complex conjugated poles $z_{3,4} = \alpha \pm j\omega$, where $\alpha^2 + \omega^2 < 1$. The parameter α influences the speed of the control-loop transient response and influences controller output changes, too. By changing parameter ω it is possible to influence the desired overshoot.

PID - B controller structure. The structure of the control loop with controller PID-B is shown in Fig. 1b). The characteristic polynomial has in this case form

$$A(z^{-1})P(z^{-1}) + B(z^{-1})\left[Q(z^{-1}) + \beta\right] = D(z^{-1}) \qquad (11)$$

where polynomial $P(z^{-1})$ has the same form as in equation (7) and second polynomial in equation (11) is given by

$$Q(z^{-1}) = (1 - z^{-1})(q_0 - q_2 z^{-1}) \qquad (12)$$

Using polznomial (9) it is possible to derived controller B-1, using equation (10) PID B-2.

3. TOOLBOX DESCRIPTION

The first modification of this toolbox used the Graphical User Interface - GUI (Bobál, et al., 1999, Bobál and Böhm, 2000). The modification using GUI is depended on the individual MATLAB version. On that account the SIMULINK version of this toolbox (so called Self-Tuning Controllers Simulink Library - STCSL) has been designed.

3.1 Self-Tuning Controllers Simulink Library (STCSL)

The purpose of the STCSL was to create an environment suitable for creating and testing of self-tuning controllers. The library is available free of charge at Internet site (see Bobál and Chalupa, 2002). The library was created using MATLAB version 6.0 (Release 12), but it can be ported with some changes to lower MATLAB versions. Controllers are implemented in the library as standalone SIMULINK blocks, which allow an easy incorporation into existing simulation schemes and an easy creation of new simulation circuits. Only standard techniques of SIMULINK environment were used when creating the controller blocks and thus just basic knowledge of this environment is required for the start of a work with the library. Controllers can be implemented into simulation schemes just by the copy or drag & drop operation and their parameters are set using dialog windows. Another advantage of the used approach is a relatively easy implementation of user-defined controllers by modifying some suitable controller in the library. Nowadays the library contains 29 simple single input single output discrete self-tuning controllers, which use discrete models of second and third order for the on-line process identification. The library package contains not only the controllers but also reference manual with simple description of the algorithm and the internal structure of each controller.

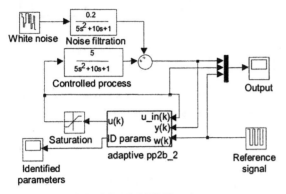

Fig. 2. Control circuit in SIMULINK environment

The typical wiring of any library controller is shown in Fig. 2. Each self-tuning controller from the library uses 3 input signals and provides 2 outputs. The inputs are the reference signal (w) and the actual output of controlled process (y). The last controller input is the current input of controlled process – the control signal (u_in). The value of this signal does not have to be the same as controller output e.g. due to the saturation of controller output. The main controller output is, of course, control signal – the input signal of the controlled process. The second controller output consists of the current parameter estimates of the controlled process model. The number of parameters this output consists of depends on the model used by on-line identification.

A scheme analogous to the scheme in Fig. 2 can be used to simulate the control process of both a discrete and a continuous controlled process with much more complicated structure. It is possible to implement processes with time variable parameters, processes described by non-linear differential equations etc.

3.2 Overview of library controllers

The Self-Tuning Controllers Library is started by opening the SIMULINK file *cntrl_lib.mdl*, which contains block schemes of all controllers. The name of the controller always corresponds with the name of the file, which provides the calculation of controller parameters and has the following structure: **xx(n)yyyy**. First two characters (**xx**) indicate controller type – **zn** stand for the controller synthesis based on Ziegler-Nichols method, **pp** represent controller with a pole placement synthesis, **mv** denote minimum variance controller, etc. The third character (**n**) in a controller name is a digit 2 or 3 corresponding to the order of the model used by on-line identification part of the controller. Following characters (**yyyy**) serve to cover other controller details. The survey of the individual controllers with the name of the m-files and requisite input parameters is introduced in Bobál *et al.* (2001). The

detailed description of most of the controller algorithms is introduced in Bobál, *et al.* (1999a).

3.3 Internal controller structure

Each library controller is constructed as a mask of a subsystem, which consists of SIMULINK blocks and has inputs, outputs and parameters. Internal controller structure consists of SIMULINK blocks which provide, among others, the possibility of easy creation of a new controller by a modification of some suitable library controller. The structure of controller *pp2a_1* is presented in Fig 3 as an example. Each library controller consists of three basic parts:

- on-line identification block,
- block computing controller parameters,
- block computing controller output.

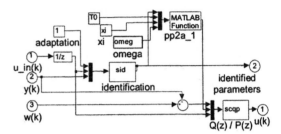

Fig. 3. Block scheme of PP2A_1 controller

3.4 Reference guide and help

Besides the files implementing the functionality of all library controllers the library package includes reference guide with a short description of all controller algorithms, of an operating with the controllers and of the internal structure of the controllers. The guide is provided in two versions:

- *pdf* format suitable for printing,
- *html* format used for the context help

Moreover the help for each function included in the library can be invoked by entering `help function_name` at MATLAB command line.

4. PRACTICAL USAGE OF STCL IN REAL - TIME APPLICATIONS

When using the library to control some real-time model or equipment the first phase consists of selecting appropriate controller and some simulations are performed to tune controller parameters. In the next phase the testing the controller using real model is performed. In the last phase the SIMULINK environment is used to generate source codes of program working outside MATLAB - SIMULINK environment. This code can be compiled, linked and loaded into industrial controllers.

Practical verification of library controllers has been performed using several laboratory models. One of them, the air heating system, is shown in Fig. 4. This system has two inputs (rotations of ventilator and a power of resistance heating) and two outputs (the flow of an air measured by the rotations of aircrew and a temperature inside the tunnel measured by resistance thermometer). The system was divided into two input-output pairs, where the first pair consists of the rotations of ventilator as an input and the flow of an air as an output. The second pair consists of the power of the resistance heating as an input and the temperature as an output. The aim is to control a two inputs two outputs system using two standalone single input single output controllers. This approach is known as the decentralized control.

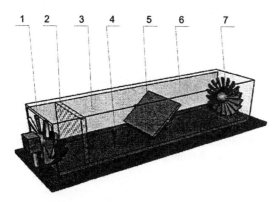

Fig. 4. Laboratory air-heating model (1- ventilator, 2-resistance heating, 3- pressure sensor, 4-resistance thermometer, 5-shutter, 6-shield, 7-air flow measurement)

The connection of the laboratory model and the SIMULINK environment has been realized through control and measurement PC card Advantech PCL-812 PG. Blocks for the reading analogue inputs and for writing to the analogue outputs on the PC card were used to communicate with the model. These blocks are implemented as s-functions written in C language, which allows low-level access to the ports of the PC computer. This mechanism allows the connection of SIMULINK and any PC compatible equipment designed to collect external data.

The sample of control course using a pole-placement controller is plot in Fig. 5, where reference signal is green, controlled signal is red and control signal is blue. The time constants of the pair ventilator-flow are relatively small when comparing with time constants of the heating-temperature pair and thus the output of the first pair becomes steady earlier. The courses demonstrate that the change of the resistance heating power does not affect the flow of an air, but the influence of ventilator rotation to the temperature is substantial. The parameters of the heating-temperature controlled system are thus changing in time and on-line identification method used should assign greater weight to newer data then to older data. Time courses of parameter estimates are presented in Fig 6. It is obvious from this picture

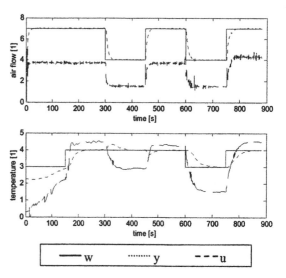

Fig. 5. Control courses using PP2B_1 controllers

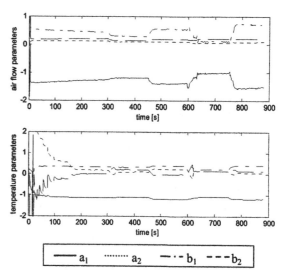

Fig. 6. Courses of model parameter estimates using PP2B_1 controllers

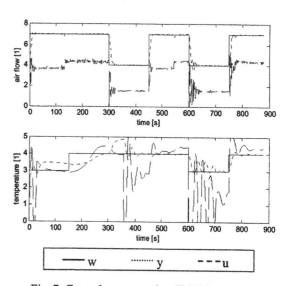

Fig. 7. Control courses using ZN2BR controllers

That the changes of parameters are greater in first phase – this is caused by inaccurate initial parameters estimates. Further changes of parameters correspond to changes of control signal what indicates the presence of a non-linearity in the system.

The control course is different when controller of another type was used as shown in Fig. 7. The course of reference signal is the same as in previous case, but the *zn2br* controllers were used. The *zn2br* controller uses Ziegler-Nichols algorithm of computation the value of control signal with the backward rectangular method of discretization integration component. In this case the oscillations of controlled signals are significantly greater and the quality of control process is lower when comparing with the pole placement controllers.

5. CREATING APPLICATIONS WITH REAL - TIME WORKSHOP

The MATLAB - SIMULINK environment can also be used to generate code to be used in controllers in industrial practice. *Real-Time Workshop*, one of the toolboxes shipped with MATLAB, allows generating of source code and programs to be used outside the MATLAB environment. The process of generating the source code is controlled by special compiler files that are interpreted by *Target Language Compiler*. These files are identified by the *.tlc* (target language compiler) extension and describe how to convert SIMULINK schemes to target language. Thereby source code is generated and after compiling and linking the resulting application is created. Applications for various microprocessors and operating systems can be created by selecting corresponding target language compiler files.

The target language compiler can create applications to be used under the Windows environment, which perform control algorithm and save results in a binary file with the structure acceptable by MATLAB. An analysis of the control process can then be performed using advantages of MATLAB functions and commands. Selecting another *.tlc* file leads to creation of a MS-DOS application or an application to be used on PC based industrial computers without a requirement of any operating system.

Many manufactures of industrial computers and controllers has created their own target language compiler files used to create applications for equipment they produce. *Real-Time Workshop* provides a relatively opened environment for the conversion of block schemes to various platforms where users can create their own target language compiler files for converting the block scheme to a source code and hence reach the compatibility with any hardware.

An application creation consists only of selecting appropriate target language compiler file, eventually setting compiler parameters and then start-up of compilation process.

6. CONCLUSIONS

The Self-Tuning Controllers Simulink Library is used in university course of adaptive control systems. Its architecture enables an easy user orientation in SIMULINK block schemes and source code of controllers' functions. The controllers provided are suitable for modification and thereby implementation of user-defined controllers. The compatibility with *Real-Time Workshop* ensures not only the possibility of laboratory testing using real time models but also the possibility of creating applications for industrial controllers.

ACKNOWLEDGEMENTS

This work was supported in part by the Grant Agency of the Czech Republic under grants No.102/02/0204, 102/00/0526 and in part by the Ministry of Education of the Czech Republic under grant MSM 281100001.

REFERENCES

Bobál, V., J. Böhm, J. Prokop and J. Fessl (1999a). *Practical Aspects of Self-Tuning Controllers: Algorithms and Implementation*. VUTIUM Press, Brno University of Technology, Brno (in Czech).

Bobál, V., J. Böhm, and R. Prokop (1999b). Practical aspects of self-tuning controllers. *International Journal of Adaptive Control and Signal Processing*, **13**, 1999, 671-690.

Bobál, V. and J. Böhm (2000). MATLAB-Toolbox for design and verification of self-tuning controllers. In: *Prep. of IFAC Symposium on Advances in Control Education*, Gold Coast, Australia, Paper No. 10S-1.

Bobál, V., J. Böhm, and P. Chalupa (2001). MATLAB-toolbox for CAD of simple self-tuning controllers. In: *Proc. 7th IFAC Workshop on Adaptation and Learning in Control and Signal Processing ALCOSP 2001*, Cernobbio-Como, Italy, 2001, 273-278.

Bobál, V. and P. Chalupa (2002). Self-Tuning Controllers Simulink Library. http://www.utb.cz/stctool/.

Kulhavý, R. (1987). Restricted exponential forgetting in real time identification. *Automatica*, **23**, 586-600.

Ziegler, J. G. and N. B. Nichols (1942). Optimum settings for automatic controllers, *Trans. ASME* **64**, 1942, 759 – 768.

MODEL AND SOFTWARE DESIGN FOR HEALTH CARE AND LIFE SUPPORT

Tsutomu Matsumoto* Shigeyasu Kawaji**

*Kumamoto National College of Technology
2659-2 Suya, Nishigoshi, Kikuchi, Kumamoto 861-1102, Japan
matumoto@ec.knct.ac.jp
** Graduate School of Science and Tecnology
Kumamoto University
2-39-1 Kurokami, Kumamoto 860-8555, Japan
kawaji@cs.kumamoto-u.ac.jp

Abstract:
The importance of developing a intelligent control system has been pointed out for ill structure which includes human such as health care support, life support, education and so on. In order to develop the software system which targets these field, analyzing task and information based on cybernetics are very useful for modeling and software design. Unfortunately, design methodology education is not given to student. Because it is very difficult to derive mathematical equation for the behaviour of object.
In this paper, modeling and software design of health care life support system as an typical example of ill structure are described. Main part of this paper has been used as an educational material to final project students who belong to our research group. They learn the design method of the system including human via the tutorial and the educational material. Copyright © 2003 IFAC

Keywords: Ill structure, Modeling, Patient Model, Disease Model, Clinical Diagnosis Decision Support, Health Care Modeling, Design Method

1. INTRODUCTION

When control system is desgined, the first thing, the behavior of controlled object is expressed by mathematical equation based on the physical law. If the controlled object is complex or large scale, the mathematical equation that expresses the behaviour of controlled object could be determined by the input output relationship and/or the assumption of internal structure. In the case of controlled object being nonlinear and/or obtaining no mathematical equation for expressing the behaviour of controlled object, the behaviour of object by Neural Networks or Fuzzy may be expressed. Like these objects which are formulated are well known as good structure issue.

On the other hand, issues such as medical, education and emotion and so on which include human in itself or are difficulty or impossible formulated. These are well known as ill structure. Real world has many issues such ill structure.

Medical diagnosis is regarded as one of control system since clinician identifics client by obtained client information and gives medical treatment as control input to client. Education is also regarded as one of control system since educator identifies the understanding of student and gives an appropriate teaching material.

Development of modeling and control law to ill structured object should be created by analyzing controlled object based on cybernetic. Unfortunately design method of control system by cybernetic has not educated to student.

In this paper, first a design method of information support system to ill structure object including human is described using clinical diagnosis support system and health care life support system as actual examples. Second how educates student for design method of ill structured information support system is discussed. Finally evaluation of proposed method is described.

2. HEALTH CARE AND LIFE SUPPORT MODEL

Development of a health care and life support model including medical diagnosis by analyzing task and information is discussed in this section as a typical example of ill structure problem.

2.1 *Task of Health Care and Life Support*

In order to construct a medical consultation model, first we discuss the medical task of physician's diagnosis and giving therapy to patient. Physician interviews a new patient for acquiring various personal information, disease history of patient and family, life style, main complaint and so on. And physician does inquiry, inspection, auscultation to find abnormal states of patient. Physician orders blood check and/or bio-chemical medical check as occasion demands to clinical laboratory.

Physician estimates the state of patient using the data from patient and clinical laboratory data and inspection. Finally physician identifies the disease of client. Then physician makes a medical treatment plan after discussion on the subject with the client and/or family. After that, the physician makes a detail treatment plan. The treatment has two types. The one is to provide directly treatment to the client. The other is to provide indirectly to the client's treatment, for example through nursing staff. In this case physician gives an instruction to nursing staff then nurse staff carries out the instruction. There may be a case that physician gives a description to client and/or gives some advice to family. The family then cares for the client. Physician should confirm that treatment is given to the client.

In the case of health care and life support which include nursing elderly people, comedical who are family and nursing staff finds client's problem as client identification. For this decision making, Minimum Data Set is used. Then care plan is developed, is implemented.

Some time later physician and comedical evaluates the effect of treatment and/or care plan given to client and improves the plan. Above-mentioned tasks of physician and comedical are similar to an feedback control system configuration, a client as a controlled object, healthy as target and manipulated variables are medical treatment and care plan with cybernetics.

2.2 *Information for Health Care and Life Support*

Client information can be classified into two categories types of static and dynamic. The static information is not to change or even if change, time interval of changing is ignored, e.g. sex, age, etc. The dynamic information indicates variation with time and other causes, e.g. body temperature, auscultation sound, index of clinical laboratory data. In this paper client information is classified as follows. The zero order information is static information. And first - third order informations are dynamic.

(1) Zero order information :
    Personal information(sex, age, disease history, career), disease history of family, life style (smoking, drinking, meal custom etc).
(2) First order information : Obtaining directly or using simple detector
    Main compliant, inspection, auscultation, blood pressure, pulse count, body temperature etc.
(3) Secondary information : Blood check and biochemical check up data
    Information from analyzing available sample of living body, ex blood, urine, stool, pus, snivel.
(4) Third order information : using electro and information technology
    ECG, electromyogram, sphygmogram, X-ray image, MRI, X-ray CT etc.

Note that higher order information is also advanced health care information.

2.3 *Health Care and Life Support Model*

Analyzing the physician's and comedical's task in the health care life support and structuring the client information will yield a scheme as shown in Fig. 1. In the figure disturbance shown means inputs for client to take bad turn. Fig. 2 is simplified by detector parts and treatment parts becoming each one black box. We call it "Health Care and Life Support Model".

Physician plays role of a controller in the Medical Consultation Model, i.e. he diagnoses client using the output y of detector which is originated from client, makes a treatment plan and gives directly and/or indirectly the medical treatment u to the client. The feedback from treatment block to physician block is required to confirm whether or not the client receives treatment. The r is normal health situation. The treatment is recognized as same as an operation in a control system. At this time physician verifies whether or not treatment is given.

3. DIAGNOSIS ALGORITHM OF PHYSICIAN

In this section, we propose a medical diagnosis algorithm based on the medical consultation model described in Section 2. Concepts of client disease

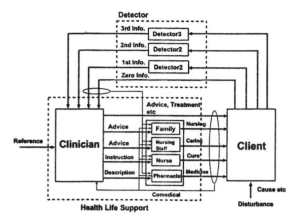

Fig. 1. Information from client and task

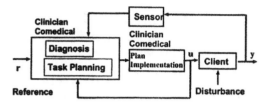

Fig. 2. Medical Consultation Model

model are also introduced. Developing diagnosis algorithm in the medical consultation model is equivalent to designing a controller. The disease model and client model corresponds to respective represented knowledge on disease and modeling of controlled object.

Medical diagnosis algorithm is generally expressed to compare evaluated value of $p = wy$ using client information y with relevant sample P based on medical knowledge (Furukawa, 1982). From the view of control theory, physician develops a real client model using client information y as a output of controlled object and evaluates p of each disease with medical knowledge. Finally a medical decision making is carried out by comparing p with normal value P which is output of disease model developed by medical knowledge.

3.1 Disease Model

The model which gives normal value P to physician is called disease model. Let disease be P_i and information of client be Y_j, then physician's diagnosis algorithm is expressed by equation (1).

$$\begin{pmatrix} P_1 \\ P_2 \\ \vdots \\ P_i \end{pmatrix} = \begin{pmatrix} W_{11} & W_{12} & \cdots & W_{1j} \\ W_{21} & W_{22} & \cdots & W_{2j} \\ \vdots & \vdots & \ddots & \vdots \\ W_{i1} & W_{i2} & \cdots & W_{ij} \end{pmatrix} \begin{pmatrix} Y_1 \\ Y_2 \\ \vdots \\ Y_j \end{pmatrix} \quad (1)$$

Next a method of determining weighted coefficients W_{ij} is discussed. Human living body system generally consists of the central nervous system, the metabolism system, the regulation system and generation system that affects each other. If one

or more system have some problems, output of the system includes abnormal state. Diseases depend on this level of abnormal state. When the abnormal exceeds a threshold, human recognizes some subjective symptoms. Subjective symptom becomes one of first order information in Section 2. Output of the metabolism and regulation system causes the second order information.

Mostly physician decides kinds of required secondary information to diagnose based on zero and first order client. This paper deal with symptoms and remarks as first order client information, clinical laboratory data as secondary client information.

3.2 Client Model

Client model is expressed by equation (2), in similar way of disease model constructed by medical knowledge.

$$\begin{pmatrix} p_1 \\ p_2 \\ \vdots \\ p_l \end{pmatrix} = \begin{pmatrix} w_{11} & w_{12} & \cdots & w_{1m} \\ w_{21} & w_{22} & \cdots & w_{2m} \\ \vdots & \vdots & \ddots & \vdots \\ w_{l1} & w_{l2} & \cdots & w_{lm} \end{pmatrix} \begin{pmatrix} y_1 \\ y_2 \\ \vdots \\ y_m \end{pmatrix} \quad (2)$$

Let y_m be the client information and p_l be pathological state and disease. In order to solve individuality, client model has individual weighted coefficients w_{lm}. Next a method of determining w_{lm} is explained.

Individual normal range of each client clinical laboratory data are set up by adopting Tango's method (Tango, 1982). This is to set a normal range of each clinical laboratory data with past no problem data $(\phi_1, \phi_2, \ldots, \phi_{(n-1)})$. Thus this range is used for finding whether or not current data ϕ_n in the normal range. Furthermore distribution of clinical laboratory data general index has been studied in medical field. Individual normal range could be set up by obtaining a interval estimation of population mean μ. This interval estimation would be individual normal range.

3.3 Diagnosis Algorithm

Physician estimates disease name from zero and/or first order information of client or makes medical decision. In order to improve the accuracy estimation, physician uses the secondary and/or the third information. Otherwise physician may identify the name of disease with them. Consequently, this system searches the name of disease with the remarks and the symptoms and estimates the pathology of client by clinical laboratory data.

Searching the name of disease is carried out with the zero and/or the first order information, and comes to the conclusion of distinction using multivariate analysis. The clinical laboratory data belongs to the secondary data but its futures can be expressed by words, e.g. GOT increasing.

As clinical laboratory data shows the output of the metabolizing system and the regulatory system, these data are used for estimating the pathological state of client. When the number of sampled data is short, general index is used. The number of sampled data exceeding specified number, individual index is placed by equation (3). Furthermore as those data have different units and index, then the index of p_l is normalized.

4. TELE-HEALTH CARE LIFE SUPPORT SYSTEM

This section describes the modeling method of telehealth care information system based upon above the health care and life support model.

4.1 System Design

The telehealth care and life support system is regarded as setting up feedback system to share/transfer the information among client, clinician and comedical in the case of some problem of client happening. This paper defines the telehealth care information system as follows. When an irregular habitual life behavior and abnormal health condition is detected, the system confirms safety of client. And the system sends a message to clinician, nursing staff and/or neighbors depending on answer of client. In addition the system provides health advice from clinician to client.

Consequently the system has following specifications in order to perform above mentioned.

(1) Obtaining the information of client
(2) Transferring the information of client
(3) Finding problem of client
(4) Confirming client safety
(5) Informing urgency to comedical, clinician
(6) Providing health care advice

The fourth and the fifth mean the collaboration among client, comedical and clinician. Finally the telehealth care life support system can be designed using layer structure shown by figure3.

4.2 Layer Structure

This section describes the relations among layers and functions of each layers.

4.2.1. Real World
The object of this system are clients, clinicians and comedical including client's family and/or neighbors.

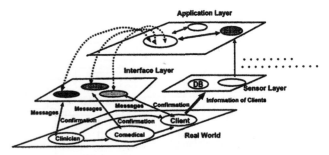

Fig. 3. Layer Structure for tele-health care

4.2.2. Sensor Layer
The sensor layer is to gather information of client and to find urgency of client(Matsumoto et al., 2000). If sensor layer finds an urgency of client, the sensor layer sends a message to application layer. The message has ID of client and urgency level that client needs clinician or not, needs nursing staff or not and so on. In addition the sensor layer provides health care record to clinician via application layer. This system focuses life information (Zero order), pulse and body temperature as the third order.

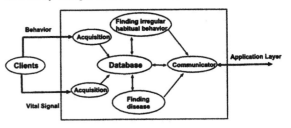

Fig. 4. Sensor layer

4.2.3. Application Layer
Application layer searches contacts of family doctor, comedical and/or client corresponding to urgency level for sending a message. With the search results application layer sends a message to interface layer. Application layer continues to send a message in order to the reply from receiver. If nobody replies, the system call an ambulance.

The application layer requests information of client to sensor layer after receiving a message of "Request : providing information of client". The application layer transfers the data to appropriate interface layer.

4.2.4. User Interface Layer

(1) Clinician Interface
Clinician interface informs urgency of client and transfers reply from clinician to the application layer. Clinician inputs the process into the system via the clinician interface. Clinician is able to retrieve health record of client.

(2) Comedical Interface
Urgency of client is informed to registered

comedical via the comedical-medical interface. Comedical who received a message has to inform the condition of client and/or result of treatment to the system via the comedical interface. Also comedical retrieves health advice from the system that clinician has provided, and comedical guides client to keep health condition.

(3) Client Interface(Interactive Response System)

Client interface sends a message to confirm safety. Client can reply "No problem", "need help", "need clinician" or no answer. The client interface informs content of reply to the system.

(4) System Administrator Interface

The system administrator interface provides functions of system administrating such as registration, removing, modifying user(client, clinician, nursing staff, neighbors).

5. DESIGN OF SOFTWARE SYSTEM

In this section design and implementation of user interface as an actual example are described based on system design described in section 4 with Object Oriented Method. Software development by Object Oriented Method has some benefits as reuse, flexibility. Furthermore UML (Unified Modeling Language) can be used to design and implement the software for Object Oriented method.

5.1 *Scenario*

UML method suggests us to develop scenario to find use case. Because an instantiation of a use case is called a scenario, and it represents an actual usage of the system; that is a specific execution path through the system. An example for the sensor layer finding the problem of client is shown as follows.

When the sensor layer finds pathological state transition. The sensor layer sends a message to the application layer. Then the application layer decode it, simultaneously sends a message to client for confirming whether or not the client has problem. If no reply from the client, the application layer sends a message for sending a comedical to the client. Also the application layer sends a message to appropriate staff according to reply from the client.

5.2 *Use Case*

After developing scenario, use case could be found. Figure 5, figure ?? and figure ?? show use cases

for each interface. A use case represents a complete functionality as perceived by an actor. The actions can involve communicating with a number of actors (users and other systems) as well as performing calculations and work inside the system. The functionality of use case is described as Use Case Description.

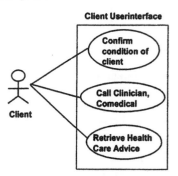

Fig. 5. Use Case for Client Interface

5.3 *Use Case Description*

Normally Use Case Description defines a functionality of use case. This is a simple and consistent specification about how the actor and the use case (the system) interact. Some of use case description for explaining briefly the system behaviour.

5.4 *Diagrams*

UML recommends the development of some kinds of diagrams to express the relation among objects and sequence of messages among objects.

A sequence diagram shows a dynamic collaboration between a number of objects. The importance aspect of this diagram is to show a sequence of message sent between the objects. Here a sequence diagrams for showing a set messages arranged in time sequence are shown as figure6.

A collaboration diagram shows a dynamic collaboration and shows the objects and their relationships. The collaboration diagram is drawn

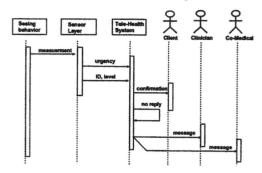

Fig. 6. Sequence Diagram : No response of client

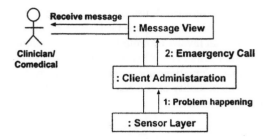

Fig. 7. Problem Happening

as an object diagram, where a number of objects are shown along with their relationships using the notation in the class/object diagram. Message arrows are drawn between the objects to show the flow of message between the objects.

For convenience, a collaboration diagrams is shown in figure 7. Figure 7 shows that when sensor layer finds client problem, how the message flows between objects.

6. APPLYING TO EDUCATION

In this chapter, how educates developing of information support software system including human to students is sequentially described.

(1)Developing structure of the object

When an information support system is developed, finding a structure of the object is a very importance. This is educated by giving actual example shown by the introduction of chapter two at the tutorial class. Simultaneously teaching material let student also clear what is supported by the software system at developing a structure.

(2)Analyzing information

Second thing to educate is to analyze information which flows feedback loop from the object to controller. Information could be categorized by their attribute. This analysis clears what information can be used for making decision and feed back loop should be either continuous or discrete.

(3) Modeling of controlled object

The third thing is to show some modeling of controlled object. As the target of the software system is to support human, this modeling has strongly related to aim of support.

(4) Formulated of making decision

In case of controller being human which is concerned, formulating of human's acting and thinking way is required. An example is shown by chapter 3.3.

(5) Layer structure

Layering system is useful for software design. This is to educate with an example shown by chapter 4.1.

This flow shows how educate students to develop an information support system including human based on cybernetics. Also students is expected to understand that control theory and cybernetics is useful not only developing traditional control system but design of information support system.

7. DISCUSSION

At the class, more details on software design documents are distributed to students. The students try to develop whole required software design documents. The software system can be developed on Java programming environment according to whole software design documents. Student can capture the thinking way of modeling and software design through the lecture and practical development of software.

This program is carried out on the beginning of final project. Then students obtain the fundamental modeling and software design method by analyzing task and information based on cybernetics.

8. REFERENCES

Furukawa, T. (1982). *Computer-based Medical Decision Support (J)*. Kyouritsu Co., Ltd.

Matsumoto, T., Y. Shimada, K. Shibasato, H. Ohtsuka and S. Kawaji (2000). Creating behavior model of a senior solitary life and detecting an urgency. *Proceedings of the International ICSC Congress on Intelligent System and Applications* pp. 234–238.

Tango, T. (1982). A new statistical method for the determination of individual normal range. *JAMHTS* 9, 241–246.

PNEUMATIC INVERTED WEDGE

Joško Petrić
Željko Šitum

University of Zagreb
Faculty of Mechanical Engineering and Naval Architecture
Ivana Lucica 5, HR-10000 Zagreb, Croatia
josko.petric@fsb.hr Phone: +385 1 6168-385
zeljko.situm@fsb.hr Phone: +385 1 6168-437

Abstract: The paper describes the inverted wedge actuated by the pneumatic cylinder. It can be considered as a two degrees-of-freedom planar robot, which is controlled by a single control input (movement of the slider) in order to keep the frame of the wedge in balance. Only one measured variable (angle of the frame) was anticipated, hence an observer was necessary to control this model using state controller. The nonminimum-phase characteristic and pneumatics make control of described system even more challenging. Thus, this model offers great possibilities as an experimental device in control education. *Copyright © 2003 IFAC*

Keywords: Control education; Experiments; Laboratory model; Mechatronics; Nonminimum-phase system; Observer; State controller

1. INTRODUCTION

The course Mechatronics at the Faculty of Mechanical Engineering and Naval Architecture on the University of Zagreb has been taught through the projects that students should make through the semester. One of these projects is a laboratory model of the inverted wedge driven by pneumatics, which is described in this paper. Laboratory models are used later in control education for practical exercises and for individual research of students in order to improve performances or make alternative design that is more suitable to control. The first laboratory model, which has been made through the course Mechatronics, was the pneumatic inverted pendulum. The comments on mechatronic education, description of the pendulum and some obtained results are given in (Petrić and Šitum, to appear). The possibilities to learn design in a mechatronic way, and to learn

control by making and using such laboratory models are really great. The reasons for the choice of aforesaid laboratory models has been in their acceptable complexity – modeling, control and hardware setting skill is necessary in order to complete the project, while in the same time they can be realized relatively simple and they are not expensive (regarding the already existing laboratory infrastructure). Important is that the physics of balancing mechanisms is intuitively clear and their motion is very attractive, as well. However, balancing mechanisms in control and mechatronic education are usual, and what distinguishes described mechanisms from common practice is their drive. They are driven by pneumatics. Pneumatics by itself is very interesting in modeling and control tasks due to air compressibility and significant friction effects in cylinders, and a challenge to make usual models with pneumatic drive vanquished over common practice. Other drive possibilities were considered and compared during the course of Mechatronics. The extensive description of the conducted research on pneumatics modeling and control is given in (Šitum, 2001) and (Šitum and Petrić, 2001).

In this paper the pneumatic inverted wedge is described. Its description, together with pneumatic and electronic infrastructure is given in Chapter 2. The control of the wedge is described in Chapter 3, while the experimental results are discussed in Chapter 4. The mathematical model, employed for the control of the wedge is given in Appendix.

2. DESCRIPTION OF THE PNEUMATIC INVERTED WEDGE

The schematic diagram of the laboratory model of the pneumatic inverted wedge is given in Fig. 1, while the photography of the model is given in Fig. 2.

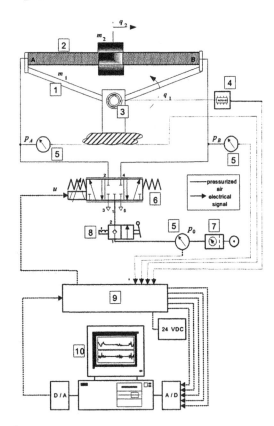

1. Frame
2. Rodless cylinder
3. Rotat. potentiometer
4. Electron. reference card
5. Pressure transducer
6. Proportional valve
7. Pressure valve & filter
8. Air supply valve
9. Electronic interface
10. PC & D/A A/D card

Fig. 1. Schematic diagram of pneumatic inverted wedge

The pneumatic rodless cylinder (in Fig. 1 marked by number 2) is FESTO DGO-12-600-P-A-B, with stroke 600 mm and diameter 12 mm. A linear motion q_2 of the slider (the piston of the cylinder, with mass m_2) is controlled by the proportional directional 5/3 valve (6), FESTO type MPYE-5 1/8 HF-010B. There has not been any sensor for measurement of linear position of the slider so far, but it will be added. The slider position can be measured by rotational potentiometer, which are connected together by a string. The angle of the pendulum q_1 is measured by rotational servo-potentiometer by Spectrol (3). The electronic reference voltage card (4) with ref-chip for potentiometer reduces the measured signal to required values. Pressure transducers (5) are SMC ISE4-01-26. The controller is implemented on the PC computer (10) via PCL-812PG acquisition card with 12-bit A/D and D/A converter. The pneumatic equipment consists additionally of some more components, like a pair of proportional pressure valves SMC VY1A00-M5 and a pair of simple on-off 3/2 valves SMC EVT307-5D0-01F. Their application (combining some other control methods, like for example Pulse Width Modulation control method) can be also considered as a cheaper alternative of the proportional directional valve employment.

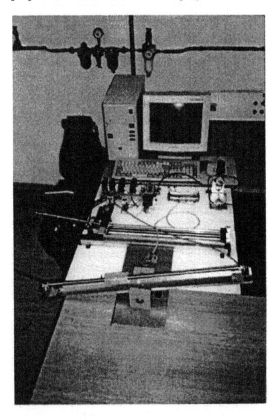

Fig. 2. Photography of inverted wedge

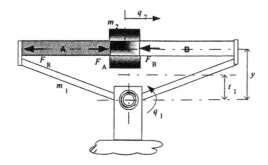

Fig. 3. Inverted wedge mechanism

The frame and the stand of the inverted wedge were designed and made by students in the Faculty's workshop. The mechanism is given in Fig. 3, while the measures can be found in List of symbols in Appendix.

The slider of mass m_2 moves in order to keep the frame of mass m_1 in balance. The slider moves by acting of force F, which is proportional to the pressure difference Δp between two chambers of the rodless cylinder:

$$F = F_A - F_B = (p_A - p_B)A = \Delta p\, A \qquad (1)$$

The proportional directional valve controls air supply and air pressure in the chambers via voltage signal u (0-10 V) on the proportional solenoid. The complex, nonlinear relation, between the input voltage and the air mass flow into the chambers is given in (Situm, 2001) and (Situm and Petric, 2001). Here, that dynamics is neglected because of significantly slower dynamics of the wedge mechanism. The mathematical model of dynamics (nonlinear and linear) of the inverted wedge mechanism is given in Appendix. The mathematical model (both nonlinear and linear) is employed for simulation while the linear model is employed in controller design. The gravity center of the frame t_1, and viscous frictions of joint b_1 and of the cylinder b_2 were determined experimentally.

In Fig. 3 can be seen an effect of reactive force F_R. Namely, pressure in a chamber does not only push the slider creating an useful moment, but it also creates a reactive moment T_R in opposite direction with harmful effect on the balance of the mechanism:

$$T_R = F_R \cdot y \qquad (2)$$

Amount of F_R can be approximated as equal as force on slider F since the area at the cylinder's end is similar to the piston area, and pressure in unpressurized chamber is almost zero in very short time (hence the difference $\Delta p = p_A - p_B$ depends mainly on the pressure in pressurized chamber). The effect of reactive force can be seen in pole-zero map of the system as a positive zero. This nonminimum-phase characteristic of the pneumatic inverted wedge makes its control even more interesting and challenging task. That characteristic could be avoided using counterbalances. Other improvements of control characteristics applying some similar, but more imaginative design of the inverted wedge can be obtained. That is, however, left to attendants of the course Mechatronics.

3. CONTROL

The wedge can be considered as a planar robot with two degrees of freedom and a single control input (it is "underactuated"). The open-loop inverted wedge is an unstable system. In (Moore et al., 2001) very advanced control strategy (genetic adaptive control) is proposed and experimentally tested on the example of inverted wedge. In this case, the problem of control of the wedge is even more complex. The actuator is pneumatic cylinder, which is interesting to control by itself. Also, measurement of the slider position has been avoided so far, and hence the observer is required to realize a state control.

The control scheme of the pneumatic inverted wedge is given in Fig. 4. System, input and output matrixes and state vector (**A**, **B**, **C** and **x** respectively) of the inverted wedge are given in eq. (4) and (5) in Appendix. **K** is a regulator gain matrix, while **L** is an observer gain matrix. The regulator gain was designed using linear quadratic regulator (LQR) optimal design, while the pole placement method was employed for getting the gain of the full-order observer.

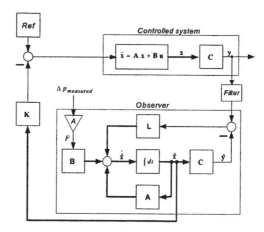

Fig. 4. Control scheme

The 2nd order Butterworth filter with cut-off frequency of 10 Hz was applied in order to eliminate noise in the measured signal. The reference was set to zero. The pressure difference, as an input value has been measured, too. The pressure difference in cylinder chambers multiplied by the area of the piston gives force F that drives the slider.

An integral action of the angle of the frame q_1 was introduced in order to improve behaviour of the closed loop system. Enlarged matrixes and vectors in that case are given in eq. (6) and (7) in Appendix. The state space controller and observer design was according to (Friedland, 1987) and (Petrić and Kecman, 1989).

4. EXPERIMENTAL RESULTS AND DISCUSSION

The laboratory model make possible to do experiments in order to test different control strategies and different controller parameters, and to obviously see their influences on the closed-loop system. The usage of the Real Time Workshop makes implementation of different controllers straightforward and fast. This Chapter presents results of the first experiments using "classical" control. It is expected that some more advanced control strategies can achieve better behaviour of the closed-loop system, hence they will be tested on this laboratory model.

The first and the simplest attempt to control the wedge were using the proportional-plus-derivative (PD) compensator. According to expectations, a simple analysis of the closed-loop mathematical model demonstrated that it can not stabilize the inverted wedge, and state controller was introduced. Since only one state has been measured (the angle of the frame of the wedge q_1) it was necessary to introduce an observer in order to estimate other states. The observability test of the **AC** pair indicated that the system is observable. For the beginning, the weighting matrix **R** was kept equal to identity matrix, while initial values for weighting matrix **Q** was equal to **C'C**. They were changed after analysis of closed loop poles and simulations, and after experiments. The poles of the observer were kept left from the system poles, that are 2 to about 10 times faster than the closed loop system. The pole-placement technique was used to design the observer gain matrix **L**. The measurement signal has been filtered in order to eliminate noise, and there were no problem to keep observer gain stronger. It was noticed that with too large regulator gains valve switches too fast, and the slider could not even

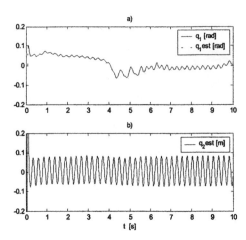

Fig. 5. Responses of inverted wedge
a)   Measured and estimated frame angle
b)   Estimated slider position

move. With too small gains the system is too slow to be stable. Also, too small observer gains cannot catch actual states, while too strong observer gains can deteriorate stability. However, regardless of different combination of regulator and observer gains, the experimental results were not satisfactory.

The integration of the angle of the frame q_1 was introduced in order to improve the results. The adequate enlargements of the matrixes and vectors from eq. (4) and (5) for one state are given in eq. (6) and (7) in Appendix. At enlarged weighting matrix **Q'** only the weight acting on the integral state was relevant. The responses of the inverted wedge using optimal state controller with integral action and with observer are given in Fig. 5. The observer poles were set up cca. 10 times faster than the poles of the closed-loop system. The initial slope of the wedge was 0.1 rad, while the observer initial states were set up to zero. The controller could hold the wedge in balance. However, the wedge could not be calmed down completely, and the slider always "worked", as that can be seen in Fig. 5. Note that the slider position in figures was estimated. The pneumatic valve opened one side after other with frequency of about 4 Hz filling the chambers alternately with pressurized air (max. pressure is set up to 7 bar), and the slider moved a few centimeters left and right. The improvements can be expected using some more advanced controllers. According to experiences and excellent results obtained with pneumatic inverted pendulum (Petrić and Šitum, to appear), better results might be expected with the introduction of the slider position measurement (for example with a rotational potentiometer connected with a string to the slider), and avoiding thus estimation of q_2.

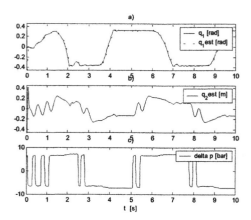

Fig. 6. Responses using stronger integral action
   a) Measured and estimated frame angle
   b) Estimated slider position
   c) Pressure difference in cylinder chambers

Intensifying integral action, the wedge became unstable. It bounced from one limit position to another (± 0.3 rad), and after while the slider went from one limit position to another (± 0.3 m). The responses in that case can be seen in Fig. 6.

If the observer gains were weak (approximately equal poles to the slowest poles of closed-loop system), the estimated states became unrealistic large, and the inverted wedge were unstable, and it also bounced from one limit position to another. That case can be seen in Fig. 7.

An influence of different initial conditions of the system and of the observer can also be experimentally examined on this model of the inverted wedge.

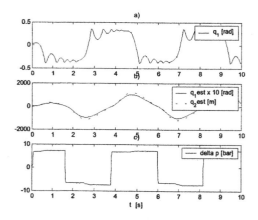

Fig. 7. Responses using weak observer gains
   a) Measured frame angle
   b) Estimated frame angle and slider posit.
   c) Pressure difference in cylinder chambers

5. CONCLUSIONS

The inverted wedge is a two degrees-of-freedom mechanism, which is controlled by a single control input (movement of the slider) in order to keep the frame of the wedge in balance. Only one measured variable (angle of the frame) has been applied, hence an observer was required to control the laboratory model. Also, the wedge has nonminimum-phase characteristic. The pneumatic actuation makes it pretty unique. Considering already mentioned attributes, the control of described laboratory model is very challenging task. It offers really great opportunities in practical control education. Modelling, control design and practical implementation skill are necessary in order to control the wedge. There are many alternative control methods, or possible design modifications whose practical effectiveness can be verified on this laboratory model.

The results using state controller with integral action are described in the paper. The full-order observer is employed to estimate the system states. The controller can keep the wedge in balance. However, advanced control strategies or additional measurement should be applied in order to make behaviour of the closed-loop system better.

ACKNOWLEDGEMENT

Many thanks to Mr. Zeljko Vukelic from FESTO d.o.o., Zagreb, Croatia, and Mr. Arman Mulic from A.M. Hidraulika d.o.o. (the representative of SMC), Zagreb, Croatia, for pneumatic equipment donations and support. The appreciations go to students, particularly to Mr. Mario Štimac, that took great part in modeling, designing and building of this laboratory model.

REFERENCES

Friedland, B. (1987). *Control System Design*. McGraw-Hill, New York.

Moore, M.L., J.T. Musacchio and K.M. Passino (2001). Genetic adaptive control for an inverted wedge: experiments and comparative analysis. *Engineering Applications of Artificial Intelligence*, **14**, 1-14.

Petrić, J. and V. Kecman (1989). Design of optimal PI controller on example of circulation system of steam generator (in Croatian). *Conference JUREMA'89*, Zagreb, **1**, 75-79.

Petrić, J. and Ž. Šitum (to appear in 2003). Inverted pendulum driven by pneumatics. *International Journal of Engineering Education*.

Šitum, Ž. (2001). Control of pneumatic servosystems using fuzzy controller (in Croatian). *Ph.D. thesis*, University of Zagreb.

Šitum, Ž. and J. Petrić (2001). Modeling and Control of Servopneumatic Drive. *Strojarstvo*, **43**, 1-3, 29-39.

APPENDIX

Mathematical model

Nonlinear mathematical model of the inverted wedge dynamics is given by:

$$m_1 t_1^2 \ddot{q}_1 + m_2 y^2 \ddot{q}_1 + m_2 q_2^2 \ddot{q}_1 + m_2 y^2 \ddot{q}_2 + 2m_2 q_2 \dot{q}_1 \dot{q}_2 +$$
$$+ b_1 \dot{q}_1 + m_2 g q_2 \cos q_1 - m_1 g t_1 \sin q_1 - m_2 g y \sin q_1 = F y$$
$$(3)$$
$$m_2 y \ddot{q}_1 + m_2 \ddot{q}_2 + b_2 \dot{q}_2 - m_2 q_2 \dot{q}_1^2 + m_2 g \sin q_1 = F$$

The linearization is obtained near the equilibrium point assuming:

$$\cos q_1 \cong 1; \quad \sin q_1 \cong q_1; \quad \dot{q}_1^2 \cong q_2^2 \cong \dot{q}_1 \dot{q}_2 \cong 0$$

The state space description of linearized system is:

$$\dot{\mathbf{x}} = \mathbf{A}\mathbf{x} + \mathbf{B}\mathbf{u}$$
$$\mathbf{y} = \mathbf{C}\mathbf{x} \qquad (4)$$

Where the matrixes and vectors are:

$$\mathbf{A} = \begin{bmatrix} 0 & 1 & 0 & 0 \\ \dfrac{m_1 g t_1}{N} & \dfrac{b_1}{N} & -\dfrac{m_2 g}{N} & -\dfrac{b_2 y}{N} \\ 0 & 0 & 0 & 1 \\ -\left(\dfrac{m_1 y g t_1}{N} + g\right) & -\dfrac{b_1 y}{N} & \dfrac{m_2 y g}{N} & \left(\dfrac{y^2 b_2}{N} - \dfrac{b_2}{m_2}\right) \end{bmatrix}$$

$$\mathbf{B} = \begin{bmatrix} 0 & \dfrac{2y}{N} & 0 & \left(\dfrac{1}{m_2} - \dfrac{2y^2}{N}\right) \end{bmatrix}^{\mathrm{T}}$$

$$(5)$$

$$\mathbf{C} = \begin{bmatrix} 1 & 0 & 0 & 0 \end{bmatrix}$$

$$\mathbf{x} = \begin{bmatrix} q_1 & \dot{q}_1 & q_2 & \dot{q}_2 \end{bmatrix}^{\mathrm{T}}$$

$$\mathbf{u} = F$$

The input variable F is a force given in eq. (1).

The denominator N from matrixes $\mathbf{A}$ and $\mathbf{B}$ is:

$$N = m_1 t_1^2 + 2 m_2 y^2$$

When the integral action at the angle of the frame is introduced, matrixes and vectors from eq. (4) and (5) are enlarged by one state:

$$\dot{\mathbf{x}}' = \mathbf{A}'\mathbf{x}' + \mathbf{B}'\mathbf{u}$$
$$\mathbf{y}' = \mathbf{C}'\mathbf{x}' \qquad (6)$$

The matrixes and vectors in this case are following:

$$\mathbf{A}' = \begin{bmatrix} \mathbf{A} & \mathbf{0} \\ \mathbf{I} & 0 \end{bmatrix}$$

$$\mathbf{I} = \begin{bmatrix} -1 & 0 & 0 & 0 \end{bmatrix}$$

$$\mathbf{B}' = \begin{bmatrix} \mathbf{B}^{\mathrm{T}} & 0 \end{bmatrix}^{\mathrm{T}} \qquad (7)$$

$$\mathbf{C}' = \begin{bmatrix} 1 & 0 & 0 & 0 & 0 \\ 0 & 0 & 0 & 0 & 1 \end{bmatrix}$$

$$\mathbf{x}' = \begin{bmatrix} \mathbf{x}^{\mathrm{T}} & \int q_1 \end{bmatrix}^{\mathrm{T}}$$

List of symbols

A	area of the piston, 0.000113 m^2
b_1	coef. of visc. frict. of frame, 0.05 Nms
b_2	coef. of visc. friction of slider, 45 Ns/m
F	force, [N]
F_A	force on side A of piston, [N]
F_B	force on side B of piston, [N]
F_R	reactive force, [N]
g	gravity acceleration, 9.81 m/s^2
m_1	mass of frame (+ cyl.; - slider), 1.1 kg
m_2	mass of slider, 0.25 kg
p_A	pressure in chamber A of cylinder, [Pa]
p_B	pressure in chamber B of cylinder, [Pa]
Δp	pressure difference in chambers, [Pa]
q_1	frame angle, [rad]
q_2	slider position, [m]
T_R	reactive moment, [Nm]
t_1	gravity centre of frame, 0.03 m
y	height of slider, 0.062 m

SMALL GAS FLOW MEASUREMENT - MICROCOMPUTER APPLICATION IN EDUCATION

Milan Adámek, Petr Neumann, Miroslav Matýsek

Department of Automatic Control, Institute of Information Technologies
Tomas Bata University
Mostní 5139, 762 72 Zlín, Czech Republic
Phone, Fax: +420 67 754 3103, E – mail:adamek@ft.utb.cz

Abstract: This paper is focused on education in the field of small gas flow measurement. Students obtain theoretical knowledge in flow and they acquaint with different types of flowmeters. The Department of Automatic Control of Tomas Bata University is equipped with the thermal mass flowmeter (time – of – flight sensor). The theoretical education is on a basis of the mathematical model of energy and temperature balance simulated in the FEMLAB. Students use various constructional materials for design of the thermal mass flowmeter and they verify the properties of designed flowmeter on laboratory models. *Copyright © 2003 IFAC*

Keywords: Microcomputers, sensors, simulation, thermistors, temperature profiles, thermal diffusivity.

1. INTRODUCTION

The necessity of extract an accuracy measuring of fluid flows is a classical and standard task in industrial process control. However, there is a wide range of gas and liquid flows and the utilization of unified measuring principle is impossible. The biogas flow measurement produced in biodegradation reactions of plastics gives the stimuli for the micro - flow measurement study. The produced biogas at the biodegradation reactions consists of methane CH_4 and carbon dioxide CO_2. The reaction kinetics of these reactions are studied by gas flow measurement, the gas flow range is $(20 - 50)$ *ml/hr*. This article describes the flowmeter design based on the principles of a time - of - flight sensor in operating flow range $(20 - 200)$ *ml/hr*.

One of control engineering courses at the Department of Automatic Control of Tomas Bata University contains a part aimed at biogas flow measurement. It is focused especially in thermal mass flowmeter (time – of – flight sensor), which is suitable for small flow range. Theoretical essentials are combined with

practical laboratory work. In the theoretical part students get knowledge of fluid dynamics together with a heat transfer. They acquaint with principles of a thermal mass flowmeter, especially with a time - of - flight sensor. Students use various constructional materials for design of flowmeter and they study properties of designed device in the FEMLAB theoretically. They verify dependence of sensor output mainly on the diffusivity of the fluid medium at the relatively low flow rates theoretically too. For evaluation of sensor output students use the microcomputer ADuC 812. Students have to create a flow diagram and a control program in C language to evaluate an output of designed sensor. Finally they verify properties of designed flowmeter on laboratory models. Laboratories of the department are equipped with two flowmeter models (time - of - flight sensor) which are made of different constructive materials.

This paper is organized as follows: Section 2 presents general concepts of flow measurement; Section 3 describes model for time – of – flight sensor; Section 4 gives description of measurement system; Section 5 describes control program; Section 6 presents experimental validation of modeling results; finally, Section 7 concludes the paper.

2. GENERAL CONCEPTS OF FLOW MEASUREMENTS

The thermal mass flowmeter is one of promising principles for low gas flows in the range (20 – 50) *ml/hr*. This flowmeter type utilizes a heated sensitive element and thermodynamic heat conduction principles to determine the real mass flow rate. A mass flow rate is determined by observing the effects of heat energy added to the flow stream as governed by the equation of the heat transfer:

$$Q = mc_p \Delta T \qquad (1)$$

where Q is a heat, m is a flow mass, c_p is a heat capacity of the fluid and ΔT is a temperature difference. For measurement of small flow in range (20 – 200) *ml/hr* was used a thermal mass flowmeter that measures the passage time of a heat pulse over a known distance. This sensor consists of a heater and one or more temperature sensors downstream, see (Fig. 1).

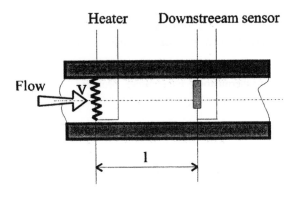

Fig. 1. Principle of a time-of-flight sensor

The heater is activated by periodical equidistant current pulses. The transport of the generated heat is a combination of diffusion and forced convection. The resulting temperature field is detected by a temperature sensor located downstream. The detected temperature output signal of the temperature sensor is a function of time and flow velocity. The sensor output is the time difference between the starting point of the generated heat pulse and the point in time at which a maximum temperature at the downstream sensor is reached, (Fig. 2). At the relatively low flow rates, the time difference depends mainly on the diffusivity of the fluid medium . At the relatively high flow rates, the time difference tends to relate to the ratio of the heater – sensor distance and the average flow velocity, see (Lammering, *et al.* 1993).

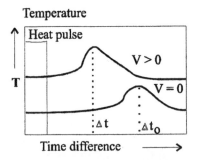

Fig. 2. Temperature at downstream sensor

3. MODEL FOR THE TIME – OF – FLIGHT SENSOR

The mathematical model of the designed flowmeter can be successfully simulated by the FEMLAB program. The simulated time – of – flight sensor type is a multiphysics model. It involves more than one kind of physics, see (Webster, *et al.* 1999). In this case, there are an Incompressible Navier – Stokes equations from fluid dynamics together with a heater transfer equation, essentially a convection – diffusion equation. There are four unknown field variables: the velocity field components u and v; the pressure p and the temperature T. They are all interrelated through bidirectional multiphysics couplings. The pure Incompressible Navier-Stokes equations consist of a momentum balance (a vector equation) and a mass conservation and incompressibility condition. The equations are

$$\rho \frac{\partial u}{\partial t} + \rho(u.\nabla)u = -\nabla p + \eta \nabla^2 u + F \qquad (2)$$

$$\nabla.u = 0 \qquad (3)$$

where F is a volume force, ρ is a fluid density and η is a dynamic viscosity.

The heat equation (4) is an energy conservation equation, that says only that the change in energy is equal to the heat source minus the divergence of the diffusive heat flux

$$\rho c_p \frac{\partial T}{\partial t} + \nabla.(-k\nabla T + \rho c_p Tu) = Q \qquad (4)$$

where c_p is the heat capacity of the fluid, and ρ is fluid density as before. The expression in the brackets is the heat flux vector, and Q represents a source term. The heat flux vector contains diffusive and convective terms, where the latter is proportional to the velocity field u (Hardy, *et al.*, 1999).

In this model, the above equations are coupled through the F and Q terms. Free convection is added to the momentum balance with the Boussinesq approximation. In this approximation, variations in density with temperature are ignored, except insofar as they give rise to a buoyancy force lifting the fluid. This force is put in the F-term in the Navier-Stokes equations.

3.1 Geometry of model and boundary conditions

For design of the flowmeter students use different constructive materials (steel, cooper, aluminium and plastic). They find suitable material considering good properties of the flowmeter – minimum time constant (Chudy, 1999). 2D geometry of the designed flowmeter is depicted in Fig. 3. For better visualization of heat transfer in the flow sensor tube students use 3D geometry in the FEMLAB too.

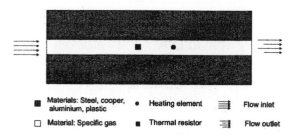

Fig. 3. 2D geometry of the designed flowmeter

The boundary conditions of the flowmeter model are shown in Fig. 4. The gas flow is modeled by no – slip, zero – velocity condition on all solid wall surfaces in inner wall of the sensor tube. The heating element is driven by pulse current, length of the current pulses is 20 *ms*.

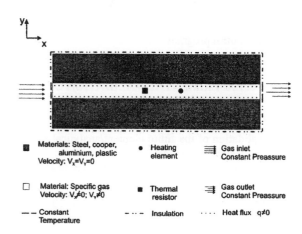

Fig. 4. Boundary conditions of the flowmeter model

3.2 Modeling results

The simulation results of a heat transport are shown in Fig. 5 and Fig. 6. The distance between the heating element and the temperature sensor is 15 *mm* and the initial temperature of running gas is 20 ^0C. Fig. 5 shows the temperature profiles of agitated air in the sensor tube (constructive material is steel). The heat transfer coefficient between the heater and air is

$5,2\ Wm^{-2}K^{-1}$ and the heat transfer coefficient between air and the temperature sensor is $3,8\ W\ m^{-2}K^{-1}$.

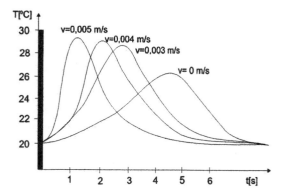

Fig. 5. Temperature profiles of agitated air in sensor tube

The temperature profiles of agitated CO_2 are depicted in Fig. 6 (constructive material of the sensor tube is plastic). The heat transfer coefficient between the heater and CO_2 is $4,1\ W\ m^{-2}K^{-1}$ and the heat transfer coefficient between CO_2 and the temperature sensor is $3,1\ W\ m^{-2}K^{-1}$. The presented heat transfer coefficients were computed by the Nusselt number (Fraden, 1996).

The static characteristics of the modeled flowmeter are depicted in Fig. 7. Students have to verify these regained characteristics on real models.

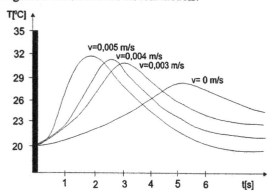

Fig. 6. Temperature profiles of agitated CO_2

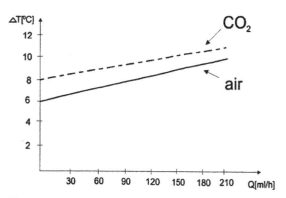

Fig. 7. Static characteristics of the modelated flowmeter

4. DESCRIPTION OF THE MEASUREMENT SYSTEM

The designed device has only one sensitive sensor (thermal resistor). The current pulses that activate the heater have a rectangular shape and their length are 20 *ms*. These excitation pulses are generated periodically with a period of generation 500 *ms*. The thermal resistor that scans temperature field downstream is connected to the Wheatstone bridge. In this circuit, the second temperature sensor is virtual and it is simulated by microcomputer. The simulation of the second temperature sensor allows raised sensitivity of the flow sensor (Sedra, A. *et al.*, 1991). The measurement principles of the flowmeter are based on the evaluation of the thermal waves propagation velocity.

The measurement system is shown in Fig. 8. The expansive thermal waves are registered by the thermal resistor R_T which is connected to a current source. The current thermal resistor supply is 50 μA. The excitation thermal waves are regulated by microcomputer (switch S), the heating circuit power is 0,028 *W*. A voltage drop on the thermal resistor is evaluated via the differential operational amplifier (DA$_1$). This measured voltage U_l is brought to the A/D converter (measuring channel 1). Voltage U_l is evaluated by the microcomputer ADuC812, the equivalent voltage is brought to the D/A converter and the differential operational amplifier (DA$_2$). This differential amplifier evaluates the same voltages, a drop voltage is amplified via amplifier (A). The thermal waves are recorded via thermal resistor, a voltage drop ΔU on the thermal resistor is measured by measuring channel 2. The circuit RS 232 is used for PC communication, see (Stepanik, 2001).

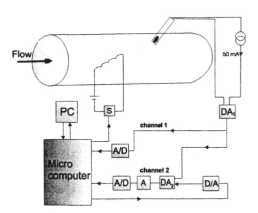

Fig. 8. The block diagram of the designed flowmeter

5. CONTROL PROGRAM

Students have to create a flow diagram and a control program in C language to evaluate an output of designed sensor. The control program philosophy is based on the algorithm shown in Fig. 9. The

diagnostic test is executed at the beginning of the control program. If the electrical circuit is disturbed, an error is written subsequently. A differential voltage U_l is measured before bracing of the heating circuit. Voltage drops are measured in very short time periods, a period of measuring is 5 *ms*.

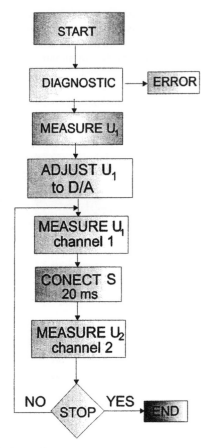

Fig. 9. Control program

6. EXPERIMENTAL VALIDATION OF MODELING RESULTS

The validation of the finite-element/analytical model of the designed flowmeter is to directly compare model generated results with experimental data.

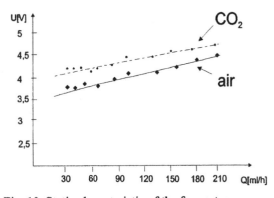

Fig. 10. Static characteristic of the fowmeter

For the calibration of the designed flowmeter students use the peristaltic pump (Dado, 1999). The designed flowmeter is calibrated for agitated air and CO_2. The static characteristics for different gases are depicted in Fig. 10.

7. CONCLUSIONS

This paper presents the education in the field of small gas flow measurement. Students use obtained theoretical knowledge for design of measurement device. They have to combine knowledge out of different study fields for verification of the properties of the designed flowmeter. Theoretical knowledge together with practical experience in control engineering is the basement of the future success of graduates in practice.

ACKNOWLEDGMENTS

The work has been supported by the Grant Agency of the Czech Republic under grant No. 102/02/D019/A and by the Ministry of Education of the Czech Republic under grant MSM 281100001. This support is very gratefully acknowledged.

REFERENCES

Chudy, V. (1999). *Measurement of technological variables*. STU, Bratislava.

Webster, J. (1999). *The Measurement, Instrumentation and Sensor*. CRC Press LLC, USA..

Lammerink, T.S. and R. Niels (1993). Micro - liquid flow sensor. *Sensors and Actuators A*, **37/38**, 45-50.

Stepanik, P. (2001*). The measurement of small fluid flow*. UTB, Zlin.

Dado, P. (1999). *Sensors*. CVUT, Praha.

Hardy, J.E. and J. O. Hylton (1999). *Flow measurement methods and applications*. Wiley-Interscience, U.S.A.

Sedra, A. and K. Smith (1991). *Microelectronic Circuits*. Oxford University Press, Oxford.

Fraden, P. (1996). *Handbook of Modern Sensors. Physics, Designs and Applications*. AIP Press Springer-Verlag, New York.

ELSEVIER

IFAC
PUBLICATIONS
www.elsevier.com/locate/ifac

Development of a Tank System for Control Studies

R. T. Pena, B. S. Torres, R. S. Assis, N. L. Carvalho, F. R. Caldeira, C. C. Penna

Department of Electronic Engineering, UFMG, Av. Antônio Carlos, 6627, Caixa Postal 209, 30161-970, Belo Horizonte, MG, Brazil, +55-31-3499-4088, rpena@cpdee.ufmg.br

Abstract: This paper describes a tank system that was built to allow students and researchers to study control strategies on a system that is as close as possible to an actual industrial plant. Its instrumentation uses Fieldbus technology and it has five control loops: two temperature control loops, two level control loops and one flow control loop. The physical aspects of the tank system and a multiloop control strategy are discussed in the paper. Some control lab experiments are suggested. The tank system can be used for the study of instrumentation, programmable logic controller and supervisory configuration, adaptive control, distributed control, expert systems, neural systems, fault detection, mathematical modeling, etc. *Copyright © 2003 IFAC*

Key Words: Interacting Tank System, Process Control, Pilot-Scale Plant, PID control.

1. INTRODUCTION

The tank system for control studies, TSCS, was built to allow students and researchers to test control strategies on a system that is as close as possible to an actual industrial plant (Carvalho, 1998). The plant has five control loops: two temperature control loops, two level control loops and one flow control loop. Figure 1 presents a picture of the system and figure 2 presents its engineering diagram.

As it can be seen in figure 2, the TSCS presents high coupling between some variables and therefore it is an excellent test bed for studying multivariable and multiloop control strategies. The attempt to control the temperature of the product tank, for

instance, requires the manipulation of TCV1, which controls the hot water input flow. As this water comes from the heating tank, the manipulation will cause a change in the level of the heating tank. To compensate this level change, the cold water control valve (LCV1) has to be also manipulated which in turns will cause a temperature change in the heating tank. This temperature change will affect indirectly the product tank temperature control that will change again the temperature control valve (TCV1) position.

Another interaction example in this system is related to the product tank outflow control valve (FCV1). If the output flow rate is increased, the system will compensate the resulting change in the product tank level by adding more cold water to this tank.

Fig. 1 – System Physical Overview

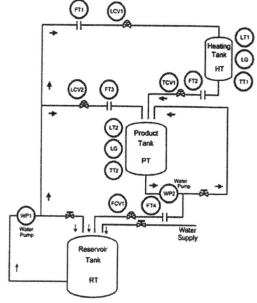

Fig. 2 – System Engineering diagram

This will cause a water temperature reduction that has to be compensated through the temperature control valve (TCV1) manipulation. Changes in this valve position will then cause variations on the heating tank level and its temperature, as described before.

Besides studying control strategies, the TSCS can also be used for the study of instrumentation, CLP and supervisory configuration, adaptive control, distributed control, expert systems, neural systems, fault detection, mathematical modeling, etc.

2. PLANT PHYSICAL DESCRIPTION

The system has 3 tanks: a 500 l reservoir tank (RT), a 50 l heating tank (HT) and a 75 l product tank (PT). The reservoir tank is located outdoors, near the plant, and is responsible for supplying water for the other tanks. As the capacity of the reservoir tank is higher than the other tanks, it is less susceptible to temperature changes than the other tanks.

The heating tank has the main task of supplying the product tank with hot water. Its cross section area is 0,113m² and it is 0,44m high. The product tank is located below the heating tank and receives hot water from HT, by gravity, through TCV1, and cold water from RT. This cold water is pumped by WP1 and

goes through LCV2. Its cross section area is 0,114m² and it is 0,66m high. Both tanks were built with stainless steel. The HT has a polypropylene external cover for thermal insulation. It also has a removable thermal insulation top cover that can be used to minimize environmental heat exchanges.

A 12 kW three-phase resistor installed at the bottom of the HT heats the water. The power dissipation in the resistor is controlled through three tyristorized bridges, one for each phase. The resistor was specified to heat the water in the tank from 20°C to 60°C in 9 minutes when the tank is full. The water level in the heating tank is controlled through LCV1.

The HT outflow rate is the heating element for PT. It has 4 manual outlet valves that can be used to add disturbances to its level control loop. The level control is similar to the heating tank level control and is accomplished by manipulation of LCV2.

The product tank output flow rate is controlled by FCV1. This valve is located downstream of a water pump that is connected to the product tank, decoupling the output flow rate and the PT level.

The instrumentation installed in the plant uses Fieldbus Foundation technology. The system is controlled through a Fieldbus network and a Programmable Logic Controller (PLC). There is also a supervisory program that runs in a personal computer.

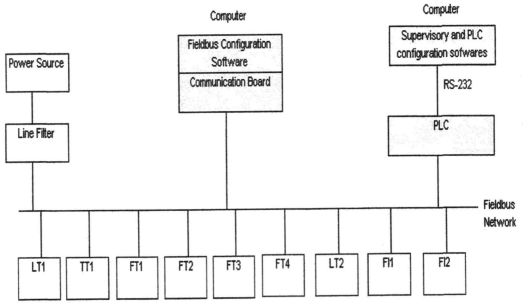

Fig. 3 – System Network Configuration.
(LT=Level Transmitter, FT=Flow Transmitter, FI=Fieldbus to 4-20 mA current transducer)

The Fieldbus network has 2 level transmitters, 4 flow transmitters, 1 temperature transmitter that is connected to 2 temperature measuring elements, installed in the tanks, and 2 Fieldbus code to 4-20 mA current transducers. Four common industrial pneumatic spherical control valves are installed in the plant.

The Fieldbus network interacts with the PLC through a Fieldbus scanner device that is attached to the PLC rack. The Fieldbus configuration is accomplished in the computer through a communication board and a configuration program. The PLC is responsible for system safety interlocking, alarms and Fieldbus devices data acquisition. The communication between the computer and the PLC is established through a RS232 serial port and using Modbus communication protocol (SMAR, 1999). The system network configuration can be observed in figure 3.

3. SYSTEM CONTROL LOOPS

3.1 Multiloop control strategy

The plant has 5 inputs (positions for four control valves and shooting angle for the tyristorized bridges) and 5 outputs (HT level and temperature, PT level, temperature and output flow). If a multiloop control strategy is used, there are many possibilities of pairing control and controlled variables.

In fact, choosing the five pairs of control/controlled variables can be a first exercise in a multivariable control lab class. The analysis of the system shows that a possible multiloop control strategy is the one shown in figure 4.

This strategy has been checked by the determination of the Relative Gain Matrix (Seborg et al, 1989; Torres, 2002).

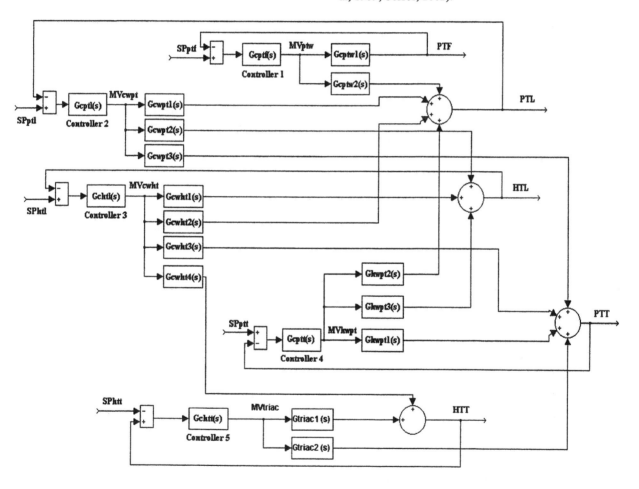

Fig.4 - TSCS Block diagram in a multiloop control strategy

3.2 Input-Output Plant modeling

Step response tests in the system yielded the transfer functions matrix, shown in equation 1 (Torres et al, 2002),

$$Y(s)= \begin{bmatrix} 1.210e^{-6s} & 0 & 0 & 0 & 0 \\ \dfrac{-0.180}{s} & \dfrac{0.110}{s} & \dfrac{-0.016}{s} & \dfrac{0.025}{s} & 0 \\ 0 & 0 & \dfrac{0.192}{s} & \dfrac{-0.030}{s} & 0 \\ 0 & \dfrac{-0.124e^{-31s}}{85s+1} & 0 & \dfrac{0.175e^{-28s}}{175s+1} & \dfrac{-0.185e^{-200s}}{692s+1} \\ 0 & 0 & \dfrac{-0.407e^{-5s}}{15s+1} & 0 & \dfrac{-0.590e^{-49s}}{489s+1} \end{bmatrix} \quad (1)$$

$$\text{Where } Y(s) = \begin{bmatrix} w_{PT} \\ l_{PT} \\ l_{HT} \\ t_{PT} \\ t_{HT} \end{bmatrix} \text{ and the inputs are} \begin{bmatrix} FCV1 \\ LCV2 \\ LCV1 \\ TCV1 \\ S \end{bmatrix}$$

where S is the shooting angle for the tyristorized bridges, w stands for flow rate, l for level and t for temperature of the water in HT (heating tank) and PT (product tank).

The transfer functions located out of the main diagonal in the transfer matrix, equation 1, represent coupling among different control loops. Both figures 5 and 6 show this coupling effect.

Figure 5 shows a step in the product tank temperature set point. The actuador, TCV1, changes its position, increasing the hot water flow rate to the PT. This causes a level change in HT, which is compensated by the manipulation of LCV1. Therefore, HT temperature changes. In figure 6, a step in the PT level set point from 250mm to 500mm caused oscillations in the temperature of this tank.

3.3 Valves Modeling

Assis (2000), based upon actual valves response measurements, derived polynomial functions relating the normalized position command signal (u) and the corresponding normalized flow rate (w).

An active region and a dead zone were also characterized for each valve. During the experimental tests, different flow rate behaviors were observed when the valves move opening and closing.

This may be due to hysteresis in the valves and can be studied on subsequent tests. The mean values for two ascending and descending flow measurement were used for the polynomial modeling of the valves. The results can be observed in table 2.The flow rate through TCV1 depends on the control signal for this valve and the level on the heating tank, as expected. Many flow curves were obtained for different levels on the heating tank (Assis, 2000).

Therefore, the relation between the control signal and the maximum flow rate through TCV1 was practically constant and a function can be adjusted to represent this relation, which does not depend on the heating tank level. The maximum flow is the variable that depends on the tank level and then the procedure to characterize TCV1 flow can be accomplished based on the following steps:

- Calculate f(uFCV1) which represents the relation between the control signal and the percentage of the maximum flow rate.
- Calculate f(hht) which represents the maximum flow rate for a specific value of the HT level
- Multiply f(uFCV1) and f(hht) to obtain the flow rate.

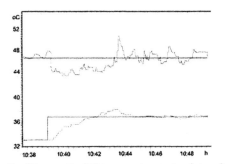

Fig. 5 – Step on the PT temperature set point and its influence on the HT temperature control

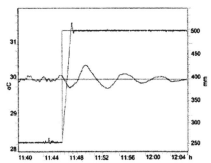

Fig.6 – Step on the PT level set point and its influence on the PT temperature control

Another characteristic of TCV1 is that its flow can be laminar or turbulent. Therefore, the maximum heating tank output flow can be calculated using one expression when h_{ht}/w_{FCV1} is linear and another expression when h_{ht}/w_{FCV1} is quadratic, as show on table 2.

3.4 Physical Model of the TSCS

The HT water is heated by a three-phase resistor. Equation 2 describes the transformation of electrical energy into heat, in this case.

$$Q(u_{heater}) = \frac{3V^2}{R}\left(1 - \frac{u_{heater}}{u_{\max heater}} + \frac{sen(2\pi u_{heater}/u_{\max heater})}{2\pi}\right)$$
(2)

Where, $Q(u_{heater})$ is the thermal power (in kW), u_{heater} is the control signal (in volts) to the tyristors bridge, V is the line voltage (220 V), R is the resistance and $u_{maxheater}$ is the maximum value of the control signal.

The shooting angle can be calculated as

$$\theta = \frac{u_{heater}}{u_{\max heater}}\pi$$
(3)

The heating tank temperature can be modeled using energy conservation law, as shown in equation 4.

$$\frac{dT_{ht}}{dt} = \frac{1}{A_{ht}}\frac{w_{in}}{h_{ht}}(T_{rt} - T_{ht}) + \frac{1}{\rho A_{ht}C}\frac{Q_{ef}}{h_{ht}}$$
(4)

Where, T_{ht} is the heating tank temperature, A_{ht} is the HT cross section area, w_{in} is the input flow, h_{ht} is the HT level, T_{rt} is the RT temperature, ρ is the fluid density, C is the fluid specific heat and Q_{ef} is the heat rate supplied by the resistor.

The PT temperature can also be modeled using the energy conservation law, as shown in equation 5.

$$\frac{dT_{pt}}{dt} = \frac{1}{A_{pt}}\frac{1}{h_{pt}}\left[w_{in}(T_{rt} - T_{pt}) + w_{out}(T_{ht} - T_{pt})\right] -$$
$$\frac{Q_{lost}}{h_{pt}\rho A_{pt}C} + \frac{Q_{pump}}{h_{pt}\rho A_{p_t}C}$$
(5)

Valve	Maximum flow rates (*l/min*)
LCV1	51,50
LCV2	36,00
TCV1	9,00
FCV1	52,00

Table 1 – Valve's maximum flow rates

LCV1 – Heating tank cold water inflow
$w(u_{LCV1}) = \begin{cases} 0, & 0 \leq u_{LCV1} \leq 0.12 \\ -4.9237u_{LCV1}^5 + 20.42744u_{LCV1}^4 - 29.8403u_{LCV1}^3 + 17.4963u_{LCV1}^2 - 2.2002u_{LCV1} + 0.0503 & 0.12 < u_{LCV1} \leq 1 \end{cases}$
LCV2 – Product tank cold water inflow
$w(u_{LCV2}) = \begin{cases} 0, & 0 \leq u_{LCV2} \leq 0.12 \\ -38.5695u_{LCV2}^5 + 102.4796u_{LCV2}^4 - 94.8061u_{LCV2}^3 + 33.059u_{LCV2}^2 - 1.268u_{LCV2} - 0.088 & 0.12 < u_{LCV2} \leq 1 \end{cases}$
FCV1 – Product tank outflow
$w(u_{FCV1}) = \begin{cases} 0, & 0 \leq u_{FCV1} \leq 0.12 \\ 5.5693u_{FCV1}^5 - 5.0682u_{FCV1}^4 - 9.0489u_{FCV1}^3 + 11.1507u_{FCV1}^2 - 1.6257u_{FCV1} + 0.043 & 0.12 < u_{FCV1} \leq 1 \end{cases}$
TCV1 – Product tank hot water inflow
$w(u_{TCV1}, h_{ht}) = \begin{cases} f(u_{TCV1}).f(h_{ht}) \\ f(u_{TCV1}) = -17.4919u_{TCV1}^5 + 50.6727u_{TCV1}^4 - 51.735u_{TCV1}^3 + 20.0766u_{TCV1}^2 - 0.5602u_{TCV1} + 0.025 \\ f(h_{ht}) = -2.3865 \times 10^{-5}h_{ht}^2 + 0.0282h_{ht} - 1.222, \quad h_{ht}/w \text{ is linear} \\ f(h_{ht}) = 0.0074h_{ht} + 5.9844, \quad h_{ht}/w \text{ is quadratic} \end{cases}$

Table 2 - Polynomial functions relating the normalized command signal and the corresponding normalized flow rate through the valves. The terms u_{LCV1}, u_{LCV2}, u_{FCV1} and u_{TCV1} are the command signals and h_{ht} is the HT level.

Where, T_{ht} is the HT temperature, A_{pt} is the PT cross section area, w_{in} is the input flow rate, w_{out} is the output flow rate, h_{pt} is the PT level, T_{rt} is the RT temperature, T_{pt} is the PT temperature, ρ is the fluid density, C is the fluid specific heat, Q_{pump} is part of the input cold water pump power that is transferred to the water as heat which was calculated as 108.6 kW, Q_{lost} is the heat lost through the tank walls. Q_{lost} can be calculated through the following expression:

$$Q_{lost} = UA_{ept}(T_{air} - T_{pt})$$
(6)

Where, U is empirically determined, A_{ept} is the PT external area, T_{air} is the air temperature and T_{pt} is the PT water temperature.

The HT and PT levels can be modeled based on mass conservation laws, as shown in equations 7 and 8.

$$\frac{dh_{ht}}{dt} = \frac{1}{A_{ht}}(w_{in} - w_{out})$$
(7)

$$\frac{dh_{pt}}{dt} = \frac{1}{A_{pt}} (w_{in\_cold} + w_{in\_hot} - w_{out}) \qquad (8)$$

Where, h_{ht} is the HT level, h_{pt} is the PT level, A_{ht} and A_{pt} are the tanks cross section areas, w_{in} and w_{out} are input and output flow rates, w_{in_cold} is the input cold water flow on the PT and w_{in_hot} is the input hot water flow in the PT.

The model of the TSCS can then be used in studies and simulations of control strategies for this plant.

4. TEACHING CONTROL LAB USING TSCS

A system as the TSCS, once installed in the lab, allows a great number of lab classes, enhancing the possibilities of learning control system principles. Follows a non unique sequence of possible experiments with the TSCS:

1- *Dynamic Modeling based on physical principles* – based upon equations 2 to 8, develop tests to determine all the model parameters.

2- *Transfer function plant modeling* – determination of the five plant transfer functions by step response or by pulse testing followed by frequency response calculations.

3- *P, PI, PID control of a single loop* – as the plant is multivariable, to study the control of a single loop, one needs to consider all the other four loop at rest. Any eventual coupling coming from other loops can be approached as a disturbance source.

4- *Control of multiloop systems* - determination of the Relative Gain Array (Seeborg et al,1989).

5- *Control loop decoupling* - study the effects of loop detuning, and of the static and dynamic design of decoupling structure.

6- *Control of multivariable systems.*

7- *Studies of intelligent controllers.*

5. CONCLUSIONS

The TSCS is a valuable tool for the study and teaching of control strategies. As its behavior closely resembles that of an actual industrial plant, the results obtained in this system give important insight for their transposition to actual world plants.

The TSCS has already been used for the study of mathematical modeling (Assis, 2000), physically based. Empirical models were also developed through tests on the plant (Torres,2002).

The built plant is a challenging system for the study of multivariable and multiloop control strategies as it has five control loops that present coupling among themselves, as exemplified on this paper.

The plant is fully operational and some control researches are being developed on it. Specifically, Torres et al (2002) studied static and dynamic decoupling structures, applied to TSCS.

6. ACKNOWLEDGMENT

The authors acknowledge the donation of the control valves by the company Neles Jamesbury, now named Neles Automation, and the donation of the Fieldbus devices by the company SMAR Equipamentos Industriais Ltda.

This paper was developed under a grant from FAPEMIG – Minas Gerais State research support agency.

7. REFERENCES

Assis, R. S. (2000), Development of a simulator to the tank system for control studies – Level, Flow and Temperature - TSCS, *M.E.E. Dissertation*, Graduate Program in Electrical Engineering – PPGEE/UFMG (in Portuguese).

Carvalho, N. L. (1998), Project and implementation of a tank system for level, flow and temperature control using fieldbus technology, *M.E.E. Dissertation*, Graduate Program in Electrical Engineering – PPGEE/UFMG (in Portuguese).

Seborg, D. E., Edgar, T. F., Mellichamp, D. A., *Process Dynamics and Control*, John Wiley & Sons, 1989.

Smar (1999), Fieldbus Tutorial – Smar Equipments, São Paulo, Brazil (in Portuguese).

Torres, B. S. (2002). Tuning of PID Controllers in a Multiloop System, *M.E.E. Dissertation*, Graduate Program in Electrical Engineering – PPGEE/UFMG (in Portuguese).

Torres, B. S., Pena, R.T., Jota, F.G. (2002), Decoupling Strategies in a Mutiloop Control System of Level, Flow and Temperature. *XIV Brazilian Automatic Congress*, Natal, Brazil, September (in Portuguese).

ELSEVIER

IFAC
PUBLICATIONS
www.elsevier.com/locate/ifac

SIMPLE LEVEL CONTROL EQUIPMENT FOR VISUALISING BASIC CONTROL PRINCIPLES

Pasi Joensuu, Kauko Leiviskä and Leena Yliniemi

University of Oulu, Control Engineering Laboratory, Box 4300, FIN-90014 Oulun yliopisto, Finland
first name.surname@oulu.fi, http://ntsat.oulu.fi

Abstract: This paper describes a simple laboratory process equipment, which is used as a support tool to visualise in learning basic concepts of process control. It is used as a part of course "Introduction to Control Engineering", the aim of which is to give the idea of feedback control and the basics of process instrumentation to the students. One problem in designing the course has been the fact that the students are coming from many different study programs with different backgrounds and motivation. The laboratory process is used in the visualisation of control concepts and in moving to the direction of "Learning by Doing". *Copyright © 2003 IFAC*

Keywords: control engineering, feedback control, instrumentation, laboratory process

1. INTRODUCTION

Control Engineering Laboratory in the University of Oulu is responsible for teaching the basics of control engineering, mechanistic and data based modelling, and process optimisation. Additionally, the laboratory is offering courses from fuzzy and neural network systems, and the special courses from modelling and control of biotechnical processes and from automation in pulp and paper industry.

This paper concerns with the realisation of the course "Introduction to Control Engineering". The course introduces the basic concepts of control; mainly the idea of feedback control. It also concerns with the development of automation systems, their functions and structures and general effects of automation.

The main part of the course concentrates on giving the basic knowledge on process measurements together with their underlying physical principles and application areas. It also deals with control valves and actuators including their dimension and selection.

The students come from various study programs. For example, in addition to control and process engineering students, there are students whose main subjects are environmental technology, electrical engineering, mechanical engineering, and sometimes biophysics. Thus, the students can be a very heterogeneous group with varying educational background, and also varying motivation.

The main aim in the exercises of this course is to teach students instrumentation symbols so that they are able to read and plan PI charts after they leave the course. The most important in this is a group work where they have to design a PI chart for a real process according to the description of the process and its functions. Examples are taken from paper industry.

The exercises include also a visit to the laboratory hall at the University of Oulu. During this visit it is possible to look pilot scale processes and to study the instrumentation and automation system where the processes are connected. Commonly, the students receive an exercise for doing it during the visit. The task is to identify instruments in the processes.

To motivate students, extra points are given for the exam according to the work that they have done during the exercises. Usually there are five additional exercises from which the design part of the PI chart takes three of five. These points are added to the points they get from exam. The overall amount of points from exam is 30.

It is very difficult, and also very boring to teach the structure and operation of instruments by lecturing. Even though the lecture materials include hypermedia parts with animations and video clips, some more concrete means for visualisation are required. This led to the idea of designing the following simple laboratory process and using it to support lectures and exercises.

2. DESCRIPTION OF THE PROCESS

2.1 Level control

The purpose of the laboratory equipment is to visualise the level control of a tank. The block diagram of the level controller is shown in Fig. 1. The differential pressure transmitter measures the level by comparing hydrostatic pressure to atmospheric pressure. The measurement data of the pressure transmitter is transferred to the unit controller. The valve actuator gets a signal from the controller and adjusts the control valve.

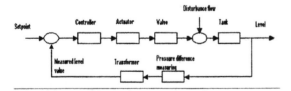

Fig. 1. The block diagram of the level control.

Omron E5CK digital process controller has been used as the unit controller. There are three possibilities to tune the controller. The controller offers a choice of either advanced PID with auto-tune capabilities or fuzzy logic self-tuning. Parameters can also be set by hand. The appropriate parameters for auto tuning are selected by step response tuning. Fuzzy logic self-tuning has three different forms of activity. The one is step response tuning (SRT) which tunes up PID parameters while the set value is changed. The other one is the disturbance tuning (DT). During disturbance the PID parameters are

tuned up so that the effect of the next disturbance is so small as possible. The last one is hunting tuning (HT). The aim of HT is to prevent the hunting of the circuit, which is done by tuning PID parameters.

At present, there is the graphical user interface (GUI) implemented to the equipment. GUI is SYS-Config v2.0, which is an advanced software package for Omron process controllers, temperature controllers and intelligent signal processors. This software has a wide range of functions to set up, configure and monitor behaviour of controller and process. SYS-Config software offers almost the same the methods for tuning a controller as the manual tuning of a controller. Omron E5CK digital process controller includes a communication unit, which is able to communicate with a computer through RS-232C. The operation of the controller can also be monitored visually.

2.2 Instruments of the equipment

The equipment contains the following parts:

- equipment rack
- process and storage tank
- pipework
- circulation pump
- unit controller
- control valve
- valve actuator
- differential pressure transmitter
- solenoid valve
- limit switch
- flow sensor and monitor
- temperature sensors and monitors
- power supply and transformer and
- manual and choke valves.

The process and PI diagram is shown in Fig. 2. The circulation pump is recycling city water (max 10 l/min), which flows to process tank through choke and hand valves. There are two mechanisms to avoid overflow from the process tank. There is a direct by-pass from the process tank. Additionally the limit switch monitors the surface of the process tank. When the water level activates the limit switch, it opens the solenoid valve. It makes the process controllable in any circumstances.

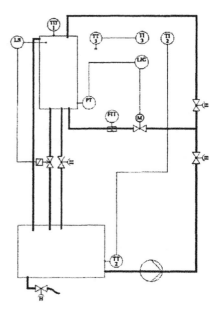

Fig. 2. The process and PI diagram.

The equipment has been installed on the equipment rack. It is equipped with wheels to make it movable. The rack itself has been made of steel pipes and faced plywood.

The process tank is constructed from transparent plastic sheet so that the level changes can be easily detected. The volume of the storage tank is bigger than the process tank, because the system failure can occur in the closed system.

The flow sensor and monitor is the compact unit (FIT). The flow sensor detects the flow rate of water and converts it into an analogous output signal. In addition a row of LEDs indicates the current flow rate. The unit was set for sensing the range 0 to 10 l/min by choosing specific T-piece.

Three kinds of sensors have been chosen for temperature measurements: Pt-1000, Pt-100 and TC K (NiCr-Ni) thermocouple. Pt-1000 sensor TT 5050 with the compact transmitter TR 2430 measures the temperature of water in the process tank (TIT-1). Pt-100 sensor measures the temperature of the storage tank (TT-2). The room temperature is measured by TC K thermocouple (TT-3). This unit has also a 7-segment LED display.

3. DOING BY HANDS

3.1 Purpose of laboratory exercises

The very common disadvantage of educational course has been great difference between the theory and practice. Practice will be learnt at work after graduating. A lot of work has been done to overcome this disadvantage. The laboratory exercise is one solution, because it gives illustrative learning experience. "Doing by hands" provides learning environment where theory and practice have been linked together.

Laboratory exercises or extensive courses have a lot of different purposes. This subject has been expanded to the conferences organized by American Society for Engineering Education (see www.asee.org). The objectives for a course can be set a variety of ways. Wankat and Oreowicz (2002) have presented the following goals:

- experimental skills
- real world
- built objects
- discovery
- equipment
- motivation
- teamwork
- network
- communication and
- independent learning.

It is easy to find these goals among any exercise, which every student may have faced up to. Students can learn different skills involved in doing experimental engineering work. Students learn to plan an experiment, which they record, analyse and so on. The opportunity of working with the modern equipment gives students the aspect of down-to-earth to theory. Also they might learn to distinguish working life's reality from theory.

To make a case worth of learning the subject has to motivate students. Active interaction between other students gives an enthusiastic environment in which students feel secure as a member of the team. Usually most of laboratory exercise works are team efforts. Networking and communication are the ways to find information from different sources, to process information and to report results. Making all these goals true, it happens independent learning all the time from the starting of the exercise to as far as for returning the report.

3.2 Use of the level control equipment

The laboratory exercise equipment introduced in Chapter 2 was originally designed to visualise the basic concepts of control engineering in the course "Introduction to Control Engineering". This educational course contains the basics of process instrumentation, and a choice of selecting instruments to a process. The exercise equipment has been built for this course, because the teachers' interest is to maximize the learning results of the students. During lectures it is possible to demonstrate the control of the water level. The students can examine instruments and tune the controller.

So far, it is become clear that the students are very satisfied with the use of the equipment. They can examine the operation of the equipment and get information about the instrument selections.

Because the equipment is quite new, there is a lack of feedback from students. The course "Introduction to Control Engineering" is lectured next time in the autumn and after that more feedback will be received from the students.

The equipment has also been used in the extra course concerning process control engineering. In this course the students get the certain problem, which they should solve. The graphical user interface (GUI) of the equipment was developed by some students as a practice exercise. This work was found very interesting, because it had connections to the real world.

3.3 Tasks carried out with the equipment

Measurements
The laboratory exercise equipment gives versatile possibilities to exemplify different measuring principles. The equipment includes two different temperature measurements i.e. two resistance thermometers and the thermocouple, the differential pressure measurement and the flow measurement. In addition to different measuring principles, several transmitter technologies and calibration systems are involved.

Actuators and valves
The system includes two actuator-valve combinations. They help in visualising the actuator operation, but have a little use in valve dimensioning.

Control loop
The equipment gives the good possibility to demonstrate different parts of the control loop e.g. to demonstrate in the concrete way, how the feedback control operates. The students have possibility to operate with a "real world"-application in the safe, but visually in clear way. This helps to understand the concept of the control loop.

Controller tuning
As mentioned earlier, the controller has versatile tuning possibilities. Although the course "Introduction to Control Engineering" does not include these topics, the equipment could be used in other courses for this purpose.

4. CONCLUSIONS

Control principles are in many cases taught as abstract concepts and mathematical expressions. In practice, control is, however, a concrete action for aiming to improve the process operation. The control principles can also be concretised with real tools and equipments. The very demanding task is to find an exercise process, which is enough simple for this use, but on the other hand enough complex to visualise large field of control applications.

This paper has introduced the simple level control exercise process, which is used as the support tool in the course "Introduction to Control Engineering". The equipment can be used for visualising different measurement and control instruments and for examining feedback control. It can be moved over to the lecture room or it can be used in the laboratory.

REFERENCES

Wankat, P.C. and F.S. Oreowicz (2002). Design and Laboratory. In: *Teaching Engineering*. https://engineering.purdue.edu/ChE/News_and_Publi cations/teaching_engineering/Book.pdf Pages 168-189.

www.elsevier.com/locate/ifac

THE MONITORING AND CONTROL SYSTEMS FOR THE REAL EQUIPMENT CONTROLLED BY PLC IN EDUCATION

Tomáš Sysala, Petr Dostál

Department of Process Control, Institute of Information Technologies
Faculty of Technology, Tomas Bata University in Zlín
nám. TGM 275, 762 72 Zlín, Czech Republic
E-mail: sysala@ft.utb.cz, dostalp@ft.utb.cz

Abstract: This paper is focused on education in the field Measurement and control of technological processes. In one part of this field students obtain theoretical knowledge in an application of commercial systems for measurement, visualisation and control of technological processes. The data are measured by a programmable controller (PLC) and as the superior systems the system InTouch, Wizcon and ContolWeb2000 were used. All systems were tested on the real equipment. The programmable controller controls the surface of two communicating vessels and at the same time is a data source for master applications that are designed for the systems InTouch, Wizcon and ContolWeb2000. The contribution also describes visualisation systems and serial communication for Windows and gives the description of the ModBus protocol. *Copyright © 2003 IFAC*

Key words: monitoring, PLC, real time, visualisation, real equipment.

1. INTODUCTION

The main task was to apply different monitoring and control systems to control of the water level in two communicating vessels model.

2. MODELS USING FOR EDUCATION

2. 1 Communicating vessels model (coupled tanks)

This model consists of these parts: two calibrated vessels, the tender, the motherboard with the water pump, two water level sensors, overfall, interconnection with water cock and water outlet. The pump power is controlled by voltage of the pump motor. It is possible to input the disturbance into the system. We can do it by the profile change of

the interconnection of the tanks. The change size depends on the water cock shift.

The water level is measured by capacitance sensor and its output is connected to the I/O (input/output) module of the PLC (10-bit D/A converter).

The scheme of the model is shown in the figure 1.

3. CONTROL OF THE WATER LEVEL

To solve that problem, three controllers designed in different ways were consequently applied. The two-position controller (Vašek, 1989), the PID-controller (Vítečková, 1996) and the pole-placement controller (Bobál & Kubalčík, 1994). We have achieved best results with the third order pole placement controller. It was incorporated into control software.

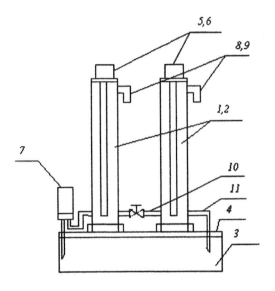

Fig. 1. The model of the communicating vessel

3. 1. The 3rd Order Controller

The controller is designed especially for process control without overshoot of the process output.

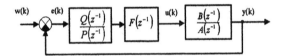

Fig. 2. Control system scheme

Let the controlled process be described by transfer functions in the form (Bobál & Kubalčík, 1994)

$$G_S(z) = \frac{Y(z)}{U(z)} = \frac{B(z^{-1})}{A(z^{-1})} \qquad (1)$$

where

$$A(z^{-1}) = 1 + a_1 z^{-1} + a_2 z^{-2}$$
$$B(z^{-1}) = b_1 z^{-1} + b_2 z^{-2} \qquad (2)$$

are second degree polynomials and Y(z),U(z), E(z)=W(z)-Y(z) are the Z-transforms of the process output, the controller output and the error.

In many cases we need to control the tank level on the reference signal with Z-transforms (W(z) is the Z-transform of the reference signal)

$$W(z^{-1}) = \frac{z^{-1}}{(1 - z^{-1})^2} \qquad (3)$$

The controller is described by transfer (Vítečková, 1996) in the form

$$G_R(z^{-1}) = \frac{Q(z^{-1})}{F(z^{-1})P(z^{-1})} \qquad (4)$$

where

$$Q(z^{-1}) = q_0 + q_1 z^{-1} + q_2 z^{-2} + q_3 z^{-3} \qquad (5)$$
$$F(z^{-1}) = (1 - z^{-1})^2 \qquad (6)$$
$$P(z^{-1}) = (1 + p_1 z^{-1}). \qquad (7)$$

We derive the following transfer function for the closed loop system

$$G_W(z^{-1}) = \frac{B(z^{-1})Q(z^{-1})}{A(z^{-1})F(z^{-1})P(z^{-1}) + B(z^{-1})Q(z^{-1})} \qquad (8)$$

In the denominator of the transfer function (8), there is the characteristic polynomial. We assume the following polynomial identity

$$A(z^{-1}) F(z^{-1}) P(z^{-1}) + B(z^{-1}) Q(z^{-1}) = D(z^{-1}) \qquad (9)$$

We desired polynomial

$$D(z^{-1}) = (1 - c z^{-1})^5 \qquad (10)$$

where c influences the speed of the closed control loop transient characteristic and controller output changes, too.

The polynomial identity (9) and polynomial (10) give a set of five linear algebraic equations

$$
\begin{aligned}
q_0 b_1 && + p_1 + a_1 - 2 &= -5c \\
q_0 b_2 + q_1 b_1 && + p_1(a_1-2) + 1 - 2a_1 + a_2 &= 10c^2 \\
q_1 b_2 + q_2 b_1 + p_1(1 - 2a_1 + a_2) + a_1 - 2a_2 &= -10c^3 \quad (11)\\
q_2 b_2 + q_3 b_1 + p_1(a_1 - 2a_2) + a_2 &= 5c^4 \\
q_3 b_2 + p_1 a_2 &= -c^5
\end{aligned}
$$

Solving the equations (11) we obtain five unknown controller parameters q_0, q_1, q_2, q_3 and p_1

The resulting transfer function of controller is

$$G_R(z^{-1}) = \frac{q_0 + q_1 z^{-1} + q_2 z^{-2} + q_3 z^{-3}}{1 + (p_1 - 2)z^{-1} + (1 - 2p_1)z^{-2} + p_1 z^{-3}} \qquad (12)$$

The control with controller based on the pole placement approach is shown on Fig. 3 and Fig. 4.

The system was controlled by programmable logic controller (PLC) Modicon TSX 3722 - Micro (Martinásková & Šmejkal, 1998). It is a product of the Schneider Electric company.

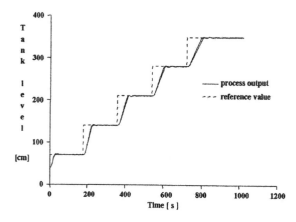

Fig. 3. Tank level control with controller based on the pole placement approach – step shape of the reference value

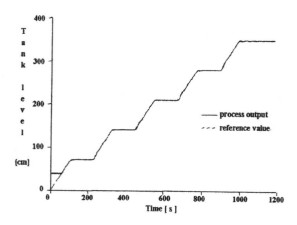

Fig. 4. Tank level control with controller based on the pole placement approach – ramp shape of the reference value

4. PLC AND MODBUS PROTOCOL DESCRIPTION

4. 1 PLC Modicon

PLC Modicon TSX Micro is the PLC of choice for smaller machines, mobile systems and vehicles where minimal space requires maximum compactness, or when environmental conditions demand reliable performance. In perfect synergy with other Schneider products, Micro enables machines to perform to their fullest potential. The expertise of machine builders is also enhanced and increases their value to end-users.

The base configuration of the Micro includes the power supply (24Vdc or 220Vac), CPU with memory, mini operator dialog, as well as the different communication ports.To meet various size needs, the Micro offers a choice of 5 modular configurations, each offering several levels of integration (I/O, analog, counting) and openness (PCMCIA for communication).

The Micro offers over 40 different specialty modules, including positioning/rapid counting, measuring and regulation (lower upper level, sensor, thermo element), safety, 24Vdc or 220Vac, single or double format, modularity (8, 16, 32, 64). All types of discrete I/O can be used within the same configuration.

PLC TSX 3722

1. 3-slot base rack
2. Centralised display block
3. integrated analogue and counter modules
4. Terminal port (TER) (Uni-Telway or Modbus Master/Slave protocol) and Man-machine interface port labeled AUX
5. Cover for accessing the power supply terminals
6. Slot for a memory extension card
7. Slot for a communication module
8. Battery housing
9. Connector for extent racks

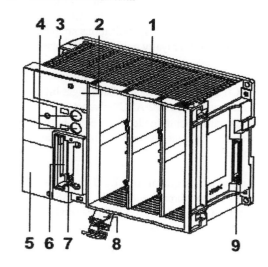

Fig. 5. The scheme of the PLC Modicon TSX Micro

To provide incredibly fast response times, the Micro uses a fast execution time of 0.15μ for Boolean, and a programmable filtering for I/O of 0.1ms to 7.5ms for event processing coming from application modules as well as rapid outputs.

For PLC programming a special software PL7 Micro is used.

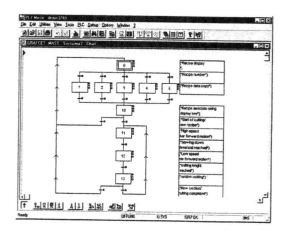

Fig. 6. Programming software PL7 Micro

4.2 MODBUS Protocol

MODBUS® Protocol is a messaging structure, widely used to establish master-slave communication between intelligent devices. A MODBUS message sent from a master to a slave contains the address of the slave, the "command" (e.g. "read register" or "write register"), the data, and a check sum (LRC or CRC).

Since Modbus protocol is just a messaging structure, it is independent of the underlying physical layer. It is traditionally implemented using RS232, RS422, or RS485 over a variety of media (e.g. fiber, radio, cellular, etc.).

MODBUS TCP/IP uses TCP/IP and Ethernet to carry the MODBUS messaging structure. MODBUS/TCP requires a license but all specifications are public and open so there is no royalty paid for this license.

The MODBUS protocol comes in 2 flavours:
- ASCII transmission mode: Each eight-bit byte in a message is sent as 2 ASCII characters.
- RTU transmission mode: Each eight-bit byte in a message is sent as two four-bit hexadecimal characters.

The main advantage of the RTU mode is that it achieves higher throughput, while the ASCII mode allows time intervals of up to 1 second to occur between characters without causing an error.

The basic structure of a MODBUS frame is shown below:

ADDRESS	FUNCTION	DATA	CHECKSUM

The address field of a message frame contains two characters (ASCII) or eight bits (RTU). Valid slave device addresses are in the range of 0 ... 247 decimal. The individual slave devices are assigned addresses in the range of 1 ... 247. A master addresses a slave by placing the slave address in the address field of the message. When the slave sends its response, it places its own address in this address field of the response to let the master know which slave is responding.

5. MONITORING AND VISUALISATION SYSTEMS

The programmable controller controls the water level of vessels and at the same time it is a data source for master applications that are designed for the systems InTouch, Wizcon and ContolWeb2000.

5.1 InTouch

InTouch is one component of -based MMI system that gives users access to a full package of automation tools - FactorySuite 2000. It is the first fully integrated suite of software for industrial automation.

System provides a single integrated view of all your control and information resources. It enables engineers, supervisors, managers and operators to view and interact with the workings of an entire operation through graphical representations of their production processes.

InTouch gives users many significant advantages and improvements over previous releases. Distributed alarm performance has been greatly improved, and the new display allows greater flexibility and analysis of alarms. The new alarm architecture in InTouch replaces NetDDE with TCP/IP for increased performance and decreased network loading. In addition, SuiteLink timestamping has been added as the alarms occur to enable better alarm analysis and the new Distributed Alarm System allows you to configure a secondary backup Alarm Provider to establish a fail-safe system.

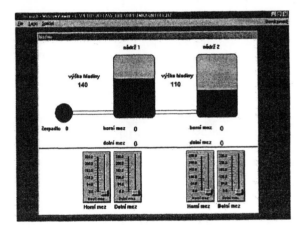

Fig. 7. The application made in system InTouch

5.2 Wizcon for Windows and Internet

Wizcon for Windows and Internet provides all of the tools users need to efficiently build effective operator interface and supervisory control applications. Productivity is further enhanced with Wizcon's online engineering which lets users design and run applications within a single environment. Users can make changes online and see the results immediately, minimizing initial configuration time and on-site installation efforts.

Wizcon's Application Studio offers easy access to all of the application's components

It provides a powerful Image Editor to create and view the Images that enable the operator to visualize part or all of a control process. An image consists of objects, which are geometric figures or text. Geometric objects can be either open or closed, and all objects can be filled with specified patterns and displayed in unlimited colors.

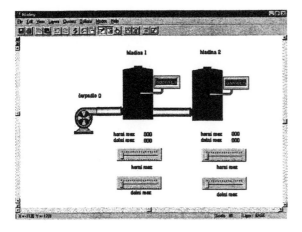

Fig. 8. The application made in system WIZCON

Using Wizcon.s Multilanguage support, tag descriptions, alarm text and image field text can be created in one language and exported to an ASCII file. This easy-to-handle file can be translated to another language and re-imported into Wizcon. The same procedure can be applied to multiple languages. During runtime, a user can choose the required language. For example, users can enter an alarm description in English when they configure the application, but display it in French during runtime. Support for different languages is dependent on the operating system and its support for that language.

5.3 Control Web 2000

Control Web is a component, object-oriented system for the development and operation of visualisation, measurement, control, regulation and communication programs and programs for the collection, archiving and processing of data. Similarly, the application can control a machine or technological line in real time and simultaneously be incorporated into a large company information system.

The Control Web environment enables you to create programs using symbolic descriptions of the designed system. If you do not want to study all the advanced features of the system you can simply develop your own application as if you were actually creating the indicator and control panel on a distribution board or control room. You are using the *virtual models* instead of real instruments which brings about incomparable means of expression and potential for creativity.

Based on *visual programming* you can very easily create your own application by arranging the objects with the mouse. All you have to do is to capture the relevant icon, which is a graphical representation of the required instrument, and insert it within the structure of the application you are developing. The

data concerning the inserted instrument are automatically transferred into the application source text.

Even programmers who are used to programming applications directly in source text can work in the text editor. The text that you enter via this editor automatically translates into the structures of the application you are developing.

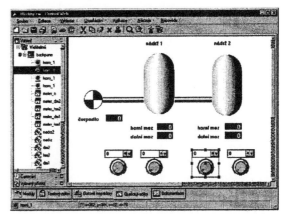

Fig. 9. The application made in system ControlWeb 2000

In education students have to create applications in described system. They are preparing for implementation of these systems in the factories, in industry, in their new job.

Examples of the projects made in different control systems are shown in figures no. 7, 8, 9.

Acknowledgements: This work was supported in part by the Grant Agency of the Czech Republic under grant No. 102/02/D020/A, grant No. 102/03/0070 and by the Ministry of Education of the Czech Republic under grant MSM 281100001.

REFERENCES

Bobál, B., Kubalčík, M. (1994). Self-Tuning Controller for Temperature Control of a Thermo-Analyser. *Proc. 3rd IEEE Conference on Control Applications.* p. 1443, Glasgow.

Martinásková, M., Šmejkal, L (1998). *Řízení programova-telnými automaty*, Vydavatelství ČVUT, Praha.

Schneider Group, (2001). PL7 Mikro, *CD-rom*, Schneider Automation Inc., Germany.

Sysala, T. (2002). The monitoring and control system for the real equipment controlled by PLC, *International Conference ICCC' 2002*, Malenovice, Czech Republic

Vašek, V. (1989). *Teorie automatického řízení II*, ES VUT, Zlín.

Vítečková, M. (1996). *Seřízení regulátorů metodou inverze dynamiky*, VŠB-TU Ostrava, Ostrava.

ELSEVIER

IFAC
PUBLICATIONS
www.elsevier.com/locate/ifac

INDUSTRIAL TOOLS AND EMULATORS IN BACHELOR EDUCATION

Jana Flochová, Iveta Zolotová, Dušan Mudrončík

Department of Automatic Control Systems FEI STU, Bratislava, Slovak republic
Department of Cybernetics and Artificial Intelligence TU Košice, Slovak republic
++421 2 60291 667, ++421 55 602 2570
flochova@kasr.elf.stuba.sk, iveta. zolotova@ tuke.sk, mudronci@mtf.stuba.sk

Abstract: The aim of this paper is to show the simulation, emulation and diagnostic possibilities of several industrial controllers and tools installed in our laboratories as well as the possibilities of exploitation of this simulation in the SCADA/HMI design, testing and in the teaching of SCADA/HMI design principles. Some software models of chemical plants have been designed with help of industrial simulation and emulation tools in our laboratories. This has been used for teaching and for SCADA/HMI applications and WEB control design and testing. *Copyright © 2003 IFAC*

Keywords: Simulation, Emulation, Programmable logic controllers, Distributed computer control systems, Human-machine interface, Control education

1. INTRODUCTION

By simulating of the plant and by emulating of the controllers as part of a complete control system, you can test and prove software for faster startups, minimize process downtime and startup costs. Early computer simulation tools were primarily analogue and oriented to small-scale simulation. Today's fully digital, large-scale simulation tools have expanded modelling capabilities to include pretty much any component of the controlled plant and its control system.

A typical industrial smaller control system consists of a controller (or of several controllers) with a communications link to a supervisory system. The controlled process would be an element of a manufacturing plant, the controller may very likely be a programmable controller or a programmable logic controller (PLC) and the supervisory system would incorporate a SCADA/HMI (Supervisory Control and Data Acquisition/Human Machine Interface) software packages for overall plant monitoring and advanced control (Mudrončík, Zolotová 2000). PLC testing in an office environment has always proven difficult. In the past, the use of the actual control panels, I/O racks and

a dexterous simulator adjusting controls has been required. Utilizing a simulation program occupying spare PLC memory or running on a PC, a more efficient test of the program could be performed (Smejkal, Vyhlídal 2000, Fedor, Perduková 2000). Actual I/O response, including analogue devices, may be simulated with adequate accuracy for the most demanding applications. A combination of a separate controller emulator and a simulation application may be used and it is often used in industrial practice too. This approach allows the testing of control programs before controller delivery and startup.

A distributed control system or industrial control system has often a control configuration tool with built in basic and advanced control strategies. Configurations of such systems could be used like process simulators; the dynamics of a process model may be modelled with help of pre-programmed control blocks.

These mechanisms of industrial practice have been used for PLC programming techniques teaching and for SCADA/HMI applications and WEB control design and testing in our laboratories (DACS FEI STU Bratislava, DCAI FEI TU Košice). Some software models of chemical plants have been designed with help of industrial simulation and emulation tools. Among others, we developed a simplified software model of an ethylene storage plant, which has been used in a more complex version in the refinery Slovnaft, batch reactor models, and a three conveyors model. Our experiences with several industrial tools and their simulation capabilities are described in the following chapters. The paper is organised as follows. Simulation and emulation capabilities of RSLogixEmulate500 are described in section 3, sections 4 and 5 are devoted to SCADA/HMI design applications (RSView32, InTouch) and to our projects on integrated control systems. In section 6 we introduce the equipment of our laboratories. Some advantages of student projects and work with industrial tools are discussed in the end of the paper.

2. DEFINITIONS

Simulation is a computer representation of a real-world system. When experimentation with the actual equipment in an entire system is too expensive, not practical or impossible from technological point of view, than a computer simulation model provides an environment where ideas and the flow of work can be tested before its actual implementation. "What-if" scenarios can be performed, determining how the system's elements work together under varying conditions.

An emulator is a program that maps and executes machine instructions intended for one machine architecture onto the host machine's architecture. In other words, emulators allow someone to run software on a machine in which it was not intended to run on. Generally one knows the term emulation in connection with a hardware, which is simulated on command level.

3. RSLOGIX EMULATE

Windows-based software packages RSLogixEmulate 5/500 (Rockwell Automation, Inc.) are troubleshooting and debugging tools that can emulate most operations of Allen-Bradley PLC-5 and SLC 500 family processors. The programs use the personal computer's CPU to scan the rungs of a ladder program. They execute the ladder logic in a computer like a real PLC processor, updating programs' data tables, allowing the users to approximate what is going to happen after the download of the programs to physical PLC-5 or SLC500 processors.

RSLogixEmulate5/500 requires the use of RSLinx, RSLinxGateway or WinteligentLinx communication drivers, ladder control and debug RSLogix5/500 files. The user needs neither a processor nor a physical model. He only determines which ladder programs he wants to run, which of them are to be used for the control and which for the plant simulation. The software emulates the plant models and the operation of one or more processor stations. Since these packages are not connected to I/O modules, I/O emulation is handled via debug files - ladder logic files that the user creates to simulate the inputs; and which can be designed like a software model of a real plant. The program rungs read input from and write output to the data table stored offline with the ladder logic project - a file containing ladder programs and SLC500 data tables. The offline data table is also active during emulation. To generate responses in the ladder program, the user will need to change the value of the desired I/O bits, storage bits, or storage words acting as inputs to his ladder program.

When RSLogixEmulate is running, the user can see the emulator as if it was a programmable logic controller PLC-5 on a DH+ network or a SLC500 on a DH-485 network. Other applications that use RSLinxGateway can also access the programs as if they were on a communication network. The engineer can program several RSLogix applications, and run several SLC500 or PLC5 emulator nodes in the same time.

The procedure of creating RSEmulate500 models (shown in the pictures below) consists of following steps:

1. The design of control ladder logic programs for the plant, a RSLogix main ladder logic file and ladder logic subroutines.
2. The design of ladder logic programs for the emulation/simulation of the plant, a RSLogix main ladder logic file and ladder logic subroutine files.
3. The download of the programs (fig.1) and the RSLogixEmulate settings (fig.2).
4. Ladder programs testing.
5. Design of the SCADA/HMI (fig. 3).
6. Steps four and five can be exchanged.

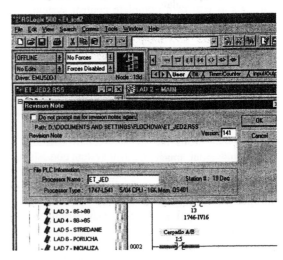

Fig. 1. The environment of RSLogix500 and the download procedure

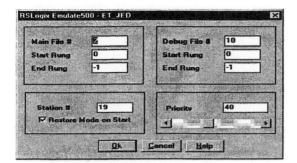

Fig. 2. The RSLogixEmulate main window

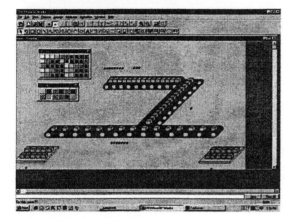

Fig. 3. Design of the SCADA/HMI - a simple three conveyors RSEmulate model

RSLogixEmulate can be used for:

Troubleshooting — the process can be stopped whenever a selected rung goes true, the program can be effectively frozen at the instant that any error occurs for simple and effective troubleshooting.

Ladder Logic Scanning Options — the ladder logic can be scanned continuously, one program scans at a time, rung-by-rung or a specific block of rungs can be selected to emulate. By selecting a block of rungs, the user is able to isolate that particular section of its program for testing purposes. The designer can set breakpoints in the ladder logic to stop execution when a specified event occurs (e.g. when an important value goes out of range), freezing the conditions that caused the breakpoint to trigger.

The basic order of events during emulation is no different from when a PLC processor runs a project. The software scans the rungs, pauses to update the output and input image tables, and scans the rungs again. Since there is no real I/O, the emulated ladder logic takes cues only from the state of the data table. To generate responses in the ladder program, the emulator changes the values of the desired I/O bits, storage bits, or storage words acting as inputs to the control ladder program. All of RSLogix Emulator's functions are selectable from a single menu bar.

One of the most powerful features of the software is the play trend feature. This feature allows the user to take data gathered from his process, and see what would happen if he changes the ladder logic or other address values. The user takes data records from a physical plant and uses these data files as input to the emulator. The play trend feature "feeds" data to the software from .DBF files, replaying the processor's actions in response to the data. The user can then monitor his ladder logic, looking for the effects of the changes.

Categories of Emulated Instructions (alphabetically sorted)

- ASCII Instructions
- Bit Shift, FIFO, and LIFO Instructions
- Branching Instructions
- Communications Instructions
- Comparison Instructions
- Control Instructions
- End Instruction
- Fault Routine
- File Manipulation Instructions
- Math Instructions and Advanced Math Instructions
- Move and Logical Instructions
- Relay-type Instructions
- Selectable Timed Interrupts
- Sequencer Instructions
- Timer and Counter Instructions

There are several instructions that are not emulated and executed by RSEmulate500. The list of these instructions is shown in the table below:

Table 1 Emulation exceptions

Instruction	Mnemonic	Execution
I/O related instructions	BTR, BTW, COR, REF, IIN, IIM, IOM, IOT,	There are no real I/O to read or write
I/O interrupts	IID, IIE, INT, RPI	ditto
MSG	Message	
PID	Proportional, Integral, Derivative	The software does not attempt to execute the PID instruction.
SVC	Service Communications	There are no processor communications to service

4. SCADA/HMI - INTOUCH, RSVIEW32

Both SCADA/HMI tools can be used for process simulation. InTouch (Wonderware – Invensys Systems, Inc.) scripting, derived tags in RSView32 (Rockwell Automation, Inc.) or Visual Basic application running in the background enable simultaneous work of a model and a SCADA application. A simple InTouch model may consist of scripts (fig. 4), RSView32 models of a collection of derived tags and Visual Basic modules attached to the objects.

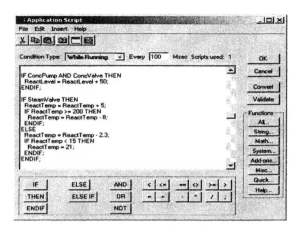

Fig. 4. An InTouch script – a simple simulation of some measurements in a reactor model.

5. INTEGRATED CONTROL SYSTEMS

Integrated control systems I/A Series Softpack (fig. 5) for Windows NT Foxboro Invensys Systems, Inc.) and CS3000 (Yokogawa Electric Corporation) have powerful control configuration tools with built in basic and advanced control blocks and strategies. Several Softpack models of simple process plants controlled by cascade loops have been developed. The dynamics of controlled plants has been modelled by parallel and serial compositions of dead time and lead-lag blocks.

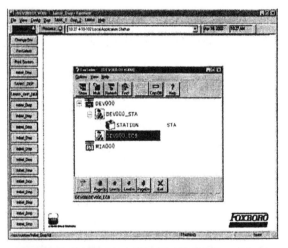

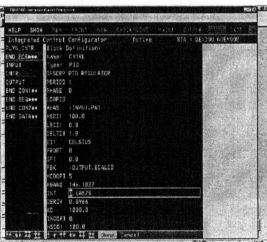

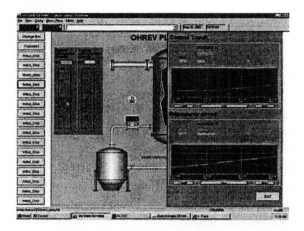

Fig. 5abc. I/A Series Softpack , a system configuration, Integrated control configurator and a students graphic display.

The control system Yokogawa CS3000 has a built in simulator of discrete and continuous systems, the works on it haven't finished yet. A more detail description of control configuration of both systems (in a table form), of their simulation possibilities and of the designed equipment and plant models will be included in the presentation

6. THE PLC LABORATORIES AND MODELS

The five and half years' study at our faculties is divided into three periods. The basic stage (2 years) gives students theoretical knowledge (Mathematics, Physics and Control Theory). The second period (2 years), which ends with the Bachelor's degree, provides a basis for a specific discipline. In this period, various laboratories among them being the Allen Bradley Laboratory, Rockwell software laboratory and Industrial control systems design laboratory support the courses. In the last study period at our departments which ends with obtaining the title of Engineer (Master's degree), students submit their diploma theses and take the final (state) examination in two main subjects.

Allen Bradley and Rockwell Automation Laboratory (fig.6) were established at our universities eight years ago. It was possible due to the Global Development Program of Rockwell Automation, Inc. and due to the help and support of the Allen Bradley product authorized distributor in the Slovak Republic Spel-Procont. Students coming to the laboratory (in 6th, 7th, 9th semester) are already familiar with control systems hardware, plant equipment, operating systems, real time programming and basic principles of the real-time control. The present laboratory equipment covers PCs Pentium, Windows NT/2000 OS, SLC500, Micrologix and PLC5 systems, PanelView operator terminals, corresponding communication drivers, control, emulation and SCADA/HMI software, RSSQL and models of real plants. There are also workplaces for elaborating of student projects and diploma theses. The DH485, DH+, ControlNet, DeviceNet and Ethernet networks have been built in the laboratories to establish the communications of all PLC and PC nodes.

An Ethernet link connects the Allen Bradley laboratory and the laboratory for industrial information systems design and IT technologies. The equipment of IT laboratory consists of Windows 2000, MSOffice2000, Visual Studio, Microsoft SQLServer, RSSQL, Sybase PowerDesigner, CAD tools, Siemens SCADA/HMI application WinCC and Yokogawa Centum CS3000.

Both laboratories have been used for teaching PLC programming techniques, the standard IEC1131-3, SCADA/HMI design, industrial information and communication technologies, the concepts of RDBS and real-time DBS; IT laboratory for teaching CAD, their use in control systems design and logistic principles.

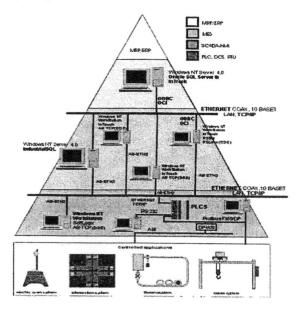

Fig. 6. Laboratories of DCAI Košice

Some software models and hardware models e.g. models of simplified chemical plants and equipments (Hrúz et al. 1997, Zolotová, Štofko 1999, Flochová, Hüber 2000 – fig.7ab, Landryová 1996, Žilková 1999) have been designed with help of industrial simulation and emulation tools in the laboratories. Among others we have developed a simplified (RSLogixEmulate) software model of an ethylene storage plant, which has been used in a more complex version in the refinery Slovnaft in Bratislava, a model of a small batch reactor, and a conveyors model.

The same models or the procedure of creating such models can be used in control programs and HMI design for HMI applications, which have built in drivers for Allen Bradley PLC 5 or SLC 500 (RSView32Studio, InTouch, Intellution Fix, WinCC, MonitorPro, etc.).

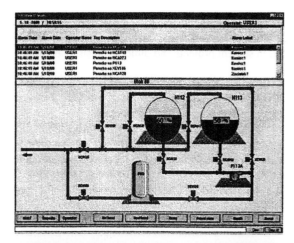

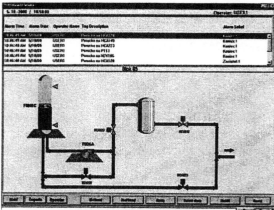

Fig. 7ab. Graphic displays for simplified real RSLogixEmulate chemical plant model – the storage tanks of ethylene

7. CONCLUSIONS

Software simulation and emulation tools will help the users to verify their ladder logic applications, HMI displays and other aspects of the control system before going out to the plant, reducing start-up time and improving quality. They can be used for students and operators training eliminating the need to build up costly hardware test stands and training centres, as well as for teaching of IEC 61131-3 programming techniques, and of SCADA/HMI design principles.

The software packages RSLogixEmulate5/500 emulate one or more running SLC500 or PLC5 processor stations and the process plant with corresponding control programs. The software RSEmulate models need neither real processors nor expensive and failure prone physical equipment. They are user-friendly, simply designed, robust and cost effective. They support network architectures and DH485 and DH+ access. RSLogixEmulate 5/500 can be used to create virtual discrete or part-time hybrid plant models and PLC stands that simulate discrete control systems; their use in a continuous PID control is limited (emulation of the PID instruction is missing).

Control systems Yokogawa CS3000 and I/A Series Softpack are powerful simulators, suitable for either discrete or continuous plant models. They can be used in basic (Bachelor's) and advanced (Master's) courses and in student projects and diploma thesis elaboration.

InTouch scripting and RSView32 or RSViewStudio derived tags design simple plant models. A Visual Basic code may run in the background.

The industrial tools and their simulators improve student's knowledge of physics, mathematics and other skills, which have been taught by the department staff, they increase their imagination, creativity and learn them deal with practical real world problems.

This work was supported in part by Slovak grant 7630 – Intelligent method for modelling and control, project APVT 51-011602 and by Leonardo da Vinci Project - SK/98/2/05020/PI/II.1.1.b/FPC.

REFERENCES

Fedor, P. - Perduková, D. (2000): *Programmable Controllers in Electrical drivers*. Košice, Mercury-Smékal, p. 86 , ISBN 80-968-550-0-X (in Slovak)

Flochová, J., Hüber, M. (2000): Utilization of RSEmulate500 in process simulation and in SCADA/HMI design teaching. In: *Proceeding of international conference Řip 2000*, Kouty nad Desnou, Czech republic, pp.231-236.

Hrúz, B., Ondráš, J., Flochová, J. (1997): Discrete event systems-an approach to education, In: *Proceeding of the 4th IFAC symposium on Advances in control education*, Istanbul, Turkey, pp.283-288.

Landryová L. (1996): Design of Processing Systems With the Use of Modelling in In Touch. In.: *Sborník XIX. Semináře ASŘ - Počítače v měřeni, diagnostice a řízení*. Ostrava, VŠB-TU/KAKI/DAAAM/IMEKO pp. 16/1-6. ISBN 80-02-01094-9.

Mudrončík,D., Zolotová, I. (2000): *Industrial controllers*. Elfa ltd., Košice, 163 pages.

Smejkal, L, Vyhlídal, T. (2000): Modeling, simulation and PLC. In: *Automatizace 8/2000*, pp. 526-532.

Zolotová,I., Landryová,L. (2000): SCADA/HMI Systems and Emergency Technology, In: *Proceedings Volume from the IFAC Worshop Programmable Devices and Systems*, PDS 2000, Ostrava, február, pp. 17-20, Pergamon - Elsevier Science, ISBN 0-80-043620X.

Žilková,J. (1999): State Variables Estimation of an Induction Motor. *MicroCAD'99*, Proceedings, Miskolc, 1999, pp. 137-141.

Rockwell software products, 2002,
http://www.software.rockwell.com
http://www.software.rockwell.com/navigation/products/index.cfm,
http://www.software.rockwell.com/rslogixemulate
http://www.ab.com/manuals/
FactorySuite Wonderware manuals. Wonderware Inc. 2001.
http://wonderware.com
http://www.ad.siemens.de/hmi/html_76/products/software
I/A Series ® *Software Real-time Database manager* PSS 21S-4K1 B3; Foxboro, US, 1995
http://194.184.64.99/SCADA_function.htm
http://194.184.64.99/i_a_series_scada_ntl.htm; Foxboro

www.elsevier.com/locate/ifac

ADVANCED CONTROL: SIMULATION TOOLS IN LABVIEW ENVIRONMENT

Imre BENYÓ*, György LIPOVSZKI**, Jenő KOVÁCS*

University of Oulu, Finland, Budapest University of Technology and Economics, Hungary
Imre@paju.oulu.fi, Lipovszki@rit.bme.hu, Jeno.kovacs@oulu.fi*

Abstract: A simulation package developed for advanced control education is reported in this paper. The package is aimed to assists students in acquiring the knowledge and to serve as an easy-to-use demonstration tool for lecturing. The simulation blocks, developed in LabVIEW™ environment, are listed and the effectiveness of the simulation package is demonstrated in examples. *Copyright © 2003 IFAC*

Keywords: computer aided, education, identification, predictive control,

1. INTRODUCTION

Interactive simulation instruments, for example Matlab/Simulink™ and LabVIEW™, have an important and increasing role in the engineering education. Especially in control engineering education, where the demonstration of the control-loop performance promotes the overall understanding. The simulation environments, in general, provide efficient and inexpensive manner for knowledge transport. Also, according to the authors' practice, students spend time with pleasure on simulation.

The aim of the work, reported in this paper, was to develop a simulation package to enable the demonstration of advanced control methods. The user, who may be the teacher or the students themselves, may easily build different control loops by simply selecting the necessary elements from a menu list. The list of available blocks is design to satisfy the need of an undergraduate course on advanced control engineering, part of the authors' laboratory's curriculum.

The simulation package is developed in LabVIEW™ environment – which alloys the simplicity of graphical programming with the traceability of the analogue devices –, utilising mainly the TUBSIM toolbox.

The paper is organised as it follows: Section 2 gives an overview of some educational applications using LabVIEW™. Section 2 describes the TUBSIM for LabVIEW™ simulation package, which provides the basic simulation environment. Section 3 introduces the new blocks developed for the advanced control course while Section 4 illustrates the new components via two examples. Finally Section 5 summarises the achievements and Section 6 reports the main references.

2. THE LABVIEW

The LabVIEW™ software package, product of National Instruments (NI), was originally designed for industrial data acquisition and instrument control. Beside the original task, many other applications (so-called toolboxes) have been developed since 1986.

To facilitate the use of LabVIEW™ for simulation task, the NI provided the GSIM Control and Simulation Software in 1998. Further supporting the simulation of control methods, PID Control Toolkit for G and Fuzzy Toolkit for G were developed by the NI.

Due to its convenient features, LabVIEW™ is widely used in the education as well. An educational example is presented in Aradi (1996), where the LabVIEW™ is used as a supporting tool in systems and control engineering. In Ertugul (2000), the needs of nowadays engineering education are summarised. The LabVIEW™ is claimed in the paper as a possible alternative in the engineering education in various fields from electrical to environmental engineering through the biomedical, chemical-control, etc. engineering. The author underlines the possibility of LabVIEW™ to convert the project into standalone application.

Keller (2000) presents a simulation trainer for PID and Fuzzy controller design. According to the author's observation, the trainer well supports the understanding of the methods. The powerful debugging tools of the LabVIEW™ is emphasised which allows the fast development and testing of the simulation software.

3. THE TUBISM MODELLING TOOLS

TUBSIM (Lipovszki and Aradi, 1995) is a system simulation extension of LabVIEW™. TUBSIM is the interpretation of an analogue computer in the LabVIEW™ graphical programming environment. The simulation system is based on the analogue simulation principle. Besides the typical analogue computers elements, the TUBSIM Library contains different Boolean blocks, typical system engineering elements (low order transfer function elements, continuous time controllers like PI, PID, time delay) and sampled system blocks.

The extension of the TUBSIM Library with Fuzzy blocks was reported in Aradi (1998).

The TUBSIM Library is successfully used in the control engineering education at the Budapest University of Technology and Economics (Aradi, 1996). It assisted the development of the modelling skills and the education of the control theory through the simulations supported by the TUBSIM.

4. THE NEW BLOCKS

The aim of the development of the new blocks was to facilitate the simulation of the advanced identification and control techniques. In the present state of the development, the new block library includes tools for SISO discrete-time systems. The algorithms of the implemented methods can be found in Ikonen and Najim, (2002).

The new blocks, presented in Fig.1, can be divided into four subgroups according to their task: process modelling tools, identification tools, state observers, general predictive controller. A pop-up help window provides a brief description of each block as shown in Fig.2.

4.1 Blocks for process modelling

The new set of blocks provides several possibilities for process modelling: both the input-output approach and the state-space approaches are supported.

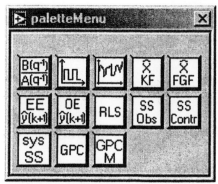

Fig. 1 The palette of the new blocks.

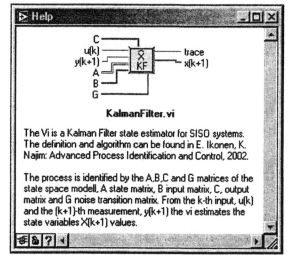

Fig. 2 The pop-up help of the Kalman Filter block.

The input-output approach defines the process by its pulse transfer function. The block "$\frac{B(q^{-1})}{A(q^{-1})}$" represents the pulse transfer function, defined by the coefficients of the B and A polynomials.

The state-space subgroup consists of three different blocks. The "sys SS" block defines the process model in state-space equation. The user shall define the state-space matrices.

In control and observer design, the controllable and observable canonical forms of the state-space equation are generally applied. Those forms can be obtained from the pulse transfer function (e.g. identification in input-output approach) by using the "SS Contr" (State-Space controllable) and the "SS Obs" (State-Space observable) blocks.

4.2 Blocks for identification

The identification subgroup contains three blocks: the recursive least-square algorithm (RLS), the equation error (EE), and the output error method (OE).

The "RLS" block estimates the parameters of a given linear regression model using the recursive least-square method. The input of the block is the regression vector and the output is the correlation vector. The covariance matrix of the parameter adaptation algorithm can be maintained by applying the offered methods: forgetting method, constant trace method or the bounded information algorithm. Since, the "RLS" block requires only the regression vector as an input, the user has the freedom to choose the model structure; consequently, the user has to create that vector.

In the "EE" and "OE" blocks, the structure of the regression model is a priori defined by the equation-error method and the output-error method, respectively. The parameter adaptation algorithm is in both cases the RLS algorithm. The block's inputs are the input and output signals of the process to be identified and the order of the model. The blocks return the estimated pulse transfer functions given by the coefficients of the numerator and denominator polynomials, the output of the regression model, $\hat{y}(k+1)$ and the covariance matrix.

4.3 Blocks for State Observer

Among the state observer blocks, one can choose between the Kalman-Filter ("$\hat{x}KF$") nd the "Fixed Gain State Observer ("$\hat{x}FGF$").

The implemented Kalman-Filter estimates the state variables at the $(k+1)^{th}$ sampling instant based on k^{th} process input and $(k+1)^{th}$ output measurement: $x(k+1) = f(u(k), y(k+1))$. The outputs of the block are the state vector and the trace of the covariance matrix.

The Fixed Gain State Observer estimates the state variables in the $(k+1)^{th}$ instant based on k^{th} process input and k^{th} output measurement. The output of the block is the $(k+1)^{th}$ state variables vector: $x(k+1) = f(u(k), y(k))$.

The state-observers utilise the observable canonical form of the state-space model. The required form can be generated by the "SS obs" (State-Space Observable) block, which transforms the pulse transfer function into observable canonical state-space form.

4.4 Blocks for General Predictive Controller

There are two blocks relating to the GPC control algorithm. The "GPC M" calculates the gain matrices of the controller; the "GPC" block is the controller. The block realizes a GPC for SISO process. The controller is computed based on the state-space model of the controlled process.

The inputs of the block are the state vector, the output signal of the controlled process and the reference signal. The main output of the block is the control signal. There is another signal for graphical purpose: the desired future process outputs on the prediction horizon.

Besides the general parameters of the GPC algorithm (prediction, minimum and control horizons, weighting factors of the control error and control signal in the cost function; there is possibility to use weighting matrices for enabling weighting the terms in the appearance of time), the pulse transfer function of the desired closed-loop behaviour can be defined.

In the case of online identification, it is possible to use the new identified model of the process in every time step, otherwise the time demanding gain matrices computation is done only in the first time step (Adaptive button on/off).

5. EXAMPLES

The outlook and the utilisation of the developed blocks are demonstrated in the following two examples. The first example is an on-line identification combined with state observation, while the second example presents a GPC control structure.

5.1 Identification and state estimation

In this example, simple parameter estimation and state estimation problems are solved. The "unknown" process is on-line identified using the input-output approach (output error method). Then the identified model is transformed to state-space model that is used for the estimation of the state variable (Kalman-filter). The LabVIEW^{TM} realisation, shown on Fig. 3., clearly demonstrates the simplicity of programming.

The process was modelled by the output error method as

$$y(k) = \frac{B(q^{-1})}{A(q^{-1})}u(k) + e(k) \qquad (1)$$

The pulse transfer function was changed at the 75th sampling instants as

$$\frac{B(q^{-1})}{A(q^{-1})} = \frac{bq^{-1}}{1+aq^{-1}} = \frac{0.2q^{-1}}{1-0.8q^{-1}} \Rightarrow \frac{0.2q^{-1}}{1-0.6q^{-1}} \quad (2)$$

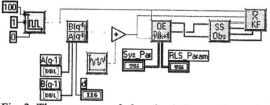

Fig. 3 The program of the simulation without the time setting and graphical blocks.

The e(k) noise is a Gaussian white noise with the variance 0.1. Within the RLS algorithm of the OE parameter estimation, the forgetting factor $\lambda = 0.98$ was applied to maintain the covariance matrix.

The state-space model is simple since the process is a first order,

$$x(k+1) = \alpha x(k) + \beta u(k)$$
$$y(k) = x(k)$$
$$(3)$$

The initial values of the parameters were set $[a\ b]_{t=0} = [-1\ \ 0.6]$. Fig. 4 demonstrates the parameter convergence while Fig. 5 shows the process and model state variables, $x(k)$ and $\hat{x}(k)$, respectively.

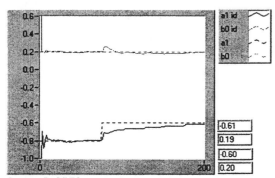

Fig. 4 LabVIEW graph: parameter estimates; true parameters (dashed lines), estimated ones (solidlines)

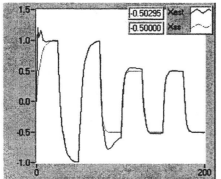

Fig. 5 LabVIEW graph: process and model state variables "Xss" and "Xest".

5.2 General Predictive Control

This example demonstrates the application of the General Predictive Control block. The scheme of the control loop is on Fig. 7. The process output (y) is controlled by the GPC to follow the reference signal (r) with a prescribed tracking behaviour. The reference signal is a square wave signal, and the process has Gaussian white noise type measurement noise.

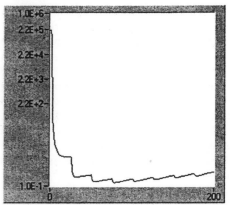

Fig. 6 LabVIEW graph: the trace of the covariance matrix of the OE estimator.

The state variables can not be measured, only the process input and output signals are available, thus state observer must be used, in this case for the sake of simplicity the Fixed Gain Feedback Observer.

The Fig. 8 shows the LabVIEW program of the simulation, with all of its accessories for graphical purpose. The Control Panel of the Simulation is shown on the Fig. 9. On the control panel, the numerical control buttons can be used to change the parameters even during run-time. Furthermore, all the process and control parameters are displayed on the control panel. The results of the simulation can be followed through the graphs and the numerical displays.

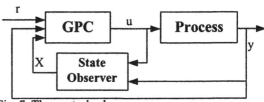

Fig. 7. The control scheme.

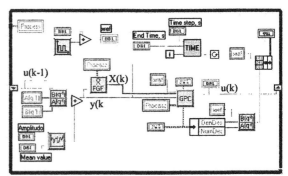

Fig. 8 The program diagram of the simulation, with the same signal notation as on the previous figure.

240

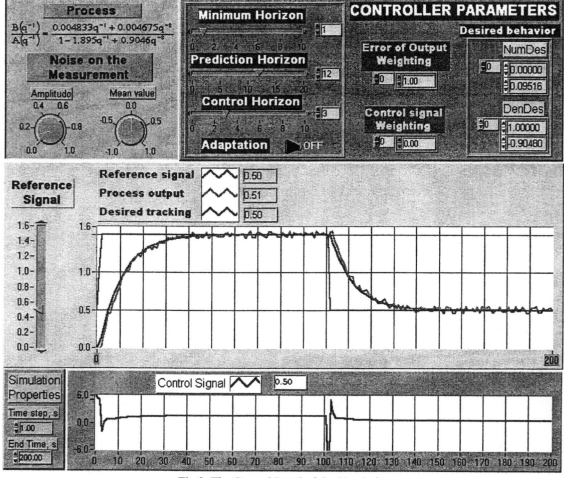

Fig 9. The Control Panel of the Simulation

6. CONCLUSIONS

A package of advanced simulation tools has been introduced in the paper. The package is designed to serve a particular course on advanced control engineering. The developed simulation elements provide an easy-to-use and easy-to-understand manner to deepen the enhanced knowledge of the course.

The presented tools support the simulation of discrete-time modelling, identification, state-observer and general predictive control design.

The programming environment was chosen to be the LabVIEW™ due to its attractiveness.

7. REFERENCES

Aradi, P. (1996), *Using LabVIEW in Education of Systems and Control Engineering,* NIWeek'96, Austin, TX, USA.

Aradi, P. (1998), *Fuzzy simulation system,* Conference

Ertugrul, N. (2000), Towards Virtual Laboratories: a survey of LabVIEW™-based Teaching/Learning Tools and Future Trends, Engineering Education Special Issue: LabVIEW Applications in Engineering Education, 2000.

Ikkonen, E. (2002). *Advanced Process Identification And Control,*

Keller, J. P. (2000), Teaching PID and Fuzzy Controllers with LabVIEW™, Engineering Education Special Issue: LabVIEW Applications in Engineering Education, 2000.

Lipovszki, Gy., Aradi, P. (1995), *General Purpose Block Oriented Simulation System Using LabVIEW,* NIWeek'95, Austin, TX, USA.

National Instrument. (1993): LabVIEW for Windows, Tutorial.

www.elsevier.com/locate/ifac

NEW DIRECTIONS IN PROCESS CONTROL EDUCATION

Sirkka-Liisa Jämsä-Jounela

Helsinki University of Technology
Laboratory of Process Control and Automation
Kemistintie 1, 02150 Espoo
FINLAND
E-mail: Sirkka-l@hut.fi

Abstract: Process automation is nowadays recognized as an interdisciplinary education activity. In the future, students will have to receive a broad education in order to cope with cross-disciplinary applications and the rapidly changing technology. In this paper the new tools available for process automation education are presented and discussed. First, a web-based study support environment is described and evaluation of the application results reported. Second, a state-of-the-art flotation cell pilot plant and a remote support system for a pressure filter for experimental training are presented. Finally the evaluation of the programme is briefly discussed. *Copyright © 2003 IFAC*

Keywords: Control Education, Process Automation

1. INTRODUCTION

During the past twenty years the field of process control has undergone a considerable transformation. Two decades ago, the typical control-engineering graduate had a course in feedback control theory, and those interested in a career in this field secured a position in the process industries. During this period, however, the number of new control engineering positions in the process industry has declined while, at the same time, increasing emphasis has been placed on positions requiring a multidisciplinary knowledge of process engineering, automation and information technology. As a result, control theory has played a very minor role.

Due to this multidisciplinary requirement for knowledge and the latest developments in the different fields, especially in Information Technology, the need has also arisen for a change in the education and courses offered at universities. In order to fulfil the demands of industry, the students will have to be broadly educated to cope with cross-disciplinary applications and the rapidly changing technology.

The Laboratory of Process Control and Automation at the Helsinki University of Technology (HUT) provides education in process automation mainly in the fields of Chemical Technology, Forest Products Technology and Materials Science and Rock Engineering. Process automation is a special program, crossing departmental lines, that enables students in different areas such as chemical, forest product technology and material science engineering to obtain the interdisciplinary training necessary for work on process automation and its applications.

In this paper the structure of the master's degree is first presented, followed by the laboratory-industry interaction, and the new applications of teaching and education tools that the laboratory is utilizing to meet the challenges set by these new requirements. Finally, the results of the program evaluations are reported and briefly discussed.

2. THE STRUCTURE OF THE MASTER'S DEGREE

The Master's degree comprises two parts. Part I consists of approximately 70 credits and constitutes a relatively uniform basis for engineering. It includes mathematics, physics, chemistry, information science and general studies, as well as foreign languages. In part II the student chooses a major subject from her/his own degree programme, and a minor one that can even be outside the degree programme. The structures of part I and II are presented in Fig. 1.

Fig. 1.Structure of the master's degree.

The compulsory and optional courses include, in addition to the traditional process control courses, also topics related to automation systems, software development and process engineering.

3. THE LABORATORY-INDUSTRY INTERACTION

The laboratory has a close interaction with industry through its involvement in programme and sponsored research. The students have a period of compulsory training during their summer vacation. This experience of the industrial environment helps them to better understand the classroom teaching, especially in the later years of their education. In addition, a number of courses have been developed in which companies have assisted with course development and staffing. Visits are organized to local industries. These visits are fully integrated into the teaching-learning strategy. External speakers from the industrial sector have made various inputs into the delivery of the course.

3.1 A State-of-the-art pilot plant for control education

A pilot-scale flotation cell was built for educational purposes in co-operation with Outokumpu Oyj. The main aim of the project was to investigate the capacity of field-distributed control, and the possibilities of using the additional field device information offered by Foundation Fieldbus for fault diagnostics.

Pilot Flotation cell. A PI diagram of the process is shown in Figure 2. The flotation cell

(V101) is a 3 m3 Outokumpu TankCell unit equipped with a real rotation mechanism. Another tank (V102) is used as a recycle supply. The flow is recycled using a slurry pump. The rotation speed of the pump can be controlled using a Vacon frequency converter. Slurry flow is measured by a Khrohne magnetic flowmeter. Output flow from the tank is controlled either by a Larox valve (303) or an Outokumpu dart valve (302). The tank level is measured using Milltronics ultrasonic radar (301). Flotation air is fed into the tank through the hollow axle of the rotating mechanism by means of an Aerzer roots compressor (201). A Vacon frequency converter controls the rotation speed of the compressor. The air flow can be measured by three different methods: a thermal massflow meter by Sierra Instruments, a temperature and pressure compensated pitot tube transmitter by Fisher-Rosemount, and an integrated control valve and flowmeter, NelFlow by Metso Automation.

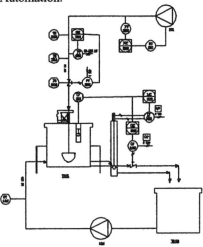

Fig. 2. PI&D of the flotation cell.

Control and information system of the pilot flotation cell. All the control actions of the flotation cell are distributed between the field devices using Foundation Fieldbus (FF) technology. Some of the instruments cannot be connected directly to a FF segment because of their "4-20 mA" technology. In these cases, Smar "Current-to-FF" converters are used. The FF segment is connected to an OPC server using a Smar DFI 302 unit via an Ethernet RJ-45 cable. Smar digital I/O units are used for controlling discrete IO.

A Smar OPC server maintains all the tag information. Information is accessed via the OPC interface by Cimplicity HMI. Tag values can be accessed either by a local OPC client or by a remote OPC client. Process data are logged into the MS SQL7 server through an

ODBC interface by a Cimplicity Data Logger application. A diagnostics tool is also connected to the system by an Ethernet cable, and diagnostics information is accessed via the OPC interface. The automation system is shown in Figure 3.

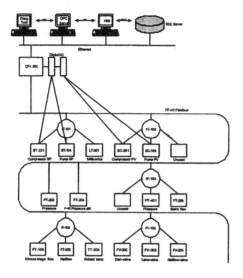

Fig. 3. Network topology of the automation system.

Wireless automation. The development of wireless technologies has been rapid during the past two decades, and this has made it possible to utilize the new technologies for process automation purposes. Wireless test environment using Bluetooth and SMS technologies was built in the laboratory recently in co-operation with NOKIA. The aim of the system was to simulate the communication structure of a flotation cell located far at a considerable distance from the backbone system. ✓

The structure of the system is presented in Fig.4. The first Bluetooth device (Starter Kit) is connected to the sensor signal and sends the information to Bluetooth device 2 (WRAP 1260). WRAP1260 is connected to the Nokia M2M terminal. The signal is then sent on to another Nokia terminal via an SMS connection. Finally, this M2M terminal is connected to the backbone system where the sensor data is presented to the operator.

In the flotation cell the measurement signal from the flow meter can be a 4-20 mA analogue signal. The A/D conversion is done in a Starter Kit where the signal is also converted to engineering units. In the SMS system the measurements are sent to the backbone system collectively and, in this system, twenty half-minute averages are sent once every ten minutes. An alarm message is

then presented in the operator's cellular phone or in the backbone system.

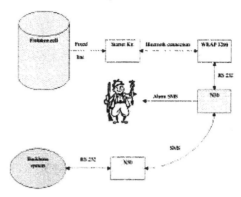

Fig.4. The structure of the wireless measurement system.

Course format. Due to the multidisciplinary nature of the process automation field, laboratory experience is particularly important with regard to the teaching of control systems. The pilot plant has been used as a teaching and research tool in undergraduate courses and in projects related to process automation. The course format of "Information Technology in Process Automation" consists of lectures and assignments with practical experience in control engineering. The topics of the invited lecturers from industry include:
* Utilization of fieldbuses in the information and control of process equipment
* Foundation Fieldbus
Integration of Industrial Automation
* The possibilities for process equipment using fieldbuses, the Metso smart valve as an example
* Com, Corba and ODBC interfaces
OPC-based data transfer in large scale process simulations
* Wireless automation in the process industries

The first practical project work includes laboratory experience with smart instruments, fieldbusses and field-distributed control. The aim of the second is to learn to set up a wireless measurement system.

3.2 Remote support system for the pressure filter

Maintenance has become a major cost factor in production plants. Increasing availability and minimising the service costs are the main objectives for the remote diagnosis of process equipment. A remote, fault-diagnosis system for a pressure filter was built for educational and research purposes at the laboratory. The aim of the project was to provide students with distance work applied to control when utilizing new information and communication technologies.

Remote support system. The proposed system at a factory site consists of a database server (SQL server connected to a LAROX pressure filter), an IIS server, workstations and a local area network. At the client end there are workstations, and both ends are connected to each other either via the Internet. Data are transported using a browser and HTTP as the interface. Due to security reasons, all data transmissions on the Internet between the factory and the client are encrypted with a DES algorithm, and access to the local network is permitted only through the firewalls. The system is shown in Figure 5.

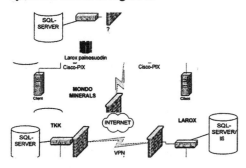

Fig. 5. Structure of the remote support system for the pressure filter.

Preliminary experience indicates that this set-up has provided a very effective means for students to gain experimental experience. The remote lab project seems to have increased the students' enthusiasm and motivation for learning the course topics.

3.3 Process monitoring

On-line process monitoring with fault detection can provide stability and efficiency for a wide range of processes. In the process monitoring course the student carries out a carefully selected specialised project which enables integration of information covered in the course: data collection, data pre-processing, selection of the monitoring variables, off-line testing and data analysis. The projects are industry oriented and are supervised by both internal and industrial supervisors. Last year the aim of the project was to model the fibre line of the Kaskinen pulp mill, starting from cooking, in order to find and analyze the most important variables that have an effect on the chosen quality variables of Birch- ECF pulp.

3.4 Production planning and control

Industrial automation is undergoing an unprecedented rate of change throughout its history. Where earlier systems mainly followed process developments, the new methods are now revolutionising the entire field of process management. The implementation of advanced technology and the timely use of information have the potential to optimise overall company productivity and, furthermore, to reduce costs under variable processing and economic conditions. The emphasis will be on plant-wide control, integrated business and production control, rather than on unit process control.

SAP R/3, along with SAP Advanced Planner & Optimizer (SAP APO), provide a flexible platform to manage the business processes, while assuring integrated execution in a complex supply chain or global customer relationship strategy. The system will be used in production planning and control courses, especially as a tool for global production planning.

4. WEB-BASED EDUCATIONAL ENVIRONMENTS

4.1 Study support system for process control courses

The aim of the web-based, study support environment was to increase study effectiveness in process automation courses. The purpose of the system is to inform students and to communicate with them while, at the same time, offering frequent advice on their progress in mastering the course content. The system was first tested in an introductory undergraduate course "Basics in Process Automation", which is a course in which the transfer of insights from the teacher to the students is often limited due to the large number of students (e.g. 200). Since then, this format has also been adopted in other courses provided by the laboratory.

Course format The format of the course is based on conventional lectures, exercises and assignments. The students can download the transparencies of the lecture notes and exercises in advance before each lecture. The students are also given the option of carrying out voluntary on-line self-test exercises, which follow the learning paths of the course. These exercises consist of several questions, alternated with small assignments. After submitting the solution to an exercise, the student immediately receives feedback. The feedback consists of the grade result, as well as an evaluation of each individual question in the exercise and a model of the right answer. This study guide has been developed to help the student master the main topics of the course. Communication between the teacher, course assistant, and the other students, is arranged through the message box, the electronic bulletin board.

Realization of the web-based environment. The system runs on a Dell PowerEdge 2400 server that has a Microsoft Internet Information Server 5.0 and a Microsoft SQL Database server 7.0. The web based education tool has been built using the Microsoft ASP (Active Server Pages) language. All computation is done on the server, and the end user utilizes a browser.

The ASP-based software development has been implemented in two parts. The laboratory manager has programmed the engine, which makes all the necessary calculations. The contents of the courses is stored in their own *.txt files or on the SQL server. This means that the lab personnel can easily participate in making the teaching material available on the www-site.

The web-based system consists of several components. One component is the navigation tool such as the Microsoft Windows Explorer shown in Fig. 6. (Kämpe, 2000) A navigation tool is used for the laboratory main site, as well as for each course given by the laboratory. A bulletin board is shown in Fig. 7.

Benefits were obtained in increased student participation during the entire course and improved final marks were obtained. Especially the implemented study guide with on-line, self-test exercises improved the grades considerably. The distributions of the results before and after the system implementation are presented in Fig. 8.

4.2 Knowpap

The Laboratory of Process Control and Automation also provides courses to students from the Department of Forest Products Technology. A learning environment for self-study and the teaching of paper making, process control and mill automation, was developed in the KnowPap project. The KnowPap system is based on multimedia, hypertext and simulation models, and the system is used through a www browser.
KnowPap includes operations in paper and paperboard mills, and their interaction with the operating environment. The system also features simulation tasks and question pages, and includes an extensive glossary of paper

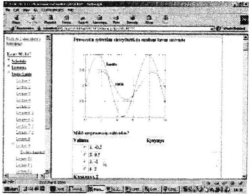

Fig. 6. Study guide with on-line, self-test exercises.

Fig. 7. A bulletin board

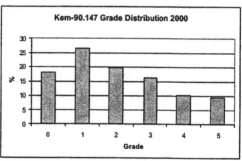

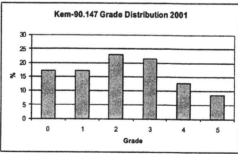

Fig. 8. The distributions of the results before and after the system implementation.

technology and automation. The learning environment was developed together with VTT Automation, the Laboratory of Papermaking, the Laboratory of Process Control and Automation and the Finnish pulp and papermaking industries.

247

5. EVALUATION

There have been several approaches to the evaluation of the degree, e.g. student questionnaires. Generally the feedback from the students has been both positive and encouraging.An evaluation was conducted to assure the quality of courses and outcomes of the programme in summer 2002.Web-based questionnaires were distributed to 100 MSc engineers that graduated from the laboratory.

The education program is adapted to the actual demands of the Finnish process industry, the main emphasis being on industrial applications. Employment of the students after graduation is presented in Fig. 9. 35% of the graduates are employed in the process industries – chemical-pulp&paper and metal processing,- 17% in the automation suppliers, and 13% in the research and development field. Based on these results and evaluations by the external experts Finland is experiencing a shortage of IT-skilled process engineers.

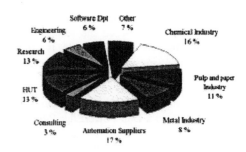

Fig. 9. Employment of the students after graduation.

Various evaluations were made on the basis of the questionnaire results, one of which was on technical knowledge related to process automation. The results of the survey before and after the introduction of the new programme are presented in Fig. 10.Five years' experience of operating the programme has indicated that the strategy is successful.

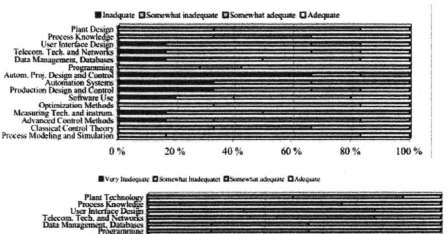

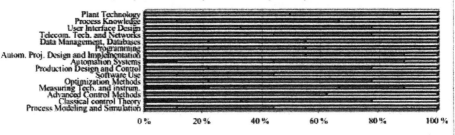

Fig. 10. The results of the survey before and after the introduction of the new programme

6. CONCLUSIONS

This paper reported the new tools available for process automation education: a web-based study support environment, a flotation cell pilot plant and a remote support system for a pressure filter. The students have found that the systems provide stimulating environments in which to develop their knowledge and experience. Increased student participation was observed during the courses, and the final marks improved. The education of students has benefited through the involvement of industry's in the course design, thus ensuring its relevance to future careers.

REFERENCES

Kämpe, J., http://kepo.hut.fi/FoldrTree/, 1.12.2000.
Kämpe,J.,http://kepo.hut.fi/FolderTree/Messag eboard/, 15.4.2002.
Jämsä-Jounela, S-L., Current status and future trends in the automation of mineral and metal processing, *Control Engineering Practice* 9 (9) 2001) pp. 1021-1035.

ELSEVIER

IFAC

PUBLICATIONS
www.elsevier.com/locate/ifac

SYSTEMS WITH VARIABLE PARAMETERS; CLASSICAL CONTROL EXTENSIONS FOR UNDERGRADUATES

Nusret Tan[1], Derek P. Atherton[2] and S. Dormido[3]

[1]Inonu University, Engineering Faculty, Department of Electrical and Electronics Engineering,
44069, Malatya, Turkey. ntan@inonu.edu.tr

[2]University of Sussex, School of Eng. and Information Technology,
Falmer, Brighton BN1 9QT UK. d.p.atherton@sussex.ac.uk

[3]Dept. Informatica y Automatica. Facultad de Ciencias, U. N. E. D.,
Avenida Senda del Rey, 28040, Madrid, Spain. sdormido@dia.uned.es

Abstract: Recently a method has been introduced for finding the minimum bounds of the frequency response of a system with variable parameters on Bode gain/phase diagrams. The results are easily derived from classical control frequency response concepts and provide a simple way of introducing the concepts of control system design under parameter uncertainty to undergraduates. This paper describes the method and shows how it can be used for investigating stability and the design of simple classical compensators. Use of the method for interactive education is also discussed.
Copyright © 2003 IFAC

Keywords: Education; Uncertain linear systems; Kharitonov theorem; Bode diagram; Interactivity.

1. INTRODUCTION

Much of the theory being taught to undergraduates in a first course involving classical control has not changed significantly during the last four decades. The major difference is that students are shown how to obtain, root loci, Bode plots etc. using computer software, such as Matlab. The goal of this paper is to suggest how an additional topic, in this case aspects of analysis and design relating to parameter uncertainty, can be easily taught using these tools. This is an important topic since most mathematical models of plants to be controlled represent an approximation of real behaviour and the values of the parameters which constitute the assumed model will not be known exactly. Therefore, from an educational viewpoint, it is important to teach simple methods to calculate the effect of parameter changes on control system performance.

Much recent work on systems with uncertain parameters has been based on Kharitonov's result [1] on the stability of interval polynomials. Kharitonov showed that for the interval polynomial

$$P(s) = a_0 + a_1 s + a_2 s^2 + a_3 s^3 + a_4 s^4 + \cdots + a_n s^n \quad (1)$$

where $a_i \in [\underline{a_i}, \overline{a_i}]$, $i = 1,2,.....,n$, the stability of the set could be found by applying the Routh criterion to the following four polynomials

$$p_1(s) = \underline{a_0} + \underline{a_1} s + \overline{a_2} s^2 + \overline{a_3} s^3 + \underline{a_4} s^4 + \cdots$$
$$p_2(s) = \underline{a_0} + \overline{a_1} s + \overline{a_2} s^2 + \underline{a_3} s^3 + \underline{a_4} s^4 + \cdots$$
$$p_3(s) = \overline{a_0} + \underline{a_1} s + \underline{a_2} s^2 + \overline{a_3} s^3 + \overline{a_4} s^4 + \cdots \quad (2)$$
$$p_4(s) = \overline{a_0} + \overline{a_1} s + \underline{a_2} s^2 + \underline{a_3} s^3 + \overline{a_4} s^4 + \cdots$$

Although this may seem a surprising result it is easily proved from the Mikhailov criterion- a graphical interpretation for the Routh criterion- which was published in the Russian literature over 50 years ago. The criterion states that the polynomial $p(s)$ has no roots with positive real parts if $p(j\omega)$ cuts successively n axes in a counterclockwise direction starting from the positive real axes. It can be easily shown that the value set of an interval polynomial at a fixed frequency is a rectangle (Kharitonov rectangle) as shown in Fig. 1, that is the value of every polynomial of the family at that frequency lies within or on the rectangle, whose sides are parallel to the real and imaginary axes. Since the sides of the rectangular value set are parallel to the real and imaginary axes, it can be easily shown that the inclusion or the exclusion of the origin from the rectangular value set can be checked by using the corner points which correspond to the Kharitonov polynomials. The Kharitonov theorem is only applicable to interval uncertain parameters but the characteristic equations of even simple control systems do not have an interval uncertainty structure. For example, to take a simple case, a plant transfer function model will typically be of the form $G(s) = K / s(T_1 s + 1)(T_2 s + 1)$ where uncertainty will exist in K, T_1 and T_2. If the plant is placed in a feedback loop with unit negative gain the characteristic equation for assessing stability is $\delta(s) = T_1 T_2 s^3 + (T_1 + T_2)s^2 + s + K$ which is not an interval polynomial. The only simple way to use the Kharitonov result is to overbound and underbound the parameters of s^3 and s^2, which produces a very conservative result.

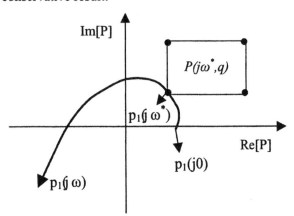

Fig. 1: Kharitonov box and the Mikhailov locus for $p_1(s)$

The purpose of this paper is therefore to look at some properties of transfer functions expressed in factored form as their parameters are varied. The parameters which are assumed to vary within lower and upper bounds are time constants, gains, natural frequencies and damping factors which are directly related to the true physical parameters describing uncertainties in plant models.

The layout of the paper is as follows. In the next section the construction of the banded Bode diagram is given. Section 3 gives some application examples for linear systems including the design of classical controllers for robust performance. Implementation of the method in *SysQuake* for interactive education is discussed in Section 4. Finally some conclusions are given in Section 5.

2. CONSTRUCTION OF THE BANDED BODE DIAGRAM

It is well known that some process dynamics can be approximated by simple first order models or by a standard second order system. In a more general form the open-loop transfer function of a real physical model can normally be written as

$$G(s) = \frac{\prod_{i=1}^{a}(s^2 + 2\varsigma_{ni}\omega_{ni}s + \omega_{ni}^2)/\omega_{ni}^2}{\prod_{j=1}^{b}(s^2 + 2\varsigma_{dj}\omega_{dj}s + \omega_{dj}^2)/\omega_{dj}^2} \cdot \frac{K\prod_{k=1}^{m}(1 + sL_k)}{s^N \prod_{l=1}^{n}(1 + sT_l)}e^{-\tau s} \quad (3)$$

where s^N in the denominator represents a pole of multiplicity N at the origin and $2a + m \leq N + n + 2b$. It is assumed that the parameters K, L_k, ς_{ni}, ω_{ni}, ς_{dj}, ω_{dj}, T_l and τ are not known exactly but vary within intervals as follows

$$K \in [\underline{K}, \overline{K}], L_k \in [\underline{L_k}, \overline{L_k}], \varsigma_{ni} \in [\underline{\varsigma_{ni}}, \overline{\varsigma_{ni}}],$$

$$\omega_{ni} \in [\underline{\omega_{ni}}, \overline{\omega_{ni}}], \varsigma_{dj} \in [\underline{\varsigma_{dj}}, \overline{\varsigma_{dj}}], \omega_{dj} \in [\underline{\omega_{dj}}, \overline{\omega_{dj}}]$$

$$T_l \in [\underline{T_l}, \overline{T_l}] \text{ and } \tau \in [\underline{\tau}, \overline{\tau}] \quad (4)$$

Although frequency response computations for uncertain systems such as the Nyquist, Nichols and Bode envelopes have been extensively studied in the literature [2-5] and references therein, the computation of the Bode envelopes of the uncertain transfer functions of the form of Eq. (3), which is the form of most real process systems and leads to simpler results than using the more general formulations usually considered, has been recently proposed in [6]. It is of course important to point out that computation of the exact boundary of the Nyquist and Nichols envelopes of an uncertain transfer function of the form of Eq. (3) is a difficult and challenging problem. However, for the computation of the Bode envelope the scenario is different since for construction of the Bode envelope one needs to find the magnitude and phase extremums at each frequency.

Although detailed mathematics can be used to get the required results it is much simpler to understand from basic knowledge of Bode diagrams. First, it is obvious that the maximum gain and maximum phase lag at a particular frequency will be obtained from the product of the maximum gains and the sum of the maximum phase lags at that frequency of the individual elements, with a similar result for the minimum values. For the time delay, since its gain is always unity, the maximum (minimum) phase lag is obtained with the maximum (minimum) value of τ. Sketching Bode gain and phase diagrams for a single time constant transfer function it is immediately obvious that the curves for $\underline{T}$ give the maximum gain and minimum phase shift and those for $\overline{T}$ give the minimum gain and maximum phase shift, respectively. Curves for all other values of T lie within these boundaries. It is well known that the Bode gain diagram for the second order complex pole transfer function

$$G(s) = \frac{\omega_0^2}{s^2 + 2\varsigma\omega_0 s + \omega_0^2} \quad (5)$$

only has a peak in the response if ς is less than 0.707. Thus, if the Bode gain and phase diagrams are considered for this transfer function with $\varsigma \in [\underline{\varsigma}, \overline{\varsigma}]$ and $\omega_0 \in [\underline{\omega_0}, \overline{\omega_0}]$, and are drawn for the four cases of $\overline{\omega_0}$ with $\underline{\varsigma}$ and $\overline{\varsigma}$ and $\underline{\omega_0}$ with $\underline{\varsigma}$ and $\overline{\varsigma}$ it can easily be seen that:

1. The minimum magnitude for all ω if $\overline{\varsigma} > 0.707$ is

$$|G(j\omega)| = \left| \frac{\omega_0^2}{-\omega^2 + 2\overline{\varsigma}\,\underline{\omega_0}\,j\omega + \underline{\omega_0^2}} \right| \quad (6)$$

and if $\overline{\varsigma} < 0.707$ then from $\omega = 0$ to $\omega_x = \sqrt{(2\overline{\omega_0^2}\,\underline{\omega_0^2}(1 - 2\overline{\varsigma}^2))/(\underline{\omega_0^2} + \overline{\omega_0^2})}$

$$|G(j\omega)| = \left| \frac{\omega_0^2}{-\omega^2 + 2\overline{\varsigma}\,\overline{\omega_0}\,j\omega + \overline{\omega_0^2}} \right| \quad (7)$$

and from ω_x to ∞ is

$$|G(j\omega)| = \left| \frac{\omega_0^2}{-\omega^2 + 2\overline{\varsigma}\,\underline{\omega_0}\,j\omega + \underline{\omega_0^2}} \right| \quad (8)$$

2. The maximum gain can be computed as follows: If $\underline{\varsigma} > 0.707$ then

$$|G(j\omega)| = \left| \frac{\omega_0^2}{-\omega^2 + 2\underline{\varsigma}\,\overline{\omega_0}\,j\omega + \overline{\omega_0^2}} \right| \quad (9)$$

and if $\underline{\varsigma} < 0.707$ then from $\omega = 0$ to $\omega_{p\min} = \underline{\omega_0}\sqrt{1 - 2\underline{\varsigma}^2}$ it is given by

$$|G(j\omega)| = \left| \frac{\omega_0^2}{-\omega^2 + 2\underline{\varsigma}\,\omega_0\,j\omega + \omega_0^2} \right| \quad (10)$$

The maximum value of the gain at $\omega_{p\min}$ is

$$\frac{1}{2\underline{\varsigma}\sqrt{1 - 2\underline{\varsigma}^2}} \quad (11)$$

and the maximum possible gain remains constant at this value until $\omega_{p\max} = \overline{\omega_0}\sqrt{1 - 2\underline{\varsigma}^2}$ and then for $\omega \in [\omega_{p\max}, \infty)$ it is given by

$$|G(j\omega)| = \left| \frac{\overline{\omega_0^2}}{-\omega^2 + 2\underline{\varsigma}\,\overline{\omega_0}\,j\omega + \overline{\omega_0^2}} \right| \quad (12)$$

3. The maximum phase for $\omega \in [0, \underline{\omega_0})$ is given by

$$\arg[G(j\omega)] = \arg[\frac{\omega_0^2}{-\omega^2 + 2\underline{\varsigma}\,\underline{\omega_0}\,j\omega + \underline{\omega_0^2}}] \quad (13)$$

and for $\omega \in [\underline{\omega_0}, \infty)$ by

$$\arg[G(j\omega)] = \arg[\frac{\omega_0^2}{-\omega^2 + 2\underline{\varsigma}\,\overline{\omega_0}\,j\omega + \overline{\omega_0^2}}] \quad (14)$$

4. The minimum phase for $\omega \in [0, \overline{\omega_0})$ is given by

$$\arg[G(j\omega)] = \arg[\frac{\omega_0^2}{-\omega^2 + 2\overline{\varsigma}\,\overline{\omega_0}\,j\omega + \overline{\omega_0^2}}] \quad (15)$$

and for $\omega \in [\overline{\omega_0}, \infty)$ by

$$\arg[G(j\omega)] = \arg[\frac{\overline{\omega_0^2}}{-\omega^2 + 2\overline{\varsigma}\,\underline{\omega_0}\,j\omega + \underline{\omega_0^2}}] \quad (16)$$

For example, for $\omega_0 \in [0.8, 1.4]$ and $\varsigma \in [0.2, 0.6]$, the Bode envelopes of 100 transfer functions using the gridding technique(solid line) and the method presented(+ line) are shown in Fig. 2. From Fig.2 it can be seen that the method gives exact results.

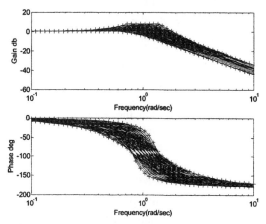

Figure 2: Bode envelopes using the method presented(+ line) and 100 transfer functions using the gridding technique(solid line)

3. STABILITY ANALYSIS AND DESIGN OF CONTROLLERS

A control system with an uncertain transfer function of the form of Eq. (3) has a characteristic equation with a polynomial uncertainty structure and thus much of the mathematical theory developed has failed to address the real problem. In the previous section, it has been shown that the Bode envelopes for such systems can be easily constructed, thus providing a very good tool for analysing the stability problem for these types of systems. The stability of a control system can be determined using the Bode plots by calculating the phase at 0 db and the magnitude at -180. Based on the values at these points, the following conclusions, provided the plant transfer function is open loop stable, can be made for the stability:

- A system is stable if the magnitude is negative decibels when the phase is -180.
- A system is said to be marginally stable if the magnitude is 0 db when the phase angle is -180.
- A system is unstable if the magnitude is positive decibels when the phase angle is equal to -180.

Thus, once the Bode envelope of an uncertain process has been constructed, it can then be used to check the robust stability and to design robust controllers by classical methods as shown in the following examples:

Example 1: In this example the application of the proposed method for determining the maximum possible gain of the transfer function

$$G(s) = \frac{K\omega_0^2}{(Ts+1)(s^2 + 2\varsigma\omega_0 s + \omega_0^2)} \quad (17)$$

to maintain stability for $\omega_0 \in [0.75, 0.95]$, $\varsigma \in [0.64, 0.86]$ and $T \in [0.25, 0.45]$ is illustrated. One can use the Routh stability criterion and get $K < ((2\varsigma\omega_0 T + 1)(2\varsigma\omega_0 + \omega_0^2 T)/\omega_0^2 T) - 1$ which gives $K < 2.3$ after over and under bounding the parameters. However, using the Bode envelopes it has been computed that the approach of this paper gives a maximum value of K of 3.3 which is significantly greater than 2.3 although it is still conservative. One can, of course, grid the parameters and using 11 values within each interval of T, ς and ω_0 1331 Nyquist plots have been drawn for $K = 1$ and the maximum allowable gain for stability has been calculated to be 5.26.

Example 2:

Let the plant transfer function be

$$G(s) = \frac{K}{s(T_1 s + 1)(T_2 s + 1)(T_3 s + 1)} \quad (18)$$

where $K \in [2.5, 2.8]$, $T_1 \in [0.25, 0.35]$, $T_2 \in [0.15, 0.2]$ and $T_3 \in [0.01, 0.05]$. The objective is to design a lead controller of the form

$$C(s) = \frac{Ts + 1}{\alpha Ts + 1}, 0 < \alpha < 1 \quad (19)$$

for which the compensated overall system has a phase margin of 40. The Bode envelope of the uncompensated system is shown in Figure 3 where the minimum phase margin is equal to 22. In order to obtain the required phase margin then

$$40 = 22 + \phi_m - 15 \quad (20)$$

where 15 is the additional phase estimate. Eq. (20) gives $\phi_m = 33$ and α is found from $\alpha = (1 - \sin\phi_m)/(1 + \sin\phi_m)$ giving $\alpha = 0.3$. The 'centre frequency' of the phase lead, ω_m, is then found from $|G(j\omega_m)| = -10\log_{10}(1/\alpha) = -5.23db$ which gives $\omega_m \approx 3.55 rad/\sec$. Since $\omega_m = 1/T\sqrt{\alpha}$, substituting for ω_m and α gives $T = 0.51$ and the lead controller is

$$C(s) = \frac{0.51s + 1}{0.15s + 1} \quad (21)$$

From Figure 3 it was calculated that the minimum phase margin of the compensated system is about 36. Gridding the parameters, 1296 Nyquist plots have been drawn and the minimum phase margin has been computed as 41.

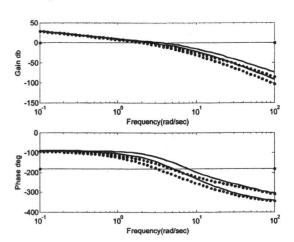

Figure 3: Bode envelopes (... uncompensated, — compensated)

Example 3: Consider

$$G(s) = \frac{K(L_1 s + 1)}{(T_1 s + 1)(T_2 s + 1)(T_3 s + 1)^2} \quad (22)$$

where $K \in [10, 12]$, $L_1 \in [0.7, 1.1]$, $T_1 \in [2, 2.8]$ and $T_2 \in [0.4, 0.65]$ and $T_3 \in [0.2, 0.3]$. The aim is to design a PI controller of the form

$$C(s) = \frac{K_p s + K_i}{s} \quad (23)$$

which guarantees that the entire family has a phase margin of at least 45. From Figure 4 it was found that the new maximum gain crossover frequency of the family should be moved to $\omega_1 = 1.66 rad/sec$ where $180° + \min \arg[G(j\omega_1)] \approx 50°$. The maximum magnitude of $G(s)$ at $s = j\omega_1$ is 14 db. Thus, from $-20\log_{10} K_p = \max |G(j\omega_1)| = 14db$, $K_p = 0.2$. One can choose the corner frequency K_i / K_p to be one decade below ω_1 in order to ensure that the phase lag of the PI compensator only affects the phase of the compensated system at ω_1 by approximately $5°$. Therefore, from $K_i = (\omega_1 / 10)K_p = (2.21/10)0.18 = 0.033$. Thus, the designed PI controller is

$$C(s) = \frac{0.2s + 0.033}{s} \tag{24}$$

From Figure 4 it was computed that the minimum phase margin of the compensated system is greater than $45°$. Gridding the parameters, 15625 Nyquist plots have been obtained and it has been calculated that the minimum phase margin is $71°$.

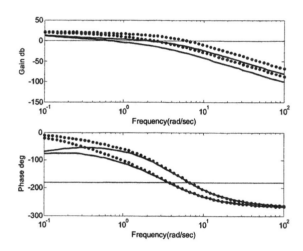

Figure 4: Bode envelopes (... uncompensated, — compensated)

Example 4: Consider a plant with

$$G(s) = \frac{K}{(T_1 s + 1)(T_2 s + 1)} e^{-\tau s} \tag{25}$$

where $K \in [1, 1.5]$, $T_1 \in [0.5, 0.8]$, $T_2 \in [1.4, 2]$ and $\tau \in [0.2, 0.5]$. The objective is to find the parameters of a PID controller of the form

$$C(s) = K_p (1 + \frac{1}{sT_i} + sT_d) \tag{26}$$

using the critical point information, critical frequency ω_c and critical gain K_c, only. There are several approaches to designing(tuning) a controller given ω_c and K_c for a process. One of the simplest approaches, although not recommended if the set

point response has to have small overshoot, is the Ziegler-Nichols approach which takes for a PID controller of the form of Eq. (26), $K_p = 0.6K_c$, $T_i = 0.5T_c$ and $T_i = 4T_d$ where $T_c = 2\pi / \omega_c$.

From the uncompensated Bode envelope shown in Figure 5, it can be seen that the frequency where the minimum gain margin is achieved is equal to 1.8 rad/sec and at this frequency the minimum gain margin of the family is equal to 7.46 db. From, $20\log_{10} K_c = 7.46db$, one gets $K_c = 2.36$. Thus, using the Z-N approach, the following PID controller is designed

$$C(s) = \frac{1.1s^2 + 2.47s + 1.42}{1.74s} \tag{27}$$

From the Bode envelopes of Figure 5 it was found that the phase margin of the compensated system is greater than $45°$ and the gain margin is greater than 6.85 db.

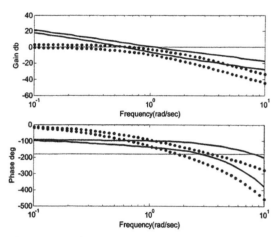

Figure 5: Bode envelopes (... uncompensated, — compensated)

4. AN INTERACTIVE TOOL FOR ROBUST ANALYSIS

Changes in computer technology now allow us to use new and high-quality educational techniques such as interactive tools, virtual and remote laboratories to make use of the the World Wide Web etc. [7]. For classical control and robust analysis and design of control systems with uncertain parameters, interactive tools are extremely useful. They enable students to explore changes in system performance as parameters are varied, and to do so looking at several diagrams simultaneously. This can be done without any programming by just using a mouse to adjust any parameters and the effects can be immediately seen. *SysQuake* provides an excellent environment for such studies. Figure 6 shows implementation of the method presented in this paper in an interactive environment for the analysis of uncertain transfer functions in factored form. When a user manipulates the parameters in the graphical window of Figure 6 the new results are automatically produced clearly showing the effects of the parameter variations.

5. CONCLUSION

The objective of this paper has been to draw attention to a recently developed method for computation of the Bode envelopes of uncertain transfer functions expressed in factored form, which is the form of most physical systems, and to show how the method can be introduced into a first course on classical control. This is the case since the results are easily derived from classical control frequency response concepts, and coverage allows undergraduates to think more practically in terms of parameter uncertainty. An interactive tool has also been introduced which allows one to change the parameters by using the mouse and see the affect on various characteristics.

6.REFERENCES

[1] Kharitonov, V. L., "Asymptotic stability of an equilibrium position of a family of systems of linear differential equations." *Differential Equations*, vol. 14, pp. 1483-1485, 1979.

[2] Barmish, B. R., *New Tools for Robustness of Linear Systems*. MacMillan, NY, 1994.

[3] Djaferis, T. E., *Robust Control Design: A Polynomial Approach*. Kluwer Academic Publishers, Boston, 1995.

[4] Ackermann, J., *Robust Control: Systems with Uncertain Physical Parameters*. Springer-Verlag, 1993.

[5] Bhattacharyya, S. P., H. Chapellat and L. H. Keel, *Robust Control: The Parametric Approach*. Prentice Hall, 1995.

[6] Atherton, D. P and N. Tan, "Design of robust controllers for uncertain transfer functions in factored form," *15th IFAC World Congress on Automatic Control, Barcelona, Spain 2002*.

[7] Dormido, S. "Control learning: present and future," *15th IFAC World Congress on Automatic Control, Barcelona, Spain 2002*.

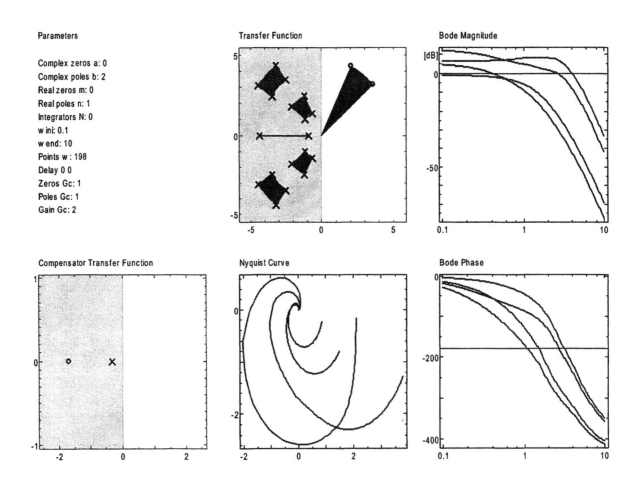

Figure 6: An interactive tool for robust analysis of uncertain transfer function in factored form

MIMO SYSTEMS IN EDUCATION – THEORY AND REAL TIME CONTROL OF LABORATORY MODELS

Marek Kubalčík, Vladimír Bobál, Petr Navrátil

Department of Control Theory, Institute of Information Technologies
Tomas Bata University
Nám. TGM 275, 762 72 Zlín, Czech Republic
Phone, Fax: +420 67 754 3103, E – mail:kubalcik@ft.utb.cz

Abstract: This paper presents an approach to education in the field of MIMO (multi input – multi output) systems practised at our department. The approach is based on putting of theoretical knowledge together with practical laboratory experience. The theoretical education is on a basis of the polynomial theory. Students use various structures of control loops in continuous, discrete and delta representations. They verify their designed algorithms by controlling of laboratory models. Laboratories of the department are equipped with two MIMO models – the coupled drives apparatus and the through – flow air heater. *Copyright © 2003 IFAC*

Keywords: MIMO system, controller synthesis, decoupling, adaptive control, laboratory education, real – time control

1. INTRODUCTION

Design of MIMO controllers represents an interesting branch of control theory and applications. Many technological processes require that several variables relating to one system are controlled simultaneously. Each input may influence all system outputs. The design of a controller able to cope with such a system must be quite sophisticated. There are many different methods of controlling multivariable systems. Several of these use decentralised PID controllers (Luyben 1986), others apply single input-single-output (SISO) methods extended to cover multiple inputs (Chien *et al.* 1987). Decoupling methods can be used to transform the multivariable system into a series of independent SISO loops (Krishnawamy *et al.* 1991; Tade *et al.* 1986).

One of process control courses at our department contains a part aimed at MIMO systems. It is focused especially in TITO (two input – two output) systems, which are the most common example of MIMO systems. There are a lot of industrial processes, which have character of TITO systems.

Theoretical essentials are combined with practical laboratory work. In the theoretical part students get knowledge of matrix description of MIMO systems, principles of controller design using pole – placement and polynomial theory extended to MIMO systems (Kučera 1980). They learn to work with various structures of closed loop systems using matrix notation, deal with problems of decoupling. They design controllers for chosen models of controlled systems both in deterministic and adaptive versions. For adaptive controllers the recursive least squares method is used. (prepared M – files for identification are used). Continuous, discrete and delta models are used in control algorithms. Students have to create programs and diagrams in Matlab + Simulink to simulate and verify their designed algorithms. Finally they use the algorithms for control of laboratory models. Laboratories of the department are equipped with two TITO models – the coupled drives apparatus and the through – flow air heater. Programs for control of the models are written in M – files using the functions of Real Time Toolbox.

The paper is organised as follows: Section 2 presents examples of models for purpose of control; Section 3 describes particular control structures; Section 4 gives simulation example; Section 5 describes the laboratory models including experimental results; finally, Section 6 concludes the paper.

2. DESCRIPTION OF TITO MODELS

The internal structure of TITO system is shown in Fig. 1

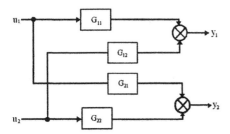

Fig. 1 A two input – two output system – the "P" structure

The transfer matrix of the system is

$$G = \begin{bmatrix} G_{11} & G_{12} \\ G_{21} & G_{22} \end{bmatrix} \qquad (1)$$

2.1 Continuous TITO model

We can assume that the system is described by the matrix fraction (s is a complex variable)

$$G(s) = A^{-1}(s)B(s) = B_1(s)A_1^{-1}(s) \qquad (2)$$

Where polynomial matrices $A \in R_{22}[s]$, $B \in R_{22}[s]$ are the left indivisible decomposition of matrix $G(s)$ and matrices $A_1 \in R_{22}[s]$, $B_1 \in R_{22}[s]$ are the right indivisible decomposition.

The matrices of a continuous model could be

$$A(s) = \begin{bmatrix} s^2 + a_1 s + a_2 & a_3 s + a_4 \\ a_5 s + a_6 & s^2 + a_7 s + a_8 \end{bmatrix} \qquad (3)$$

$$B(s) = \begin{bmatrix} b_1 s + b_2 & b_3 s + b_4 \\ b_5 s + b_6 & b_7 s + b_8 \end{bmatrix}$$

and the differential equations of the model are

$$y_1'' + a_1 y_1' + a_2 y_1 + a_3 y_2' + a_4 y_2 = b_1 u_1' + b_2 u_1 + b_3 u_2' + b_4 u_2$$
$$y_2'' + a_5 y_1' + a_6 y_1 + a_7 y_2' + a_8 y_2 = b_5 u_1' + b_6 u_1 + b_7 u_2' + b_8 u_2$$
$$(4)$$

2.2 Discrete TITO model

For purposes of computer control it is suitable to have a discrete model of the controlled system. If $G(s)$ is the transfer function of a continuous-time dynamic system, then the following expression for the discrete transfer function with the zero - order holder is valid

$$G(z) = \frac{z-1}{z} Z \left\{ L^{-1} \frac{G(s)}{s} \right\} \qquad (5)$$

The simple model structure, easy recursive identification using measurable data, suitability for the synthesis of the discrete control loop as well as for the description and expression of different types of stochastic processes, including disturbance modelling, are all advantages of the z – transform function.

We can describe a discrete MIMO system by the matrix fraction as well as a continuous system

$$G(z^{-1}) = A^{-1}(z^{-1})B(z^{-1}) = B_1(z^{-1})A_1^{-1}(z^{-1}) \qquad (6)$$

Where polynomial matrices A, B, A_1, B_1 are defined analogically to a continuous system

The matrices of a discrete model could be chosen as

$$A(z^{-1}) = \begin{bmatrix} 1 + a_1 z^{-1} + a_2 z^{-2} & a_3 z^{-1} + a_4 z^{-2} \\ a_5 z^{-1} + a_6 z^{-2} & 1 + a_7 z^{-1} + a_8 z^{-2} \end{bmatrix} \qquad (7)$$

$$B(z^{-1}) = \begin{bmatrix} b_1 z^{-1} + b_2 z^{-2} & b_3 z^{-1} + b_4 z^{-2} \\ b_5 z^{-1} + b_6 z^{-2} & b_7 z^{-1} + b_8 z^{-2} \end{bmatrix}$$

and the difference equations of the model are

$$y_1(k) = -a_1 y_1(k-1) - a_2 y_1(k-2) - a_3 y_2(k-1) - a_4 y_2(k-2) + b_1 u_1(k-1) + b_2 u_1(k-2) + b_3 u_2(k-1) + b_4 u_2(k-2) \qquad (8)$$

$$y_2(k) = -a_5 y_1(k-1) - a_6 y_1(k-2) - a_7 y_2(k-1) - a_8 y_2(k-2) + b_5 u_1(k-1) + b_6 u_1(k-2) + b_7 u_2(k-1) + b_8 u_2(k-2)$$

2.2 Delta TITO model

The Z - transformation has some disadvantages when the sampling period decreases:

Z - transformation parameters do not converge as the sampling period decreases to the Laplace – transformation continuous parameters from which they were derived,

- very small sampling periods yield very small numbers at from transfer function numerator,

- the poles of transfer function approach the unstable domain as the sampling period decreases.

The disadvantages of the discrete models can be avoided by introducing a more suitable discrete model. The δ - model, where operator δ converges with decreased sampling period T_0 to a differential operator s

$$\lim_{T_0 \to 0} \delta = s \qquad (9)$$

is best suited to this purposes.

One of approaches to the design of these new discrete δ - models were published in (Middleton and

Goodwin, 1990). If the new variable γ is introduced then it is possible to prove (Mukhopadhyay, *et al.*, 1992), that equality

$$\gamma = \frac{z-1}{\alpha T_0 z + (1-\alpha)T_0} \qquad (10)$$

holds for interval $0 \le \alpha \le 1$. By substituting α in equation (10) we obtain an infinite number of new δ- models. For controllers designs at our courses the forward δ-model (11) is mainly used.

for $a = 0$ $\qquad \gamma = \dfrac{z-1}{T_0} \qquad (11)$

The matrices of the discrete model are

$$A(\gamma) = \begin{bmatrix} \gamma^2 + \alpha_1 \gamma + \alpha_2 & \alpha_3 \gamma + \alpha_4 \\ \alpha_5 \gamma + \alpha_6 & \gamma^2 + \alpha_7 \gamma + \alpha_8 \end{bmatrix}$$

$$B(\gamma) = \begin{bmatrix} \beta_1 \gamma + \beta_2 & \beta_3 \gamma + \beta_4 \\ \beta_5 \gamma + \beta_6 & \beta_7 \gamma + \beta_8 \end{bmatrix} \qquad (12)$$

and the differential equations of the model are

$$y_{1\delta}(k) = -\alpha_1 y_{1\delta}(k-1) - \alpha_2 y_{1\delta}(k-2) -$$
$$- \alpha_3 y_{2\delta}(k-1) - \alpha_4 y_{2\delta}(k-2) + \beta_1 u_{1\delta}(k-1) +$$
$$+ \beta_2 u_{1\delta}(k-2) + \beta_3 u_{2\delta}(k-1) + \beta_4 u_{2\delta}(k-2)$$

$$y_{2\delta}(k) = -\alpha_5 y_{1\delta}(k-1) - \alpha_6 y_{1\delta}(k-2) -$$
$$- \alpha_7 y_{2\delta}(k-1) - \alpha_8 y_{2\delta}(k-2) + \beta_5 u_{1\delta}(k-1) + \qquad (13)$$
$$+ \beta_6 u_{1\delta}(k-2) + \beta_7 u_{2\delta}(k-1) + \beta_8 u_{2\delta}(k-2)$$

3. CONTROL STRUCTURES

3.1 1DOF control structure

The 1DOF control structure is depicted in Fig. 2. It contains only a feedback controller.

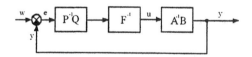

Fig. 2 Block diagram of the 1DOF control structure

Following expressions are applicable for all representations. The operators s, z^{-1} and γ will be omitted from the expressions for the sake of simplification. In the same way as the controlled system, the transfer matrix of the controller takes the form of matrix fraction

$$G_R = P^{-1}Q = Q_1 P_1^{-1} \qquad (14)$$

Generally, the vector of input reference signals W is given by

$$W = F^{-1}h \qquad (15)$$

The reference signals are mostly considered from a class of step functions.

The compensator F^1 is a component formally separated from the controller. It has to be placed ahead of the system to fulfil the requirement on the asymptotic tracking. If the reference signal is from a class of step functions, F^1 is an integrator.

The control law apparent in the block diagram has the form

$$U = F^{-1}Q_1 P_1^{-1} E \qquad (16)$$

It is possible to derive the following equation for the system output

$$Y = P_1 (AFP_1 + BQ_1)^{-1} BQ_1 P_1^{-1} W \qquad (17)$$

The closed loop system is stable when the following diophantine equation is satisfied

$$AF P_1 + BQ_1 = M \qquad (18)$$

where $M \in R_{22}$ is a stable diagonal polynomial matrix with the same diagonal elements.

The roots of this polynomial matrix are the ruling factor in the behaviour of the closed loop system.

The degree of the controller matrices' polynomials depends on the internal properness of the closed loop. The structure of matrices P_1 and Q_1 is chosen so that the number of unknown controller parameters equals the number of algebraic equations resulting from the solution of the diophantine equation using the uncertain coefficients method. The solution of the diophantine equation results in a set of algebraic equations with unknown controllers' parameters.

3.2 2DOF control structure

The 2DOF control structure contains also a feedforward part.

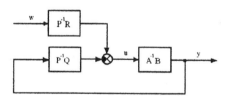

Fig. 3. Block diagram of the 2DOF control system

The control law can be described by matrix equation

$$U = P^{-1}RW - P^{-1}QY \qquad (19)$$

The block diagram leads to an equation for the system output, which takes the form

$$Y = B_1 (PA_1 + QB_1)^{-1} RW \qquad (20)$$

To achieve stability in the closed loop system the following diophantine equation must be fulfilled

$$PA_1 + QB_1 = M \qquad (21)$$

Vector of the reference signals we can assume as (15). The block diagram leads to an equation for the control error. (this equation is valid only in case of the same diagonal elements of the matrix M)

$$E = M^{-1}(M - B_1 R)^{-1} F^{-1} h \qquad (22)$$

Requirement on the asymptotic tracking is fulfilled if there exists a polynomial matrix T so that following equation is fulfilled

$$M - B_1 R = TF \qquad (23)$$

Polynomial matrices T a R are solutions of the second diophantine equation

$$TF + B_1 R = M \qquad (24)$$

The coefficients of the polynomial matrices of the left matrix fraction are given by solving matrix equation

$$BA_1 - AB_1 = 0 \qquad (25)$$

The structure of the polynomial matrices P, Q, R and T is chosen according to the same rules that are presented in the section *3.1*. The matrices P, Q and R are matrices of the controller. The matrix T resulting from the solution of the diophantine equation (24) is not useful.

3.3 Decoupling control using compensators

One possibility to control multivariable systems is the serial insertion of a compensator ahead of the system (Krishnawamy *et al.* 1991, Peng 1990, Tade *et al.* 1986). The aim here is to suppress of undesirable interactions between the input and output variables so that each input affects only one controlled variable.

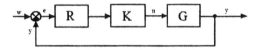

Fig. 4 Closed loop system with compensator

The resulting transfer function H is then given by

$$H = KG \qquad (26)$$

The decoupling conditions are fulfilled when matrix H is diagonal.

To the courses were incorporated controllers with two compensators. These will be referred to as C_1 and C_2.

Compensator C_1 is the inversion of the controlled system. Matrix H is, therefore, a unit matrix.

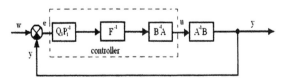

Fig. 5 Closed loop system with compensator C_1

This controller is unsuitable for non – minimum phase systems.

Compensator C_2 is adjugated matrix B. When C_2 was included in the design of the closed loop the model could be simplified by considering matrix A as diagonal. The multiplication of matrix B and adjugated matrix B results in diagonal matrix H. The determinants of matrix B represent the diagonal elements. When matrix A is nondiagonal, its inverted form must be placed ahead of the system in order to obtain diagonal matrix H, otherwise it may increase the order of the controller and sophistication of the closed loop system. Although designed for a diagonal matrix, compensator C_2 also improves the control process for non – diagonal matrix A in the controlled system.

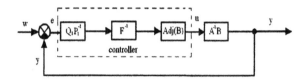

Fig. 6 Closed loop system with compensator C_2

The principles of controllers' synthesis for both cases are given in (Kubalčík *et al.* 2002).

4. SIMULATION EXAMPLE

The control of the model below is given here as an example.

$$A(z^{-1}) = \begin{bmatrix} 1 + 0,3z^{-1} + 0,1z^{-2} & 0,1z^{-1} + 0,2z^{-2} \\ 0,1z^{-1} + 0,3z^{-2} & 1 + 0,3z^{-1} + 0,1z^{-2} \end{bmatrix}$$

$$B(z^{-1}) = \begin{bmatrix} 0,1z^{-1} + 0,4z^{-2} & 0,9z^{-1} + 0,4z^{-2} \\ 0,6z^{-1} + 0,2z^{-2} & 0,3z^{-1} + 0,4z^{-2} \end{bmatrix} \qquad (27)$$

A typical Simulink scheme of the 1DOF control structure is shown in Fig. 7.

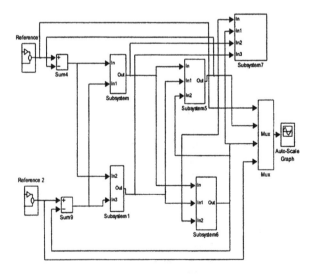

Fig. 7 Basic Simulink scheme of the 1DOF control structure

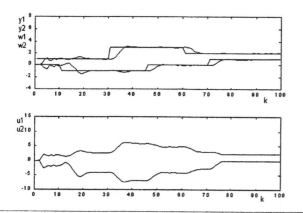

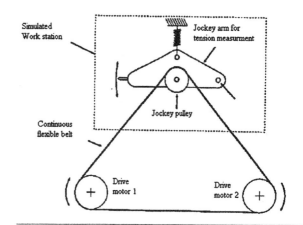

Fig. 8 Adaptive control with compensator C_2 – discrete representation

Fig. 9 Principal scheme of CE 108

5. LABORATORY MODELS

Laboratories of the department are equipped with two TITO models – the coupled drives apparatus and the through – flow air heater.

5.1 Coupled drives apparatus

This apparatus, based on experience with authentic industrial control applications, was developed in cooperation with the University of Manchester and made by a British company, TecQuipment Ltd. It allows us to investigate the ever-present difficulty of controlling the tension and speed of material in a continuous process. The process may require the material speed and tension to be controlled to within defined limits. Examples of this occur in the paper-making industry, strip metal and wire manufacture and, indeed, any process where the product is manufactured in a continuous strip.

The industrial type material strip is replaced by a continuous flexible belt. The principle scheme of the model is shown in Fig. 9. It consists of three pulleys, mounted on a vertical panel so that they form a triangle resting on its base. The two base pulleys are directly mounted on the shafts of two nominally identical servo motors and the apparatus is controlled by manipulating the drive torques to these servo motors. The third pulley, the jockey, is free to rotate and is mounted on a pivoted arm. The jockey pulley assembly, which simulates a material work station, is equipped with a special sensor and tension measuring equipment. It is the jockey pulley speed and tension which form the principle system outputs. The belt tension is measured indirectly by monitoring the angular deflection of the pivoted tension arm to which the jockey pulley is attached.

The time responses of the control example are shown in Fig. 10 and Fig. 11. The manipulated variables are the inputs to the servo - motors and the controlled variable y_1 is the tension and the controlled variable y_2 is the speed at the work station.

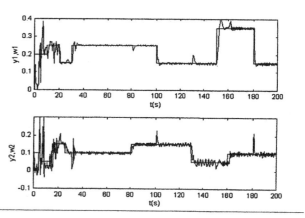

Fig. 10 Adaptive control with 1DOF controller – discrete representation

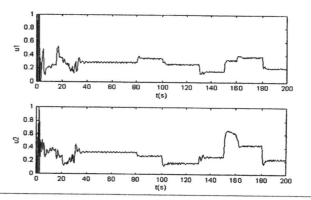

Fig. 11 Controllers' output

5.2 Through – flow air heater

This laboratory equipment is also a two input – two output system. It is shown in Fig. 12. Manipulated variables are the heat source (electric resistance heating 2) and the air flow source (ventilator 1). Controlled variables are the air temperature, measured by resistance thermometer 4 and air flow, measured by flow speed indicator (position 7). The air flow can be changed with the throttle valve (position 5).

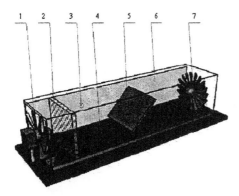

Fig. 12 Laboratory through – flow air heater
1 – ventilator, 2 – electric resistance heating , 3 – pressure sensor, 4 – resistance thermometer, 5 – throttle valve, 6 – cover of tunnel, 7 – flow indicator

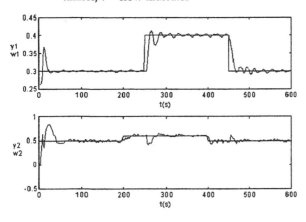

Fig. 13 Adaptive control using 2DOF controller – delta representation

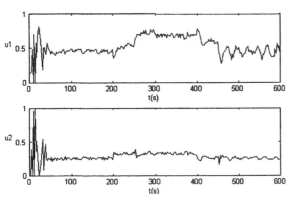

Fig. 14 Controllers' output

The time responses of the control example are shown in Fig. 13 and Fig. 14. The controlled variable y_1 is the air temperature and the controlled variable y_2 is the air - flow. The manipulated variable u_1 is the heat source and the manipulated variable u_2 is the air - flow source.

6. CONCLUSIONS

Education of process control belongs to the most difficult fields of study. Theoretical knowledge together with practical experience in process control is the basement of the future success of graduates in practice. This paper presents an approach to education in the field of MIMO systems practised at our department based on putting of theoretical knowledge together with practical laboratory experience.

ACKNOWLEDGMENTS

The work has been supported by the Grant Agency of the Czech Republic under grants No. 102/02/0204, 102/03/0070 and 102/01/P013 and by the Ministry of Education of the Czech Republic under grant MSM 281100001. This support is very gratefully acknowledged.

REFERENCES

Bobál V., Navrátil P. and Dostál P. (2002). Adaptive Control of a Two Input – Two Output System Using Delta Model. *Proceedings of the 10th Mediterranean Conference on Control and Automation*, Lisabon.

Chien, I.L., Seborg, D.E., Mellichamp, D. A. (1987). Self-Tuning Control with Decoupling. *AIChE J.*, 33, 7, 1079 – 1088.

Krishnawamy, P.R. et al. (1991). Reference System Decoupling for Multivariable Control. *Ind. Eng. Chem. Res.*, 30, 662-670.

Kubalčík, M., Bobál V. and Maca M. (2002). Polynomial Design of Controllers for Two – Variable Systems – Practical Implementation. *Preprints of the 15th Triennial World Congress of IFAC*, Barcelona.

Kučera V. (1980). Stochastic Multivariable Control: a Polynomial approach. *IEEE Trans. of Automatic Control* , 5, 913 – 919.

Luyben W. L. (1986). Simple Method for Tuning SISO Controllers in Multivariable Systems. *Ing. Eng. Chem. Process Des. Dev.*, 25, 654 – 660.

Middleton, R.H. and G.C. Goodwin (1990). *Digital Control and Estimation - A Unified Approach*, Prentice Hall, Englewood Cliffs: N. J.

Mukhopadhyay, S., A. Patra and G.P. Rao (1992). New Class of Discrete-time Models for Continuos-time Systems. *International Journal of Control*, 55, pp. 1161-1187.

Peng, Y. (1990). A General Decoupling Precompensator for Linear Multivariable Systems with Application to Adaptive Control. *IEEE Trans. Aut. Control*, AC-35, 3, 344-348.

Tade, M.O., Bayoumi, M.M., Bacon, D.W. (1986). Adaptive Decoupling of a Class of Multivariable Dynamic Systems Using Output Feedback. *IEE Proc. Pt.D*, 133, 6, 265-275.

ELSEVIER

IFAC

PUBLICATIONS
www.elsevier.com/locate/ifac

MULTIVARIABLE PI-CONTROL OF A LAMP SYSTEM

Joonas Varso and Heikki N. Koivo

Helsinki University of Technology
Control Engineering Laboratory
PL 5400 FI-02015 TKK
joonas.varso@hut.fi

Abstract: The purpose of the paper is two-fold: to present a simple, multivariable, dynamical system and to use it to demonstrate the benefits of a multivariable PI controller compared with the scalar PI controller. The two-lamp system presented in the paper is multivariable and nonlinear containing strong interactions. Multivariable PI controllers can be tuned almost as easily as scalar PI controllers, but provide significant advantages compared with scalar controllers. Demanding state-space or frequency domain design of controllers with matrix algebra are avoided, when a multivariable PI controller with familiar scalar PI subcontrollers is used. *Copyright© 2003 IFAC.*

Keywords: multivariable system, PI control; relay tuning; interaction; computer simulation; real-time control; modeling.

1. INTRODUCTION

Many of the reported laboratory exercises in the literature use state-space multivariable controllers, e.g. (Feeley, *et al.*, 1999) or fairly sophisticated frequency domain techniques, e.g. (Kocijan, *et al.*, 1997). In this paper, the emphasis is on tuning of a multivariable PI controller (e.g., Menani and Koivo, 2001; Penttinen and Koivo, 1980). Since state-space controllers are practically unknown in process industry and also in other fields, it is important to demonstrate that multivariable controllers with a simple structure, like multivariable PI controllers, can be tuned almost as easily as scalar PI controllers, but can provide significant advantages compared with scalar controllers. Demanding state-space or frequency domain design of controllers with matrix algebra, their difficult-to-understand structure and tuning are avoided, when a multivariable PI controller with familiar PI subcontrollers is used. Tuning also resembles scalar PID controller tuning.

The purpose of the proposed laboratory exercise is to familiarize the student with the behavior of multivariable, dynamical systems. The lamp system presented in the paper is also nonlinear containing quite strong interactions, and an external disturbance signal. Once the student knows the system better, a control system is designed and tuned.

The advantages of multivariable PI controller compared with scalar control are demonstrated. The lamp system is easy to understand, cheap to construct and demonstrates many properties of multivariable dynamical systems.

In Section 2 the multivariable lamp system is discussed in detail. Section 3 describes the controllers, both scalar and multivariable PI-controllers, and their tuning. Section 4 explains related computer simulations for tuning the controllers and Section 5 controller tuning for the real system. Graphical user interface of the system is briefly discussed in Section 6. The results of the tests are reviewed in Section 7. Conclusions are offered in Section 8.

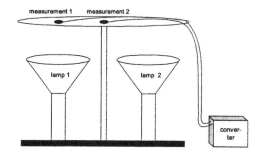

Figure 1. Schema of the lamp system.

2. MULTIVARIABLE HEATING SYSTEM

2.1. System description

The system studied is a simple two input-two output heating process (Figure 1). It consists of two lamps and a metal sheet, on which two temperature sensors are installed. The sensors are placed above the lamps asymmetrically on a thin metal sheet such that one of the sensors is directly above one of the lamps and the other is between the lamps. The metal sheet conducts heat well and the lamps emit heat to a wide area. This means strong interactions. The lamps are ordinary 250W bulbs. The temperature sensors are common Pt-100 sensors providing a resistance measurement, which is proportional to the temperature.

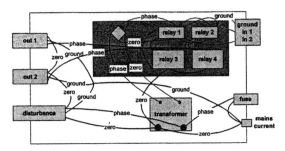

Figure 2. Physical description of the relay box

2.2. Data acquisition and implementations of control signal

The sensor resistance values are linearized and converted to a voltage signal using EM-M21A converter. This signal is sent to the computer's AD/DA converter (Analog Device RTI-815). The analog output control signals are produced using the same converter. The control signal is an on-off signal, either 0V or 10V. A relay box has been built to control the lamps shown in Figure 2.

The system uses two pairs of relays. The smaller relays (1 and 2) receive the on-off (0V or 10V) control signal from the computer. These relays control a 32V DC control signal going to the bigger relays that control the mains current. The disturbance lamp is connected straight to the mains. A manual switch to control the disturbance lamp is located on the side of the relay box (above the output connectors, not drawn in the figure).

2.3. Real Time Toolbox and Simulink

MATLAB Real-Time Toolbox is used to control and collect measurements in real-time. Real Time Toolbox is a tool that connects MATLAB and Simulink to the real world via an I/O card, which does not require knowledge on the details of the internal data transfer.

Only the scaling ranges and the channel numbers have to be known (the card has its own and RT-Toolbox its own). The control signal chopping is done in Simulink. If the signal is 50% of its maximum value, the lamp voltage is 220V for 0.5 seconds and 0V for 0.5 seconds.

3. THE TUNING ALGORITHM

The tuning of the multivariable PI-controller is described in (Menani and Koivo 2001). It is based on relay tests. The first step in the method is to find the critical point of every loop in the system (two in this exercise) using the well-known relay feedback test. Every loop is controlled separately using a relay and when the system oscillates steadily, the frequency is measured. The design frequency of the controller is chosen from the set of critical frequencies. The choice is done using the interaction indicator of the multivariable system.

In the lamp system, both critical frequencies are so close to each other that it does not matter which one is used. The gain matrices (3) and (4) are calculated using the critical frequency. Finally, the controller is tuned.

Basic PID controller is of the well-known form:

$$U(s) = K_P\left(1 + \frac{1}{T_I s} + T_D\right)E(s) \qquad (1)$$

In this study, only PI-controller is used because control is implemented with relays and it will cause slight oscillation. Derivative term would amplify the oscillations. In addition, relatively big disturbances (like temperature changes in the surrounding environment) would be difficult to handle.

Transfer function of a multivariable PI-controller is:

$$K_c(s) = K_P + K_I\frac{1}{s}, \qquad (2)$$

where K_P and K_I are the gain matrices of the proportional and integral parts of the controller, respectively. Their dimensions are determined by the number of inputs and outputs.

The tuning matrices are selected as (Menani and Koivo)

$$K_I = \rho_1 \det[G(0)]G^{-1}(0) \qquad (3)$$

$$K_P = \rho_2 \det|G(j\omega_b)|\,\|G^{-1}(j\omega_b)\| \qquad (4)$$

where G is the system transfer matrix, ρ_i, $i = 1, 2$, is a scalar fine tuning parameter, and ω_b is the frequency at which the interaction indicator reaches its minimum.

The following procedure is suggested by Menani and Koivo:

Step 1   Perform relay tests for the individual loops and record the frequency of oscillation and the amplitudes of the outputs.

Step 2   Rank the frequencies from low $?_1$ to high ω_m and define a set of critical frequencies:
$$\Omega = \{\omega_1, \omega_2, ..., \omega_m\}$$

Step 3   At each frequency, $\omega_i \in \Omega$, $i = 1, 2, ..., m$ perform an interaction test to determine the interaction indicator.

Step 4   Choose $\omega_i \in \Omega$ that corresponds to the lowest interaction indicator.

Step 5   Estimate the transfer function matrices at ω_b and at zero.

Step 6   Choose $\rho_1 = \rho_2 = \rho = 0.1$.

Step 7   Set the controller parameters using (3) and (4).

Step 8   If the results are not satisfactory, then change ρ and go back to step 7.

Advantages of the controller in (2) are that it has a multivariable structure and the interactions can be taken into account. It also gives a better performance than decentralized, scalar controllers do. A disadvantage is that there is no clear procedure to change the fine-tuning parameter ρ and stability is not guaranteed. It is also possible to use open-loop tuning (Penttinen and Koivo, 1981), but because of lack of space, only relay tuning is discussed in the rest of the paper.

The choice of tuning parameters is somewhat complicated. In the simulations, it was noticed that starting point $\rho = 0.1$ is too big. In addition, with one parameter it is not possible to find a good performance. Tuning succeeds much better with two parameters but parameters differ largely of each other, almost four decades.

3.2. Modeling

Model structure is assumed to be linear, first order, without delays around an operating point. This is analogous to Ziegler-Nichols type of models for scalar systems. The real process is not linear but, on the contrary, strongly nonlinear.

Original model is measured manually (measurements written down from the monitors of the unit controllers) using step response tests. The step response tests are implemented as follows: Both

controls are first 30% open or control voltage is 0.3s 220V and 0.7s 0V. System is given 500s to stabilize. After this, a step from 30% to 40% is performed to the one of the lamps. The measurement is completed after 500s, so the whole measurement time is 1000s. The operating point is about 70 ^{0}C. The transfer function obtained follows the real process most accurately around the operating point.

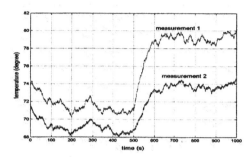

Figure 3.   Step response from the real process

Figure 3 shows a sample of the step tests, which have been performed for both lamps. Altogether seven tests were carried out. Mean values are taken from the curves to minimize the disturbances. After fitting to curves with a first order process, the resulting transfer function matrix becomes:

$$G(s) = \begin{bmatrix} \dfrac{82.7}{56.4s+1} & \dfrac{31.3}{154.9s+1} \\ \dfrac{56.6}{75.5s+1} & \dfrac{52.5}{89.6s+1} \end{bmatrix} \qquad (5)$$

Unit controllers are OMRON E5AX-A scalar controllers for temperature control. These controllers have an automatic tuning mode. Each controller is tuned individually. The control structure is presented in Figure 4.

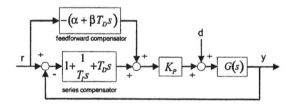

Figure 4.   Block diagram of OMRON E5AX-A temperature controller.

4.   SIMULATIONS

Simulation of the system model (5) and the tuning method was done to get indicative results on how to tune the controllers for the real process. Because the process is assumed to be a first order process with no delays, delays are added in the simulation in order to make the simulated system to act more like the real system. Delays are assumed two seconds long.

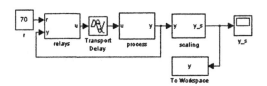

Figure 5. Simulink model for relay testing of the system model

Using relay tests (Figure 5) the critical frequency is determined to be 1.1 rad/s. Gain matrices in (3) and (4) become:

$$K_I = \begin{bmatrix} 52.5 & -31.3 \\ -56.6 & 82.7 \end{bmatrix} \qquad (6)$$

$$K_P = \begin{bmatrix} 0.53 & -0.18 \\ -0.68 & 1.33 \end{bmatrix} \qquad (7)$$

The tuning did not succeed with only one tuning parameter ρ, but both matrices had to have their own parameter $\rho_i, i = 1, 2$. With two different tuning parameters, $\rho_1 = 0.15$ and $\rho_2 = 0.000036$ good performance (Figure 6) is achieved. In the tests, four step changes are performed to reference values in turn for both of the lamps. The direction of the step (upwards or downwards) influences the result, because one can only heat the process, not cool it. Cooling is dependent very much on difference between the temperature of the operation point and the environment. The operation point is chosen so that with the minimum (lamps is off) and the maximum (lamps burn all the time) the control has equal gain. In simulations, controls are bounded corresponding to the real situation.

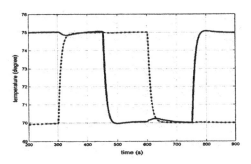

Figure 6. Typical step response of simulation

Steps are performed at 300, 450, 600 and 750 seconds. From the results (Table 1) it is seen that no overshoots exist and interactions are quite small.

Table 1 Simulation results

Step (no.)	1	2	3	4
Overshoot (%)	0	1	0	2
Interaction (%)	4	1	5	1
Rise time, 10-90% (s)	20	18	20	18

5. TUNING OF REAL SYSTEM

Relay test is done for both lamps separately (Figure 7) to determine the critical frequency. Oscillation frequency is measured from the response to be 1.11 rad/s. This is exactly the same as in the simulations. Using this value, the gain matrices of (3) and (4), are computed for the controller. The tuning parameters $\rho_1 = 0.15$ and $\rho_2 = 0.000036$ of the simulations are, however, too small, but their ratio is right. Fine tuning with this ratio (about 4167) gives good results.

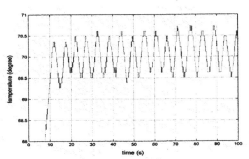

Figure 7. Relay test to lamp 1

6. GRAPHICAL USER INTERFACE

The graphical user interface is implemented using MATLAB's own GUI (Graphical User Interface). The user interface has separate windows for different stages of the laboratory exercise (modeling, finding of critical frequency, testing, old process and results). These windows are used to test the process, plot the result graphs and enter tuning parameters. The purpose of the user interface is to hide the structure of the Simulink models and the m-files used to calculate the results.

7. RESULTS OF TESTS

First, the multivariable PI controller is tested using three different tunings. The results are compared with those of unit (scalar) controllers.

Most of the important performance criteria are collected in Tables. These are overshoot (or under-shoot, depending on the direction of the step), interaction and rise time. Settling time has not been considered, because disturbances are so large. Values of the overshoot and interaction index have been calculated from the maximum deviation. These values can also be caused temporarily by the disturbance signal rather than the step input.

7.1. Tests using different tuning parameters.

Results for three different tunings are next discussed.

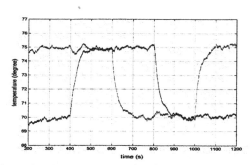

Figure 8. Step response using four steps. Tuning parameters are $\rho_1 = 12$ and $\rho_2 = 0.00288$.

Tuning 1: $\rho_1 = 12$ and $\rho_2 = 0.00288$; tuning 2: $\rho_1 = 20$ and $\rho_2 = 0.0048$; tuning 3: $\rho_1 = 41.67$ and $\rho_2 = 0.01$. The ratio of the tuning parameters is the same in all tunings.

Tests have been repeated five times to avoid the disturbance effects. The goal in this experiment is to observe the step changes in the output variables, not to test how the control compensates disturbances. This would be another interesting experiment.

Tuning 1 has been done with parameters $\rho_1 = 12$ and $\rho_2 = 0.00288$. Step responses are shown in Figure 8. The performance criteria of the step responses are given in Table 2.

Table 2. Performance, when $\rho_1 = 12$ and $\rho_2 = 0.00288$.

Step (no.)	1	2	3	4
Overshoot (%)	1	6	5	6
Interaction (%)	11	5	9	6
Rise time, 10%-90% (s)	52	52	64	59

Overshoots are relative small because of the small gain. Interactions are small too. It is easy to see that a step signal to lamp two has a larger effect to lamp one than vice versa. Control is bounded but this tuning is so slow that there are no problems. Rise times are relatively large.

Tuning 2 has been done with parameters $\rho_1 = 20$ and $\rho_2 = 0.0048$. Using this little tightening results in the responses in Figure 9. Performance indicators have been collected into Table 3. Overshoots occur now slightly more often. There is not however a significant difference. Interactions increase, especially the effect of lamp two on lamp one increases.

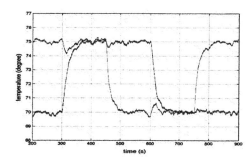

Figure 9. Step response using four steps. Tuning parameters are $\rho_1 = 20$ and $\rho_2 = 0.0048$.

Table 3. Performance when $\rho_1 = 20$ and $\rho_2 = 0.0048$.

Step (no.)	1	2	3	4
Overshoot (%)	4	7	3	4
Interaction (%)	18	3	13	6
Rise time, 10%-90% (s)	39	32	36	34

In tuning 3, parameters have been enlarged a bit more (Figure 10 and Table 4). Overshoots increase a little. Interactions on one are much bigger now. This is a consequence of increasing the gain. At the same time, rise times are very small. The control saturates and maximum control is used. Therefore, the system is not able to respond any faster.

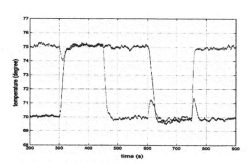

Figure 10. Step response using four steps. Tuning parameters are $\rho_1 = 41.67$ and $\rho_2 = 0.01$.

Table 4. Performance, when $\rho_1 = 41.67$ and $\rho_2 = 0.01$.

Step (no.)	1	2	3	4
Overshoot (%)	5	6	10	5
Interaction (%)	20	5	26	28
Rise time, 10%-90% (s)	15	13	19	7

7.2. The control using unit controllers – scalar case

When process is controlled with unit controllers (scalar controllers), steps are performed manually. Set points are changed from the button of unit controllers. Measurements are observed visually on the monitor. The stepping time cannot be defined exactly. Steps are performed at time steps 200, 400, 600 and 800 seconds. Because of this, rise times are not necessarily exactly right. Measurements are shown in Figure 11 and performance indicators in Table 5. Overshoots are now bigger than in any of the previous tunings. Interactions are quite strong, too. Only the most intensive tunings in the multivariable case have bigger interactions than in the case of unit controllers. From rise times, one can see that loop one is faster than loop two.

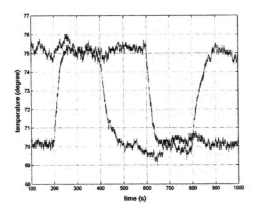

Figure 11. Step response of unit controllers.

Table 5. Performance when control with unit controllers

Step (no.)	1	2	3	4
Overshoot (%)	9	9	2	11
Interaction (%)	21	11	19	16
Rise time, 10%-90% (s)	28	66	35	54

8. CONCLUSIONS

The goal of this research was to introduce a simple lamp system, which has multivariable, dynamical characteristics. The system is easy to construct of cheap components. The other objective was to demonstrate the effectiveness of a multivariable PI controller, which has a simple structure having PI subcontrollers as elements. In 2x2 case, the cross-coupled PI subcontrollers effectively eliminate interactions between the main control loops. Multivariable PI controller is straightforward to tune based on simple (relay and step) experiments. The effectiveness of the multivariable PI controller has been compared with the system using (scalar) unit controllers.

Both simulation studies and experiments with the real system have been carried out. These effectively demonstrate the main issues of the paper. Table 6 summarizes all the measurements. The last column

(on the right-hand side) tells what kind of tuning was applied or if the test was performed using unit controllers (u.c.). The unit controllers have, when compared to multivariable control, big overshoot and a long rise time. Interaction is maybe the most important criteria, which characterizes the control performance. Interactions in scalar controller case are quite big.

When the response of unit controllers is compared to the response of multivariable controllers, one can see that, at least with this tuning algorithm, multivariable control gives better performance.

Table 6. All tests

Step (no.)	1	2	3	4	Tuning
Overshoot, (%)	1	6	5	6	1
	4	7	3	4	2
	5	6	10	5	3
	9	9	2	11	u.c.
Interaction, (%)	11	5	9	6	1
	18	3	13	6	2
	20	5	26	28	3
	21	11	19	16	u.c.
Rise time, 10%-90% (s)	52	52	64	57	1
	39	32	36	34	2
	15	13	19	7	3
	28	66	35	54	u.c.

REFERENCES

Feeley, J.J., Edwards, L.L., and R.W. Smith (1999). Optimal digital control of a laboratory-scale paper machine headbox. *IEEE Tr. on Education* **42**, 337-343.

Kocijan, J., O'Reilly, J., and W.E. Leithead (1997). An integrated undergraduate teaching laboratory approach to multivariable control. *IEEE Tr. on Education* **40**, 266-272.

Menani, S. and H.N. Koivo (2001). A comparative study of recent relay autotuning methods for multivariable systems. *Int. J. of Systems Science* **32**, 443-466.

Penttinen, J. and H.N. Koivo (1980), Multivariable Tuning Regulator for Unknown Systems. *Automatica* **16**, 393-398.

ELSEVIER

IFAC
PUBLICATIONS
www.elsevier.com/locate/ifac

TOOLBOX ENVIRONMENT FOR ANALYSIS AND DESIGN OF MULTIVARIABLE SYSTEMS

Maja Atanasijević-Kunc, Rihard Karba, Borut Zupančič

Faculty of Electrical Engineering, University of Ljubljana,
Tržaška 25, SI-1000 Ljubljana, Slovenia

Abstract: In the paper some ideas of studying multivariable control design are presented and illustrated through the usage of toolbox environment which should enable inclusion of already existing toolboxes, illustration of some interesting aspects explained by teacher and the possibility of developing, testing and presentation of solutions, prepared by students. Good experiences with graphically oriented toolbox environment usage confirm correct orientation of the discussed software. *Copyright © 2003 IFAC*

Keywords: Analysis, Design, Validation, MIMO systems, Education aids

1. INTRODUCTION

One of the lectures the students can attend in the fifth year of university study of Automatics on the Faculty of Electrical Engineering, University of Ljubljana are the lectures on Multivariable systems. They include also auditorial and laboratory exercises.

In the last few years we have gradually introduced some modifications which are now included in lectures and especially exercises and exams. With this new ideas we tried to introduce or only to emphasize the following important goals:

- exercises should be realized as illustration and an extension of the lectures with the possibility of relative simple and user friendly computer aided support as numerical treatment of multivariable systems is always complex;
- student's obligations regarding laboratory work should be to some degree simplified and guided,
- but there must also be the possibility which enables them to realize their own solutions

and to incorporate them into the same program environment;

- motivation aspects should be covered by suitably chosen and defined design problems.

As a result of mentioned goals we have prepared printed materials (Atanasijević-Kunc, 2001) with description of 15 projects and graphically oriented toolbox environment, realized inside program package Matlab (Matlab, 1999).

Each of the mentioned project bases on the realistic model of the multivariable system originating from industry practice or laboratory pilot plants. Student's tasks connected with these projects are to analyse the model and to design controllers using different approaches to meet the prescribed close-loop properties. Finally the evaluation of the solutions should point out the best result regarding all design goals.

Tasks connected with the chosen project are solved by a team of two students and are separated into four groups. By solving the first students should repeat some important aspects of system theory in general and of course a special attention is devoted also to multivariable processes which are introduced through lectures and partly repeated during auditorial exercises. This part of the project is meant to be treated during laboratory exercises and to be solved by everyone. Solving the second and third group enable to pass the written and oral part of exam if the results are successfully described and presented to auditorium consisted of the professor, laboratory staff and colleagues. It is of course also possible to pass written and oral part of exam in the classical way. In spite of the fact that each project is quite complex and somehow rounded up we wanted to show that research and design work can be continued. This ideas are indicated in the fourth group of project tasks.

Design of control systems is always related with a great number of procedures where perhaps the most often used in all design steps are different analytical operations. They are important in the modelling phase, in the phase of choosing the potentially good design algorithms, during design itself and also in the phase of result evaluation, where the solution or even a group of solutions have to be tested and compared. These were the reasons which have motivated us to build a toolbox for analytical purposes (Atanasijević-Kunc and Karba, 2002) with which we tried to fulfill the following ideas: it should remind and guide the student through important system properties in time and frequency domain; it should support open and close - loop situations; it should be used for single-input/single-output (SISO) as well as for multivariable (MIMO) systems; where possible some degree of parallelism should be established in treatment of SISO and MIMO problems; it should also be possible to estimate the degree of matching of design results with more or less exactly described design goals and so help in choosing potentially the best design result.

To enable computer aided environment of mentioned projects in combination with analytical and design operations and demonstration possibilities graphically oriented toolbox environment was designed as is described in the following section.

2. GRAPHICALLY ORIENTED TOOLBOX ENVIRONMENT

Graphically oriented toolbox environment was realized inside program package Matlab and can be started by opening the window as it is shown in Fig. 1 where we can see that each design procedure is meant to be realized in three main steps: problem definition, contoller(s) design and results validation. Pushing the *problem definition* button opens the window as shown in Fig. 2. Here we have the access to problem definitions of all described projects, further possibilities, and the user can also create his own files with corresponding prob-

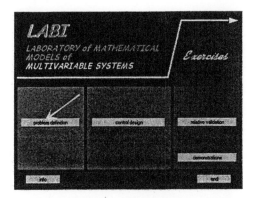

Fig. 1. Starting window of toolbox environment

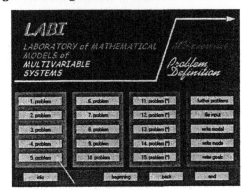

Fig. 2. Problem definition window

lem definition. Temporary information of problem definition can be displayed in Matlab workspace by pushing the buttons *write model, write mode* and *write goals.*

By pushing for example *5. problem* button the window, as shown in Fig. 3 is opened. The user can

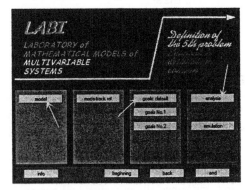

Fig. 3. Definition of the fifth problem

in this case generates linearized state - space description of semibatch distillation column (Matko *et al.*, 1992) in Matlab workspace. Corresponding matrices have in this case the following values: A = [-0.4352 0.4382 0.0172 -0.0194; -0.1229 0.1211 -0.0092 0.0104; -0.1981 0.1931 -0.3431 0.3535; 0.1017 -0.0970 -0.1698 0.1611], B = [0.1259 -0.0974; 0.1182 -0.0802; 0.2923 -0.1168; 0.2198 -0.0932], C = [1 0 0 0; 0 0 1 0], D = [0 0; 0 0]. When this information of model description is defined the user can start for example with system analysis by pushing the corresponding button which enables the call of one group of analitical functions from mentioned toolbox. The organization of available analitical functions in connection with more or less precisely specified design goals are described in the next section.

When the design problem is defined the work can proceed with design itself. Pushing the *control design* button in starting window (Fig. 1) opens the window as it is illustrated in Fig. 4. Here

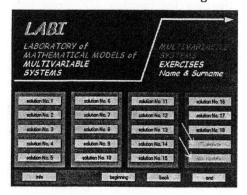

Fig. 4. Control design window

the buttons are prepared to be connected with the files created by the students when solving the tasks from the chosen project. In this way systematic approach of project solving is indicated while in the same time organized results can be used for documentation and presentation purposes. When the controller and close-loop system are defined the buttons *CL analysis* (close - loop analysis) and *absolute validation* become active and again two groups of analitical functions can be used to find out how the specified controller has influenced close - loop properties.

Relative validation (Fig. (1)) functions are also the group of analytical functions. They are meant to be used in the cases where several close - loop solutions (at least two) are available and they should help the designer to choose the best solution regarding specified design goals.

In the starting window also the *demonstrations* button is prepared. It enables the presentation of all information needed in connection with programe environment as well as with the topics of design itself.

It is also important to notice that in all windows bottom buttons enable to end the work with graphical environment, to return to the previous window or to the very beginning.

3. ORGANIZATION OF ANALYTICAL FUNCTIONS

The *Toolbox for dynamic system analysis* consists of a great number of functions which represent an extension and enlargement of the functions available in Matlab, Control System Toolbox (*Control System Toolbox, User's Guide*, 1998)

and Multivariable Frequency Domain Toolbox (Maciejowski, 1990; Maciejowski, 1989) in the way that several functions have been added, they are organized in graphical windows and can be called also only by pushing the button. In all functions explanation and where possible graphical representation of results is added. The amount of additional explanations can be controlled by the so called communication vector.

Mentioned functions can be divided in several different ways. Some of them can be used with SISO systems, some with MIMO and some are suitable for both kind of problems. Regarding different design situations both groups are organized into four levels:

Open-loop analysis functions are using only the information of linear system model. They are grouped into general, time and frequency domain properties as illustrated in Fig. 5. They should

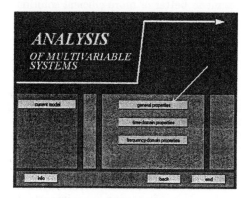

Fig. 5. Open-loop MIMO analysis

give the information of the properties of the process itself and help the user to choose between different design approaches.

Closed-loop analysis functions are organized identically however in this case also the information of used controller is needed.

Absolute validation functions are used for evaluation of matching desired design goals. In comparison with closed-loop analysis here also design goals have to be specified. Suitable and enough general specification of design goals can of course be very difficult especially in earlier design stages where needed information is not available or in situations where the user can define contradictory design goals. To avoid somehow these problems we have introduced the possibilities with which the user can define the importance of each specified design goal. In this way design goals can be specified either very precise or in a very approximate manner where perhaps all stable results are acceptable (default goals in Fig. 3). The result of absolute validation is in the range between 0 and 1. If the result of absolute validation is 0 design solution is unacceptable as one or more design goals are violated more than allowed. If the result is 1 this means that all design goals are completely satisfied. If the result is between 0 and 1 the solution is acceptable, but all design goals are not completely fulfilled. Better are of course solutions which are closer to 1. These results can also be used for some kind of relative validation. But it can occur that several solutions have the same validation result. In this case the next level can be used.

Relative validation functions tend to help the user to compare the efficacy of different design solutions and to prepare him to the real situation where also different kind of non-linearities have to be expected. One which is in practice always presented is for example limitation of control signals. Relative validation functions can simultaneously compare up to five design solutions. The validation result is calculated regarding the first solution. Result greater than 1 therefore means that solution is better than the first. Higher validation values mean better or more efficient solutions.

4. SOME ILLUSTRATIVE RESULTS

By choosing to inspect the so called general properties (Fig. 6) it is possible for example to find out the values of poles and time constants of the discussed system and the values of transmission zeros which can not be observed directly from transfer function matrix but represent the parallelism to SISO zeros. It is also possible to establish if the process can be statically or/and dynamically decoupled. This information is not important only for further design purposes but represents the structural properties of the system. The values

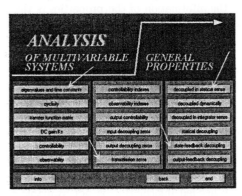

Fig. 6. General properties of MIMO system

of decoupling indexes for example give the information of the degree of the zeros at infinity for each input - output signal pair of MIMO - system. These structural properties can be connected also with the poles and transmission zeros which can lead to root - locus representation of MIMO systems in such way that it becomes very similar to SISO problems. The idea can be realized in connection with tuning procedures for MIMO P and PI controllers, which also represent some kind of parallelism to SISO tuning approaches. When using this type of design the controller of the following form is proposed:

$$G_{PI} = \gamma\, G_1 + \delta\, \frac{1}{s}\, G_2 \qquad (1)$$

where in the first step the so-called rough tuning matrices G_1 and G_2 are defined. They are usually chosen to be the inverse of the system at some desired frequency (Maciejowski, 1989). In the second step the so-called fine tuning scalars γ and δ are used in the same way as in SISO-cases when gains for P and I part are chosen to satisfy some close-loop properties.

By choosing the function *MV - root locus for P-tuning* from the group of frequency domain properties (Fig. 7) it is possible to observe the influence of parameter γ to the closed-loop pole-positions for the case, when rough tuning matrix of P-part is chosen to be the inverse of the system in steady state. This is for the discussed system

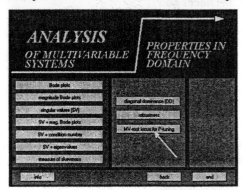

Fig. 7. Frequency domain properties of MIMO system

illustrated in Fig. 8. The system has four poles and two transmission zeros. By increasing parameter γ from zero to infinity two of close - loop poles are approaching transmission zeros while the other two are approaching the two zeros at infinity. The idea can be extended to the PI and the whole MIMO-PID structure tuning.

Suppose further that two controllers of the following form have been designed:

$$G_{PI1} = \begin{bmatrix} -0.0499 & 0.0672 \\ -0.0776 & 0.1008 \end{bmatrix} + $$
$$+ \frac{1}{s} \begin{bmatrix} -0.0150 & 0.0202 \\ -0.0233 & 0.0302 \end{bmatrix} \qquad (2)$$

$$G_{PI2} = \begin{bmatrix} -0.0899 & 0.0524 \\ -0.1899 & 0.0778 \end{bmatrix} + $$
$$+ \frac{1}{s} \begin{bmatrix} -0.0020 & 0.0028 \\ -0.0033 & 0.0042 \end{bmatrix} \qquad (3)$$

Now the close - loop properties can be inspected using the group of corresponding functions which are, as mentioned, organized very similar to open - loop analysis functions. Parallelism to SISO - problems was realized also in time domain for well known quality indicators to step responses known as delay time, rise time, settling time and maximal overshoot, only that all these parameters are now represented by matrices (Fig. 9).

In the cases where the results should match some desired design goals the expectations must or should be defined. In our case this is realized in the following manner. First the so called mode is defined (Fig. 3). This includes the definition of the

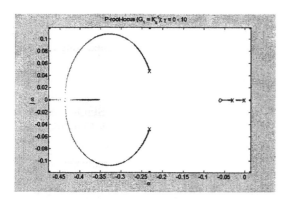

Fig. 8. Root-locus of MIMO system

variable *nacin* and (if known) the maximal values
of control, reference and disturbance signals. By
setting *nacin* = 1 we have chosen the situation
where output signals should track corresponding
references as good as possible. In the case where
nacin = 2 good disturbance rejection should be
fulfilled. When *nacin* = 3 both aspects are of the
same importance.

Design goals can be specified in time and fre-

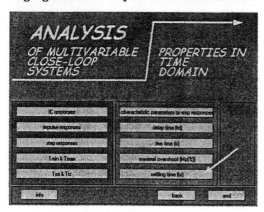

Fig. 9. Time domain properties of close-loop
MIMO system

quency domain with corresponding matrices. Suppose we have chosen *nacin* = 1 and the situation
where we want to specify design goals only in time
domain. In this case the matrix *ciljimvc* is defined. For chosen situation it consists of six rows.
Each row is connected with one of design goals:
the first with stability, the second with desired
minimal and maximal time constant, the third
with acceptable design complexity, the fourth
with desired steady-state gain, the fifth with desired settling times and the sixth with desired
overshoots. The elements of the first column are 1
if corresponding goal is of some importance or 0 if
it is of no importance. In that case it will not influence the validation process. With the subsequent
columns the so called importance and prescribed
values for each goal are defined. If for example the
element $ciljimvc(1,1) = 1$ and $ciljimvc(1,2) = 1$
this means that the design result is not acceptable if the closed-loop system is not stable. If
$ciljimvc(2,1) = 1$ and $ciljimvc(2,2) = 0.8$ and
$ciljimvc(2,3) = 100$ desired value of minimal time
constant is 100 [sec]. The importance of this goal
is 0.8 and the result is still acceptable if it is not
violated for more then 20%. In the cases when
design goal is violated more than allowed, the

result of absolute validation is 0. When certain
design goal is completely fulfilled the result is 1.
And when it is inside the allowed tolerance the
validation result value is correspondingly smaller.
The overall result is calculated by multiplying
all partial results. This kind of validation should
stimulate to improve poorly satisfied goals while
good results should not be pushed too far.
Sometimes it is of course difficult to define all
these values. In such situations design goals need
not to be specified explicitly. It is supposed that
stable solution is desired, while all other design
goals should be as good as possible and goal matrix is defined as:

$$ciljimvc = \begin{bmatrix} 1 & 1 & 1 & 1 & 1 & 1 \\ 1 & 0 & 0 & 0 & 0 & 0 \end{bmatrix}^T \quad (4)$$

where T denotes matrix transpose. Each stable
system is in this situation evaluated with 1. This
is the case also for both presented results.
When relative validation on the set of solutions
is performed also limitations on control signals
range are taken into account. In such situations
the close-loop behavior can be nonlinear if control
signals are saturated. This is the reason why all
the properties which are defined for linear systems
are omitted from relative validation. As all design
goals in time domain were defined to be of some
interest, by default complexity, settling times and
overshoots are proposed for relative validation.
The values of the last two can differ regarding
absolute validation if closed-loop system properties are nonlinear. With this information the user
can also evaluate the difference between theoretical linear and more realistic nonlinear situation.
In addition two criteria are proposed. The first
should be regarded as the measure of achieved
quality as design goal of good matching of outputs with corresponding reference signals is al-
ways presented. It is defined as: $J_1 = \sum \int |ref_i(t) - y_i(t)| dt$. On the other hand the measure of de-
manded control activity can be inspected with:
$J_2 = \sum \int |u_i(t)| dt$. We have to mention that reference signals needed for calculation of J_1 and
J_2 are generated automatically in the following
manner. First the parameter ΔT is defined as
$\Delta T = 4 * \overline{t_s}$, where $\overline{t_s}$ is the mean settling time of
compared linear systems. Then step changes are
realized in ΔT intervals for each of input signals
so that transient responses of direct paths and
cross-coupling can be observed. The situation is
illustrated in Fig. 10 for the first and in Fig. 11
for the second solution. The fist ΔT can be used
for start up in nonlinear situations.
Taken into account all presented criteria the results of relative validation are presented in Fig.
12. The first result is regarded as a norma while
the validation results of the other represent the
measure of improvement regarding the calculated
norma.
On the basis of relative validation the following

can be concluded. Both results are equal regarding complexity and approximately equal regarding control efforts. Due to considerably smaller oscillations and reduced cross couplings the control quality of the second solution is almost two times better. The overall price is obtained by multiplication of partial results and indicate the superiority of the second solution.

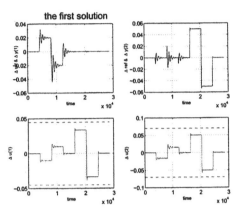

Fig. 10. Time responses of the first solution used for relative validation

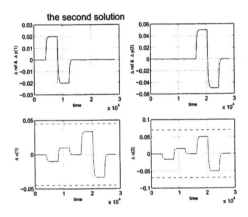

Fig. 11. Time responses of the second solution used for relative validation

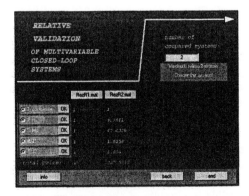

Fig. 12. Relative validation of two MIMO close-loop systems

5. CONCLUSIONS

On the basis of presented results and our own experiences the following can be concluded:
Presented graphical toolbox environment enables user friendly, simple and systematic treatment of multivariable systems in the sense of problem definition and analysis of different design stages. Students steel have all the freedom regarding controller design but some of the starting ideas can also be realized by the use of analitical possibilities.

In our opinion the analysis of parallelism and differences between SISO and MIMO systems improves students understanding of control problems.

The concept of design goals definition and evaluation is introduced in such a manner that it does not represent any kind of limitation but only stimulates the criticism of definition of what is good and why.

It also enables some further development in several directions.

Project oriented work was very good accepted by the students as it seems they have more confidence in their knowledge while approximately the same amount of time is needed for exam preparation.

In the same time the possibilities which enable further research work are indicated which also can represent a motivation for better studying results. The first experiences with toolbox environment usage confirmed correct orientation of the discussed software.

REFERENCES

Atanasijević-Kunc, M. (2001). *Multivariable systems, Collection of complex problems (in Slovene language)*. 2nd ed., Faculty of Electrical Engineering, University of Ljubljana.

Atanasijević-Kunc, M. and R. Karba (2002). Analysis toolbox stressing parallelism of SISO and MIMO problems. *Preprints of the 15th World Congress, IFAC, Barcelona, Spain*.

Control System Toolbox, User's Guide (1998). The MathWorks Inc.

Maciejowski, J. M. (1989). *Multivariable Feedback Design*. Addison - Wesley Publishers Ltd.. Oxford.

Maciejowski, J. M. (1990). *Multivariable Frequency Domain Toolbox, User's Guide*. Cambridge Control Ltd, and GEC Engineering Research Centre.

Matko, D., B. Zupančič and R. Karba (1992). *Simulation and Modelling of Continuous Systems, A Case Study Approach*. Prentice Hall Inc., Englewood Cliffs.

Matlab (1999). *The Language of Technical Computing, Version 5*. The MathWorks Inc.

ELSEVIER

IFAC
PUBLICATIONS
www.elsevier.com/locate/ifac

A MINI AERIAL VEHICLE AS A SUPPORT FOR MODELLING, CONTROL AND ESTIMATION TEACHING

Yves BRIERE, Camille PARRA, Joël BORDENEUVE GUIBE, Joan SOLA
Emails : firstname.lastname@ensica.fr

Ecole Nationale Supérieure d'Ingénieurs de Constructions Aéronautiques (ENSICA)
Avionics and Systems Laboratory
1, place E. Blouin, 31000 Toulouse, France

Abstract— At ENSICA several teams of students developed a mini aerial system that is now mature and is used for teaching. The system includes an original delta wing configuration able to bring a complete avionics system (sensors, actuators and controller) and a micro camera. We describe how a non linear model is obtained from wind tunnel tests and a linear state space model is then obtained. These models can be used for basic control theory. The board originally designed at ENSICA includes enough sensors to estimate the attitude and position of the plane. It is used at ENSICA as a basis for Kalman filtering theory teaching. *Copyright © 2003 IFAC*

Keywords: Mini Aerial Vehicle, Modelling, Attitude control, Sensors, Strapdown Inertial System, Kalman Filtering

1. INTRODUCTION

PEGASE is a Mini aerial vehicle (MAV) system which was developed by ENSICA [5]. This project is close to many others like the Dragonfly project at Stanford University [2] and the miniature acrobatic helicopter of MIT [5]. Due to its small dimensions, the plane has a very fast dynamics and therefore is hard to pilot. The payload of PEGASE consists in a miniature video camera that we use to teleoperate the plane and a microcontroller surrounded by sensors and actuators. PEGASE can be operated at three levels of autonomy: fully manual, manually controlled with an inner stabilization loop and fully autonomous.

Actually this system is the achievement of various students' projects and is now used for teaching and research.

We will describe first the short story of this project and give a detailed description of the system. In a second part we will give some results about the model of the plane and its control. In the last part we describe how this system is used at ENSICA as a support for a Kalman Filtering course.

2. WHERE PEGASE COMES FROM

Founded in Paris in 1945, ENSICA is one of the French aerospace schools. ENSICA trains multidisciplinary engineers capable, at outcome, of controlling complex projects within an international context mainly in the aerospace industry. During their three year stage at ENSICA, students have the opportunity to lead a personal project of about 150 hours. Teams of two to ten students benefit from the assistance of ENSICA's professors and technician's staff and facilities (wind tunnels, mechanical engineering, electronics, etc...)

The PEGASE project started in 1999 when a student's project led to a demonstration of live video transmission from an RC plane.

Since then a total of about "4 student-years" were dedicated to several projects related to mini and micro UAV. These efforts have now been capitalized into the PEGASE project.

3. WHAT PEGASE IS

PEGASE has a delta-wing configuration with two ailerons. The general characteristics of PEGASE are as follows:

- Wing span : $b=0.5m$
- Length : $L = 0.34m$
- Wing area : $S_{ref} = 0.0925m^2$
- Aerodynamic mean chord : $c = 0.185m$
- Speed of cruising : $V_0 = 20m.s^{-1}$

The plane is able to bring a payload of approximately 300gr. It is equipped with a video camera and a board including sensors, actuators, communication and microcontroller.

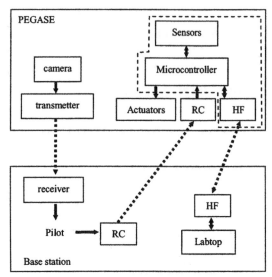

Figure 1 : PEGASE synoptic

Synoptic of the complete system is given in Figure 1. Sensors are: a set of three gyrometers, a set of three accelerometers, a three axis magnetometer, a micro GPS and a temperature sensor. Gyrometers and accelerometers are basically analogical devices and are sampled at 100 Hz. The magnetosensor is sampled at about 13Hz while the GPS is sampled at 1Hz. Future versions will include static and dynamic pressure sensors. Actuators are the two control surfaces ("ailerons") and the thrust. The plane is electrically powered. "RC" denotes the classical Radio Command system (emitter and receiver). "HF" denotes a 173MHz 19200 Bds downlink transmitter and receiver system. Next version will include a half duplex uplink and downlink system. Data is transferred and stored into a laptop via the serial link. Sensors, actuators and HF communication are connected to a single Motorola 68332 microcontroller. Total weight (included into the dashed line in Figure 1) is only 175gr.

The plane is either controlled like a classical RC plane or teleoperated via a video feedback from a miniaturised camera and video transmitter weighting only 10gr. The video feedback is displayed either on a portable video recorder or on video glasses.

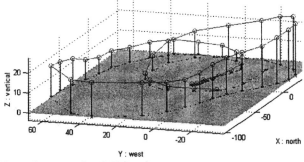

Figure 2 : example of GPS trajectory

Figure 2 shows an example of the plane trajectory during a short flight test. Position is obtained via the GPS, pre-processed into the onboard microcontroller and sent in real time to the ground station for display. All sensors output are sent to the ground station at a limited sampling rate of 13Hz due to the limited bandwidth of the transmission link.

This system is fully derived from students projects in which fundamentals of our aeronautic syllabus were involved:
- Flight dynamics,
- Sensors,
- Real time system.

4. MODELING AND CONTROL

In order to obtain a mathematic model of the MAV for the design and evaluation of guidance and control law, wind tunnel test has been carried out (Figure 3). At a fixed airspeed of 20m/s, three force and moment coefficients where measured at various angle of attack (α from $-30°$ to $+30°$), side-slip angle (β from $0°$ to $45°$) and control surfaces (elevator δ_e from -5° to +5° ; aileron δ_a from $-5°$ to $+5°$). With these data and inertial moment of the MAV, by proper interpolation, the full degree nonlinear model of PEGASE can be obtained.

Figure 3 : PEGASE during the wind tunnel test

To fulfil the mathematic model, force and moment coefficients are needed. They are multi-dimensional functions of angle of attack, side-slip angle and control surfaces, and can be obtained by interpolation of the discrete values from the wind tunnel test. From the six coefficients obtained above and the evaluation of inertial moment, a standard twelve-degree nonlinear equation can be derived.

A Matlab/Simulink full non linear model of PEGASE is made available for simulation.

It is not convenient to use the nonlinear equation to design guidance and control law; therefore, a linear state space model is computed from the full non linear model at a level straight flight condition:

$$V_x = 20 \text{ m.s}^{-1}, V_y = V_z = 0, \theta = \varphi = \Psi = 0°$$
$$\delta_e = -4°, \delta_a = -4°$$

where V_x, V_y, V_z are velocities of MAV and θ, Ψ, φ are pitch angle, heading angle and rolling angle. The linear equations of this flight state are:

$$\begin{cases} \dot{q} = -0.02V_x - 2.44V_z - 1.81\delta_e \\ \dot{\theta} = q \\ \dot{p} = -1.47\delta_a \\ \dot{\varphi} = p \end{cases}$$

where p is rolling angular rate, q is pitch angular rate. The yaw angle can only be adjusted by controlling the rolling angle because only two control variables are available.

Therefore a very simple attitude control law can be obtained:

$$\begin{cases} \delta_e = -0.55k_{11}[k_{12}(\theta_d - \theta) - q] \\ \delta_a = -0.68k_{21}[k_{22}(\varphi_d - \varphi) - p] \end{cases}$$

where k_{11}, k_{12}, k_{21}, k_{22} are control parameters to guarantee enough bandwidth, θ_d and φ_d are the desired pitch and roll of the MAV.

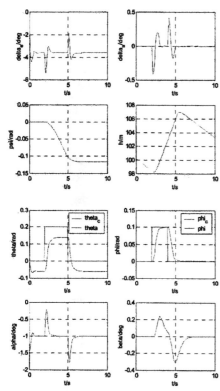

Figure 4 : Simulation results for attitude control

Nonlinear simulation has been done to verify the efficiency of attitude controller. The Figure 4 shows the step input response of pitch angle and rolling angle. At the same time, the variation of angle of attack, side-slip angle, control surfaces, heading angle and height are also given.

From the nonlinear simulation, conclusions can be drawn:
- The close loop attitude system is stable, the control law is efficient.
- Heading can be controlled by adjusting rolling angle.

For teaching purpose, the linear straight flight model can be used to design basic controllers. The controller can be then adapted to the Matlab/Simulink complete non linear model. At a basic level of control theory course, these models are a very good illustration of fundamental concepts:
- linearization,
- coupling and decoupling,
- state space control theory,
- non linearity due to actuator saturation and rate limiters,
- etc...

We plan to improve this tutorial approach with real flight tests and flight data analysis.

5. KALMAN FILTER DESIGN FOR ATTITUTE DETERMINATION

In order to control the attitude of the aircraft, three different angles (roll, pitch and yaw) must be determined. These angles are estimated from the fusion of sensors measurement into a very basic Kalman Filter as described for instance into [1]. This algorithm is used as a basis for Kalman Filtering teaching at the high level of control theory course at ENSICA.

For this course we focus in a generic attitude measurement system. Equations are derived from [5] and [3]. We only consider two frames: the earth frame, supposed to be Galilean, and the body frame. As we make no assumption on the aircraft dynamics, the system dynamic model is, in the discrete case:

$$\begin{cases} Q_{k+1} = Q_k + \dfrac{1}{2}T_e\Omega_k Q_k + T_e w_Q \\ \omega_{k+1} = \omega_k + T_e w_\omega \\ \alpha_{k+1} = \alpha_k + T_e w_\alpha \\ \varepsilon_{k+1} = \varepsilon_k + T_e w_\varepsilon \end{cases}$$

with :

$$\Omega_k = \begin{bmatrix} 0 & -p_k & -q_k & -r_k \\ p_k & 0 & r_k & -q_k \\ q_k & -r_k & 0 & p_k \\ r_k & q_k & -p_k & 0 \end{bmatrix}$$

If we consider X_k the complete state $X_k = [Q_k\ \omega_k\ \alpha_k\ \varepsilon_k]^T$ the equation above becomes :

$$X_{k+1} = f(X_k) + T_e w_k$$

275

T_e is the sample time. Q_k is the attitude (attitude of body frame relative to fixed frame, expressed in fixed frame) of the aircraft as a quaternion. $Q_k = [a_k, b_k, c_k, d_k]^T$ is therefore a 4-dimension vector. Transformation from quaternion to Euler angles is straightforward and is not described here. $\omega_k = [p_k, q_k, r_k]^T$ is the 3-dimension vector of angular rates of turn (rate of turn of body frame relatively to fixed frame, expressed in body frame). The state vector is augmented with two additional states: α is the drift of the accelerometer and ε is the drift of the gyrometers. For the design of the Kalman filter we must take into account a set of disturbance w (with subscript corresponding to each state) suppose to be Gaussian white noise with covariance Q.

The measurement model is as following:

$$\begin{cases} A_k = C_e^b(\ddot{x}_k + g) + a_k + v_A \\ G_k = \Omega_k + \varepsilon_k + v_G \\ M_k = C_e^b M_s + v_M \end{cases}$$

where:

$$C_e^b = C_b^{e\,T} =$$
$$\begin{bmatrix} a^2 + b^2 - c^2 - d^2 & 2(bc + ad) & 2(bd - ac) \\ 2(bc - ad) & a^2 - b^2 + c^2 - d^2 & 2(cd + ab) \\ 2(bd + ac) & 2(cd - ab) & a^2 - b^2 - c^2 + d^2 \end{bmatrix}$$

C_e^b is the transformation matrix (from fixed frame to body frame) and is related to the quaternion as the equation above states. A_k is the vector of three accelerometer outputs. As seen in the equation above, this measurement is related to the earth gravity $g = [0, 0, 9.81]^T$ and to the actual acceleration $\ddot{x}$ of the plane's center of gravity. We make here the assumption that this acceleration is mean null[1] and all derivatives fall into v_A. α_k is the drift of the accelerometer. G_k is the vector of three angular rates measurement. The technology of this gyrometers (cheap miniaturized piezoelectric vibrating gyrometers from Murata) makes them very sensitive to temperature: the drift ε is quite important and must be estimated online. The last measurement vector is M_k the three components of earth magnetic field. M_s is the vector of earth magnetic field in earth frame. v (with appropriate subscript) is the measurement noise supposed to be Gaussian white noise.

If we consider Y_k the complete measurement vector $Y_k = [A_k, G_k, M_k]^T$ the equation above becomes :

$$Y_{k+1} = h(X_k) + v_k$$

Let us just give the very well known Kalman Filter equations that makes the best estimate of X. (The subscripts $_k$ and $_{k+1}$ are now omitted as we describe here the algorithm, the equation $X_{k+1} = f(X_k)$ is written $X = f(X)$)

Algorithm 1

Step 1 : compute the predicted state estimate:
$$\hat{X} = f(\hat{X})$$

Step 2: compute the covariance matrix extrapolation:
$$P = F.P.F^T + Q$$

Step 3: compute the Kalman gain:
$$K = P.C^T.(C.P.C^T + R)^{-1}$$

Step 4: compute the covariance matrix update:
$$P = P - K.C.P$$

Step 5: compute the state estimate update with measurement Y:
$$\hat{X} = \hat{X} + K.(Y - h(\hat{X}))$$

F and C are respectively the jacobian matrixes of f and h.

The algorithm that we actually use is described in Figure 5 and derived from [1].

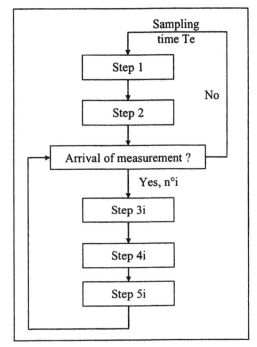

Figure 5 : Kalman filter algorithm

---

[1] This assumption leads to a good estimation of the attitude as long as the hypothesis is true. It is not true for circling flight.

Algorithm 2

Step 1 : compute the predicted state estimate:

$$\hat{X} = f(\hat{X})$$

Step 2: compute the covariance matrix extrapolation:

$$P = F.P.F^T + Q$$

Step 3i: compute the Kalman gain, corresponding to the arrival of measurement n°i

$$K = P.C_i^T.\left(C_i.P.C_i^T + R_i\right)^{-1}$$

C_i is now the i-th line of matrix C. R_i is the i-th element of matrix R. The strong assumption is that R is diagonal i.e. that measurement noises are uncorrelated

Step 4i: compute the covariance matrix update:

$$P = P - K.C_i.P$$

Step 5i: compute the state estimate update with measurement Y_i:

$$\hat{X} = \hat{X} + K.\left(Y_i - h(\hat{X})\right)$$

Y_i is the i-th measurement.

Algorithm 2 is proven to be equivalent to algorithm 1 as long as the matrix R of noise measurement covariance is diagonal.

Compared to algorithm 1, algorithm 2 as two advantages:

- Step 3i is not a matrix inversion but a scalar inversion,
- Measurement can be asynchronous.

Actually algorithm 1 cannot be used because of two reasons: huge matrix inversion cannot be processed in real time by the microcontroller and measurements are basically asynchronous.

For teaching purpose, the algorithm is done offline with Matlab routines.

Figure 7 shows the convergence of the algorithm. Only the gyrometer output (scale °/sec) is drawn but accelerometers and magnetosensors are included into the algorithm. One can see that the gyro output is not centered at 0.

About 30s of convergence is necessary to estimate the gyro drift. One can see the estimated angular rates that converge to 0 °/s after 30s. At 38s a few +/–90° rotations are made along the three axes and the attitude quaternion is estimated. After computation, the attitude is visualized in an animation (Figure 6).

For teaching purpose, students have the opportunity to work with a real attitude system. They can manage everything from the sensor output to the algorithm and drive their tests. Important notions of Kalman Filtering Theory and INS theory are illustrated in this course:

- Basis of Kalman Filtering. (For instance: filter is tuned by Q and R).

- INS errors. (Alignment errors, sensors bias and offsets, scale factors errors, nonorthogonality, temperature sensitivity, random noises)
- State augmentation (drifts estimation).
- Observability. We point out that three gyrometer and a set of three of six accelerometer and magnetosensor, properly chosen, are enough to estimate attitude.
- Synchronicity. We point out that a high sampling rate for the gyrometer and a low sampling rate for magnetosensors and gyrometers gives a good estimate of the attitude.
- Computational aspects. The two last points lead to a much liter algorithm than the basic Kalman Filter algorithm.
- At last, this attitude system is an exact copy of Strapdown Inertial Navigation Systems used in aircrafts.

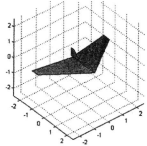

Figure 6 : 3D visualization of PEGASE attitude

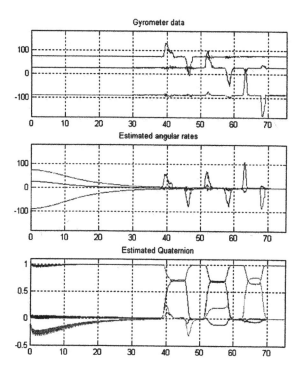

Figure 7 : attitude algorithm convergence

6. CONCLUSION

Mini UAV projects provide efficient yet inexpensive capabilities of conducting air vehicle experimental research and teaching. Energy, sensors and actuators miniaturization make now possible to reach very small dimensions and prices. In this paper we focused in the teaching aspects at an aeronautical engineer's level. The PEGASE project can be a starting base for improvements by student's projects. It is also used for academic teaching.

Of course this system is rich and versatile enough to serve research experiments. Our current research interests are teleoperation, telesupervision and telepresence in a 3D and high dynamics environment, and advanced control of very small aircrafts.

REFERENCES

[1].Alonzo, K., *Modern Inertial and Satellite Navigation Systems* (1994), CMU-RI-TR-94-15, Carnegie Mellon University, Pittsburgh

[2].Evans, J., Inalhan G., Jang J.S., Teo R., Tomlin C.J., *Dragonfly: a versatile UAV platform for the advancement of aircraft navigation and control*, Proc. Of the 20th IEEE Digital Avionics And Systems Conference, Oct 2001

[3].Grewal M.S., Weill L.R., Andrews A.P., (2001) *Global Positioning Systems, Inertial Navigation, and Integration* John Wileys and Sons, Inc, New York

[4].Grewal M.S., Andrews A.P, (1993) Kalman Filtering, Theory and Practice Prentice Hall, Englewood Cliffs, New Jersey 07632

[5].Parra C., Su B., Bordeneuve-Guibe J., Briere Y., *Development of a MAV – From theory to implementation*, Euro UVS, Bruxelles 2003

[6].Sprague K., Gavrilets V., Dugail D., Mettler B., Feron E., *Design and applications of an avionics system for a miniature acrobatic helicopter*, Proc of the 20th IEEE Digital Avionics And Systems Conference, Oct 2001

[7].Titterton D. H., Weston J.L. (1997), *Strapdown Inertial Navigation Technology*, Peter Pelegrinus Ltd, on behalf of the institution of Electrical Engineering, London, UK

ELSEVIER

IFAC
PUBLICATIONS
www.elsevier.com/locate/ifac

APPLYING KNOWLEDGE ENGINEERING TECHNIQUES IN CONTROL ENGINEERING EDUCATION

Ángel Alonso, Isaías García, José R. Villar, Carmen Benavides, and Francisco Rodríguez

[dieaaa, dieigr, diejvf, diecbc, diefrs]@unileon.es
Systems Engineering and Control Group
Dept. of Electrical and Electronic Engineering, University of León
Edificio Tecnológico, Campus de Vegazana s/n, 24071 León (SPAIN)

Abstract: The future of control engineering education will require the detailed and careful analysis of both the curricula and the methods involved in the learning process. This paper deals with the two key elements that are going to mean a revolution in the educational field: the computer and the Internet. Considering the education as a knowledge transfer process, the natural choice for formally studying the applications of computers to this process is the so-called *knowledge engineering*. If education is going to take advantage of the Internet, the necessity for constructing new, knowledge-intensive systems for collaborative work will also be shown. *Copyright © 2003 IFAC*

Keywords: Control Education, Artificial Intelligence, Knowledge Engineering, Computer-aided Instruction, Intelligent Knowledge-based Systems.

1. INTRODUCTION

This is an introductory paper that pursues to introduce a formal and engineering view on the field of control engineering education. So, no concept will be deeply treated with the aim of making a general picture of what the research efforts on this field should be. So far, no formal research has been done that covered these educational issues as a whole, and studies have been centred on the application of computer software to very concrete and restricted areas, like simulation systems, computer-based experiments, virtual laboratories, etc.

The need for a new vision in the field of education has arisen in the last years and it is nowadays seemed as a necessity in almost every subject area. Education is seen as a continuous process that doesn't finish when the university curricula is completed, but goes beyond and lasts for the rest of the engineers' professional life. But this period in the education of an engineer, being the bigger, is also the one that less

attention is paid to. The same situation is present in the case of technology transfer between university and industry, that could also be treated as a kind of educational communication.

Another important aspect to have in mind is the presence of computers, that have being helping for a long time in the more mathematical, repetitive and time consuming calculations involved in the engineering disciplines, but have being less used for another tasks, such as teaching. Time has come to change this situation. This is more notorious since the coming up of the Internet.

The aim of this research group is finding a way in what computers could help on the improvement of teaching in the field of engineering subjects (more precisely, in the field of control engineering education). Teaching, in its usual meaning, is a knowledge transfer between two human beings, and could be extended to a broader concept if we substitute the term "human being" with "entity" or "agent". This way, the term "agent" could stand for either a human being or a computer software application.

The scientific study of these knowledge transfers from an engineering point of view leads to the so-called *knowledge engineering*, a discipline rooted in Artificial intelligence research.

Knowledge engineering, when applied to education, results in a kind of knowledge-based systems called intelligent tutoring systems (ITS).

This paper is organized as follows: In section 2, some basic concepts are introduced to help understand what knowledge engineering is about and why it is so important nowadays in almost every social areas. Section 3 introduces the way knowledge engineering can be applied to communication in the field of engineering. Section 4 depicts the relations between knowledge engineering and education, focusing in the particular aspects of control engineering education, and also introduces the concept of intelligent tutoring system. In section 5, the importance of the Internet in the learning process is shown, placing special emphasis on the need for integration of both the intelligent tutoring systems and the Internet technologies. Section 6 summarizes the outline of the complete system proposed for a better control engineering education process. Finally, section 7 gives some important conclusions and sketches the future directions of research in this area.

2. KNOWLEDGE ENGINEERING

Knowledge engineering is a discipline dealing with the ways human knowledge can be represented, stored and used into computer systems. It is a research field included within Artificial intelligence, with several decades of study and research.

In its origins, Artificial intelligence had an ambitious goal: to produce machines that could completely replicate the mental processes of the human being. The first attempts were very promising. There were built *general problem solvers*, based on predicate logic, that could resolve problems dealing with some kind of automatic inferences. But it soon was clear that for real life problems, such general logic based mechanisms were not successful, even when they were used in very limited domains.

The attempts to build such general problem solvers were abandoned and the research focused its efforts in producing machines or systems which could solve some kind of problem in a very restricted domain. It was soon shown the lacks that techniques used so far when building general problem solvers had. One of the first results was the agreement that, if the system was going to be able to resolve some kind of real problem, it needed to "know" about what objects or concepts were present in the domain and what kind of particular procedures could be used to solve the problem. This way, the field of *knowledge engineering* was born.

Knowledge engineering had a very promising beginning, too. Some outstanding systems were built on the eighties, like MYCIN, a system to help in the diagnosis of infectious diseases or XCON, a system to help on the configuration of VAX computers. Those systems used the knowledge about the corresponding domain, previously extracted from a human who was considered to be an expert for that domain and implemented as a computer software application; because of this, those systems were called *expert systems*.

After these first promising years, the use and implementation of expert systems partially merged into the field of software engineering, and the research efforts became more theoretical. There was a great concern about the formal aspects of these systems and the problems of implementing them into computers. Then, a lot of research work aimed at clearing those concerns about concepts like tractability (the possibility for a computer to make the calculations and present the results in a finite amount of time) or expressiveness (the amount of information –or knowledge- the system was able to represent by using a concrete formalism). This research studied expert systems from a broader view, making less restrictions about the domain of the knowledge and trying to find some general methods to represent and reason with this knowledge. The systems studied under this point of view were called knowledge based systems (KBS).

As usual, the need for a solution to a real problem has brought to the knowledge engineering research field back to the practice. This problem is the so-called information burden generated with the advent of the Internet. Knowledge engineering is currently extensively applied to build what has been called "Semantic Web", a new idea about how the organization of knowledge in the World Wide Web should be done (see http://www.w3.org/2001/sw/ or http://www.semanticweb.org/ for details).

But it must not be blamed it only to the Internet. Nowadays, society is a very complex system, modern enterprises have to manage huge amounts of information from very different sources to be able to take some strategic action, even just to keep alive in the competitive market.

A new term, *knowledge management*, has been coined, meaning all the processes involved in the gathering, storing, analyzing and retrieving of such knowledge, implemented in the corporate computers. This term is included in a broader one which all we have heard of: Information Technology (IT).

Some of the outcomes of the research on knowledge engineering is transmitted to the enterprises and there it is built into software systems that are becoming the most valuable piece in the whole enterprise. Today, the industrial manufacturing processes are optimized (near) to their maximum. The difference, the real advantage of one enterprise over the rest, is the knowledge management it is able to implement and maintain.

3. KNOWLEDGE ENGINEERING APPLIED TO ENGINEERING FIELDS

Artificial Intelligence (AI) is not a new term for the control engineer. In fact, its techniques have been used for a long time for several different purposes. But so far, AI techniques had been exclusively used for "doing" control, that is, to improve or replace some of the traditional methods used to control some kind of physical system. Systems based on the use of neural nets, expert systems or fuzzy logic are some examples of the successful application of AI theories to control engineering.

In this paper, the techniques of Artificial intelligence applied to the control engineering field have a different approach and purpose. As was mentioned earlier, this approach is based on the use of knowledge engineering techniques. The benefits of applying knowledge engineering techniques to the field of control engineering should be evident. There are several reasons for this assertion, but the very first question one could be asked is "What can knowledge engineering do for control engineering?".

As Gruber stated early in (Gruber, 1990), the *unique* contribution of AI to science is knowledge representation, "that is, techniques for stating human knowledge in a formal, explicit, and operational manner". In this paper, Gruber pointed to the fact that knowledge engineering could help on the representational issues that engineering sciences don't implement, that is, "the human knowledge with which devices are designed, manufactured, tested, operated, diagnosed and repaired". It was also pointed out there that such a representation of engineering stuff could be used "as aides for memory and communication, where written and spoken natural language is used today". All this was said when Internet was only a network of computers devoted to scientific research. The World Wide Web, as today we know it, was only a dream.

Today, the explosive growing of computers and the Internet has led to a situation where a great amount of the communication processes is held between computers or between human beings and computers. It is easy to realize how important could be today a formal representation of engineering fields to help on these communication processes. This techniques, as

will be seen in section 4, can be directly applied to engineering education as well as to many other fields.

The communication processes between two agents (in the broader sense of the term we have seen) involves the use of expressions and concepts with a shared meaning which both the communicating agents commit to. But this commitment must be formally stated when the communicating agents are not aware of each other. This is true even when both agents are human beings. One can see, as an example, the list of keywords used by Elsevier Publishing when dealing with papers related to automatic control. This is a way to lightly formalize the terms that can be used as keywords in a technical paper with the aim of restricting the amount of different terms that can appear, facilitating the sharing of knowledge and the classification and grouping of such papers. Another interesting effort in this same direction is the thesaurus (a glossary of terms with specialization and generalization relations) built by IEEE society for the indexing of its documents.

4. KNOWLEDGE ENGINEERING FOR CONTROL EDUCATION

Control engineering education, from the standpoint of the education research field, is an activity where the constructivist view of the learning process is crearly present. As Dormido says in (Dormido, 2002), "learning by doing" is a valid expression to reflect the importance of the interactivity nature present in the learning processes in the field of control engineering (the constructivist approach can be actually extended to every education subject area). Having this premise in mind, new tools for control education are being researched and developed (Johansson *et al.*, 1998), (Dormido *et al.*, 2002). These tools offer a richer experience to the student than that of the traditional environments like MATLAB.

The approach proposed in this paper shares the point of view above mentioned, but tries to find the way of applying well-known artificial intelligence techniques to this field. As it was seen, knowledge based systems are used to store data (i.e, representing knowledge) about a domain and solve some problems or deal with some tasks. They are used also to facilitate the communication with other systems or agents. The application of such systems to education is immediate. In fact, these systems had been studied for some years under the name of *Intelligent Tutoring Systems* (ITS) (Urban-Lurain) and some systems have been built, like the one described in (Gertner and VanLehn, 2000).

Intelligent tutoring systems go beyond the traditional approaches like Computer Based Training (CBT) or Computer Aided Instruction (CAI), that only use the computer as a medium for implementing the same curricula material.

If one want to use the computer to really take advantage of the possibilities it offers, one should take a look at the ideas underlying intelligent tutoring systems. As it was shown, an ITS is a knowledge based system and so, it is composed of:

- A formal representation of the knowledge of the domain.
- A formal representation of the tasks and methods used to solve problems on this domain.

In the case of an intelligent tutoring system, the represented or formalized knowledge usually is divided into a number of different models (as shown in figure 1):

- A student model (knowledge of the learner), that stores information about the learner. As an example, this model must be able to represent and handle the performing of a student on the material he/she is being taught.
- A pedagogical model, that is, knowledge about teaching and learning strategies.
- A domain model, which consists of the knowledge about the material being taught.

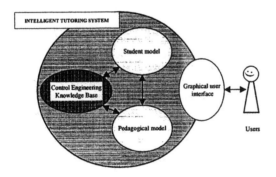

Figure 1. Intelligent Tutoring System.

The student model will help to individualize the learning processes by treating each student according to his/her progress and needs. If the computer has a formalized representation of the student's skills and evolution, it will be possible to automatically generate personalized curricula material, lab hand-outs, etc.

The pedagogical model could help, first, to clarify the important concepts involved in the education process of the control engineering field, as well as the relations among these concepts and the way these concepts may be taught.

The domain model is the base in top of which all the system is built. It consists of the concepts and relations involved in the subject at hand (control engineering). This model should not only be useful for the intelligent tutoring system, but for every kind of communication where concepts related to that subject is present. In this sense, this knowledge would be highly reusable.

The key concept to build any of these systems is the formalization of the knowledge of the domain in order to build a computer system which is able to handle it.

It is assumed in knowledge engineering that the knowledge about a given domain can be stated and studied independently of its implementation in a computer. The representation of the knowledge of a domain comprises the description of the concepts, relations and processes that occur within a given domain and is called a conceptualization; this conceptualization, when explicitly stated, is called an ontology (Gruber, 1993).

The domain ontology is the most important element when building knowledge bases and knowledge-based systems, but besides the domain ontology, which constitutes the static knowledge of the system, it is also needed to build some kind of reasoning mechanisms that are able to implement the dynamic behaviour of the system (that is, the processes involved in the system that make it change over time). This is achieved by reusing some general reasoning strategies, called problem-solving methods. Domain ontologies and problem-solving methods are the building blocks of knowledge-based systems (Musen, 1998) and intelligent tutoring systems are nothing but a special kind of knowledge-based systems.

One can realize the possibilities of a system like the one described above. But this system could no be built or used without the collaboration of every agent involved in the education process. And this fact leads to the next important issue that must be seemed as crucial for the future of education or even the human relations in a broader sense: the necessity of building collaborative systems for allowing groups of physically distant people work together with a common objective. This collaborative way of working, while not being a new necessity, has today the tool to implement it: the Internet.

5. COLLABORATIVE WORK OVER THE INTERNET

Internet has been used for learning purposes since its very early stages. In fact, its very creation was due to the necessity of sharing knowledge among the research community.

One of the most common uses of the Internet in the education field is the possibility of offering remote access to laboratories, whether real or virtual, helping in the distance learning as well as in the scheduling problems of real laboratories. This is a great area of current research and some interesting outcomes are being obtained (Dormido *et al.*, 2000).

Another use is the migration of the operation of the usual university educational structure to the Internet, building the so-called virtual campus. The material for some specific course is made available to students by accessing the web page for that course. The communications among students and teachers is implemented by forums or chat systems, including advisory boards, etc. Even administrative tasks, like matriculation or payments, are automated remotely through Internet.

The approach outlined in this paper goes far beyond those current uses of the Internet. As has been previously said, Internet is a revolution for human communications and it offers much more possibilities that the ones that have been exploited so far. The collaborative work is the key concept, but the work obtained (the knowledge obtained) with this collaboration must be properly stored and used. Here is where almost all of the existing systems fail. To be able to deal with this huge amounts of knowledge that will be got, some kind of platform or architecture is needed that has very little to do with the ones currently working (usually based on the use of a database with very little data processing effort).

The proposed approach is again based on the knowledge engineering perspective. There is a claim for the need of building big knowledge repositores (it would be better to call them knowledge based systems) and use them to automatically infer new knowledge. As an example, the results obtained after the use of a traditional Internet-based education system for an academic year are absolutely wasted. But what if we could use all this knowledge to:

- Improve the pedagogical methods, allowing the individualization of the. learning process?
- Clarify some curriculum part that has been found to be specially misunderstood?.
- Find different ways to explain the same concept and gathering the results of each one according to the outcomes of the students?
- Offering statistical data or reports based on the intensive treatment of the knowledge to the potentially new students or to the society?
- Using the students' data to keep in touch with them, offering the possibility of continuous (and individualized!) education?

To achieve these goals the current technologies must be improved. As an example, HTML documents must be substituted by semantically-enriched documents. These documents can be automatically produced by the system, dealing with such amount of information that a human being could never do. Traditional database systems must be substituted by knowledge bases and knowledge based systems.

Multimedia and virtual reality are some useful tools, but the great breakthrough will be the knowledge-intensive approach that has been shown.

6. OUTLINE OF THE SYSTEM

The whole system to be built is summarized on figure 2. The core element is the knowledge base, a place where all the knowledge on the domain of control engineering can be stored in a formalized and flexible way.

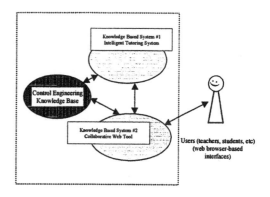

Figure 2. Web-based Tutoring System.

This knowledge base will go beyond the idea of a "centralized Internet repository" expressed on (Antsaklis *et al.*, 1999) as a recommendation for improve the use of the Internet for control engineering education. The term "repository" is a very dangerous one if we think of it as a container for the educational materials. In this paper, the term knowledge base is used instead. As it can be seen in previous sections, a knowledge base provides a way to represent and relate all the stored information in a very flexible way.

On top of the knowledge base, several knowledge based systems can be built, the particular case of an intelligent tutoring system is shown in the figure.

A collaborative web based system must be built to put the intelligent tutoring system on the Internet. Tools like forums, chats, etc will be implemented, but also a new knowledge based system for dealing all the information obtained in the interaction among the users of the system.

7. CONCLUSIONS AND FUTURE WORK

In this paper, it had been shown the necessity for a change in the way computers are being used to help in teaching of control engineering. First, the field of knowledge engineering was introduced as the engineering discipline to be used in this task and the need to take a look at the achievements from the field of intelligent tutoring systems obtained so far.

Then it was exposed the view about what the Internet-based teaching should be, that is, focusing more on the collaborative opportunities than on the possibility of tele-operation (virtual laboratories) that, being important, are not the main advantage of the Internet.

This research group is currently working on the construction of a knowledge base of control engineering stuff. The first work to deal with is the construction of a structure where all the concepts and relations among concepts of this field are formally stored. This structure is called an ontology and is the core of any knowledge base.

As it has to be a useful system, the construction of such ontology and knowledge base will require a lot of consensus among the researchers and people related to this engineering field. So, a collaborative Internet platform to be used for the work among the interested people is also in its developing process. This way, the general recommendation obtained from the 1998 NSF/CSS Workshop on New Directions in Control Engineering Education (Antsaklis *et al.*, 1999) is also achieved, that is: "It is recommended that interested organizations work to enhance cooperation among various control organizations and control disciplines throughout the world to give attention to control systems education issues and to increase the general awareness of the importance of control systems technology to society"

In a first stage, a small system will be built where the functionality of the knowledge base can be validated. This system will be an intelligent tutoring system for helping on the learning of some simple concepts of control engineering.

For the construction of the system it will be used open source software tools, and the system itself will also be open source. This is not a minor requisite for this research group, because the very knowledge based system has to be the result from a collaborative (and so opened to the community and the society) work.

The development process, discussions and project advances and outcomes will be publicly accessible at the URL: http://cekb.unileon.es.

REFERENCES

Antsaklis, P.; T. Basar; R. DeCarlo; N. H. McClamroch; M. Spong; and S. Yurkovich. (1999). Report on the NSF/CSS Workshop on New Directions in Control Engineering Education. *IEEE Control Systems Magazine*, **22**, 2, pp. 53-58.

Dormido, S. (2002). Control Learning: Present and Future. In: *Proceddings of the 15th IFAC World Congress*, Barcelona (Spain).

Dormido, S.; F. Gordillo; S. Dormido-Canto; J. Aracil (2002). An interactive tool for introductory nonlinear control systems education. In: *Proceddings of the 15th IFAC World Congress*, Barcelona (Spain).

Dormido, S.; J. Sánchez; F. Morilla (2000). Laboratorios virtuales y remotos para la práctica a distancia de la Automática. *XIX Jornadas de Automática*, Sevilla (Spain).

Gertner, A. S. and K. VanLehn (2000) Andes: A Coached Problem Solving Environment for Physics. In: *Proceedings 5th International Conference, ITS 2000*, Montreal Canada, June 2000.

Gruber, T. (1990). The Use of Formally-Represented Engineering Knowledge to Support Human Communication and Memory. *Presentation at a panel of the 1990 AAAI Spring Symposium workshop on Knowledge-Based Human-Computer Communication*, Stanford, March 27-29, 1990. Paper available as technical report KSL 89-87. http://ksl-web.stanford.edu/KSL_Abstracts/KSL-89-87.html

Gruber, T. (1993) A translation approach to portable ontology specifications. *Knowledge Adquisition*, **5**, pp. 199-220

Johansson, M.; M. Gäfvert; K. J. Åström (1998). Interactive tools for education in automatic control, *IEEE Control Systems Magazine*, 18, 3, pp. 33-40.

Musen, M. A. (1998) Modern Architectures for Intelligent Systems: Reusable Ontologies and Problem-Solving Methods. In C.G. Chute, Ed., *AMIA Annual Symposium*, Orlando, FL, pp. 46-52.

Urban-Lurain, M. Intelligent Tutoring Systems: An Historic Review in the Context of the Development of Artificial Intelligence and Educational Psychology, *[online] http://www.cse.msu.edu/~urban/ITS.htm* (checked on September, 4th, 2002)

ELSEVIER

IFAC
PUBLICATIONS
www.elsevier.com/locate/ifac

CHEMICAL MICROPLANT

Heli Pykälä[*], Mariaana Savia[**], Heikki N. Koivo[*]

[*]*Helsinki University of Technology, Control Engineering Laboratory, Finland*
[**]*Tampere University of Technology, Automation and Control Institute, Finland*

Abstract: This paper presents a chemical microplant which is a simple microfluidic system built for educational purposes. As far as the authors know there are no published reports about this kind of systems in educational use. The microplant consists mostly of commercially available components and it is designed such that varying or extending the structure is easy. The chemical reactions that are used in the system are selected such that they best serve the educational purposes of microsystem technology (MST) and control engineering. At the moment the system is mostly used as a demonstration tool in MST related courses. Later on the use will be extended to cover different kind of laboratory exercises also in the area of control engineering. *Copyright © 2003 IFAC.*

Keywords: microsystems, microfluidics, education, laboratory exercise

1. INTRODUCTION

The purpose of this paper is to introduce a simple microfluidic system in order to demonstrate students microsystem technology and system behavior in microworld. Development of microsystem technology began with miniaturization of electronic components, which resulted in integrated circuits. Especially in 1980's this led to the idea of combining mechanical and electronic components in microscale. Such systems were called micro electromechanical systems (MEMS). Many MEMS components have been developed on silicon. These have found many useful applications. For example, miniaturized pressure sensors are used in automobile airbags and microgyros, in what automobile industry calls Electronic Stabilization Programs (ESP) to prevent the car with a fast speed turning over in curves.

Success of MEMS made researchers to think of microfluidic systems. Could you produce credit card size analyzers or even a microplant? After all, when producing drugs, there is no need to have huge tanks and pipes like in oil distillation. Small size is quite sufficient.

By miniaturizing macro world components many benefits can be achieved. Components fit in smaller space and they can be used in applications, where the use of traditional components is not possible. Microcomponents have small power consumption and are often easy to replace. Production scale is typically large, so price of a single component stays quite low.

1.1. Microfluidics

Microfluidics concentrates on fluid flow in microchannels. Benefits of microfluidic systems and components are similar to benefits of any micro components: miniaturized components, lower energy consumption and reduced usage of expensive reactants. Application areas include for example chemistry, environmental technology and medical technology.

It is well known that miniaturization of fluid flow systems amplifies phenomena that can be ignored in macro scale. Theories, which work fine with macroscale flows, do not necessarily model micro flows in sufficient detail. Because of the scaling effect, surface forces become dominant in small channels. Also friction and surface tension become significant in micro channels, while they can be ignored in macro flows. In addition, pressure losses increase when channel is miniaturized. On the other hand, micro flows are practically laminar.

As far as the authors know, there are no published reports about microfluidic systems for educational purposes. The intention of this paper is to fill such a void by introducing a simple chemical microfluidic system, a microplant. The components are commercially available and reasonably priced. Considering the implementation, the word miniplant instead of microplant could also be used. The system is still under development and laboratory experiments to be carried out are currently designed.

The paper proceeds as follows. In section 2, a short description of the microplant is given. The system consists of two pumps, a mixer and a reactor, which are combined with plastic pipes. Section 3 discusses the components of the microplant. In section 4, tests that have been carried out are reported. Section 5 offers future plans for developing the laboratory experiment. The paper closes with conclusions.

2. MICROPLANT DESCRIPTION

Objective of the chemical microplant project was to construct a system, which could be used to observe chemical reactions in miniaturized scale. It can be used as a demonstration or a laboratory exercise for students. There are not many demonstrations in the field of microfluidics, so it is important to illustrate a whole microfluidic system. The system was built of commercially available components.

2.1. Structure of microplant

As can be seen in figure 1, the structure of microplant is quite simple. It consists of two micropumps, micromixer and a reactor chamber. The reactants are pumped to the mixer and then to the reactor chamber. The pumps and the mixer are commercially available components and the reactor was constructed in the workshop of Tampere University of Technology.

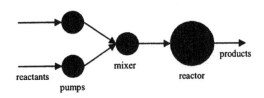

Fig 1. Structure of the microplant.

Components of the microplant are connected with 1/16" Tygon tubes. The mixer and the reactor have suitable fluidic connectors, but the pumps are assembled with polyethylene (PE) tubing. Because PE-tubes are quite hard, it was possible to connect the Tygon tube on the PE-tube. Connections seem to hold quite well and disconnecting the components is easy. However the connections have to be checked regularly, because especially connections between pumps original tube and the Tygon tube may loosen. If this becomes a problem, adding a little glue to the connection may help.

The components are placed on a 3-dimensional rack built at HUT workshop (figure 2). The rack layers are made of acrylic. Because of the structure, varying the system is easy. Order of the layers can be changed and adding new layers is possible, if new components are acquired.

2.2. Reactions

The microplant can be used to research simple chemical reactions. Because of the structure of the system it is possible to use only two reactants. Reactants, which damage the components, should not be used. Reactions should not produce any deposit, because it can clog the system. Reactions should also be easy to observe. Because in this point there are no sensors in the system, the change of color is the easiest way to observe reactions.

Most common simple reactions, which satisfy the demands, are pH and redox reactions. According to the manufacturer of the pumps, bases may damage the polycarbonate of which the pumps are made of. It was decided to use redox reaction where permanganate ion

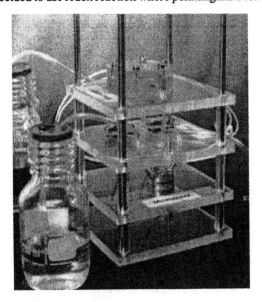

Fig 2. The microplant.

is reduced by hydrogen peroxide (equation 1). Permanganate ion is violet, but manganese in its reduced form is colorless.

$$2MnO_4^- + 5H_2O_2 + 6H^+ \rightarrow$$
$$2Mn^{2+} + 8H_2O + 5O_2 \tag{1}$$

2.3. Using the system

At this point there are no sensors in the system, so no automatic control can be implemented. Regardless, it is possible to control the pumps manually and try to find the equilibrium of the reaction. This has not yet been tested, but it should not be a problem because of the clear change of color in the reaction.

3. COMPONENTS

3.1. Micropump

The micropumps are manufactured by ThinXXS (ThinXXS Microtechnology). They are membranepumps with piezoactuator and passive check valves. The pumps are self-filling and self-priming and capable of pumping both liquids and gases. The pumps are made of polycarbonate and the housing is polypropylene.

The principle of membrane pumping is simple. When the actuator moves up, the pump chamber expands and pressure sucks fluid through the inlet valve. Then the actuator moves down and presses the fluid through outlet valve.

The pumps are operated with drive electronics purchased from IMM (Institut für Mikrotechnik Mainz). There are two units of which the other is frequency adjustable and the other amplitude adjustable.

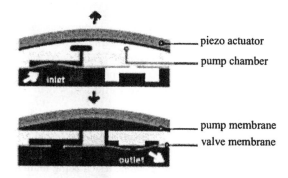

Fig 3. Function principle of the pump.

Specifications of drive-electronics:

Drive electronics, ULA

- Pulse shape: square pulse
- Output voltage: negative amplitude –90 V, positive amplitude adjustable form +100 V to +340 V
- Output frequency: 68 Hz

Drive electronics, FSB

- Pulse shape: square pulse
- Output voltage: -100 V to +340 V
- Output frequency: adjustable from 1 Hz to 100 Hz

3.2 Micromixer

The mixer is single mixer manufactured by IMM (figure 4). In macro world, mixing happens usually through turbulent flow, but flows in microfluidic systems tend to stay laminar so mixing has to happen through diffusion. The mixing element divides inflows into interdigitated channels with corrugated walls, which can be seen in figure 5.

Fig 4. The micromixer.

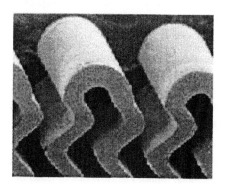

Fig 5. Channels of the mixing element.

The mixture is discharged through perpendicular slit. The configuration increases the contact surface between fluids and improves diffusion.

3.3. Reaction chamber

The reaction chamber was made of polycarbonate at TUT workshop. It is quite big, its volume is almost two milliliters, when the volume of the whole system without the chamber is 1.5 ml.

In this system reaction chamber is added only for observation purposes. Most of the suitable test reactions are very fast and happen already in the mixer. However if reactions were very slow or needed a catalyst or higher temperature, the separate reaction chamber would be necessary. Reaction chamber could also be used for placing sensors to the system. Other option for placement of sensors is the flow channels. In this case the channels are plastic tubes so it would be reasonably easy to add a flow-through chamber for sensors without any separate reactors.

4. TESTING

The pumps were tested individually without load. The tests showed that they are functioning quite differently as can be seen in figures 3 and 4. Of the six tested pumps only three were working after testing. Only one of the pumps achieved the maximum flowrate of 0.5 ml/min, which was promised by ThinXXS. Tests were performed by pumping distilled water for couple of minutes and then weighing the difference of mass in the container.

The pumps seem to function better, when they are driven with frequency adjustable FSB drive electronics. The best of the pumps reached even flow-rate of 0.6 ml/min, which is considerably higher than the maximum flow rate specified by ThinXXS.

Though the pumps function differently, the performance of each pump seems to remain quite similar in different measurements. This makes it possible to control the pumps by tabulating the functioning of the pumps.

Testing showed that the pumps are more effective to pump air than liquid. The flow rate of water in the tube seems to be even doubled when the pump chamber is filled with air. This is very interesting, because with macro world pumps the situation is normally opposite.

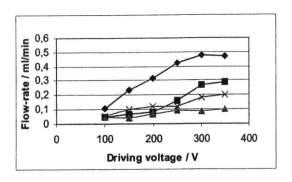

Fig 6. Flowrate of the pumps as a function of voltage.

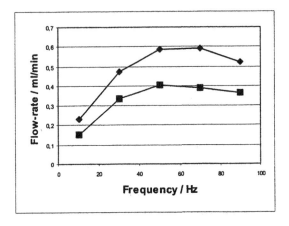

Fig 7. Flowrate of the pumps as a function of frequency.

The reason of this behavior is probably the higher friction of water in the pump valves.

5. FUTURE PLANS

The microplant is intended to be a laboratory experiment for students. Currently, it is merely a demonstration of microfluidistic system. It shows that microfluidistic components can be put together to form a complete system.

The system can fulfil its educational purposes even in its developing phase. Further testing and developing of the system can be given to students as special projects. Doing some independent research on the system under supervision is a very effective way to learn about microsystems.

The components of the system should be tested more thoroughly. There are no tests done with the mixer. The mixing performance can be tested using immiscible liquids, like oil and colored water. Also the pumps can be tested further. Currently the pumps have been tested only with water and air. Functioning of the pumps could be tested with fluids of different properties to see for example the effect of viscosity on flow rate.

After testing, the system should be implemented with sensors. Quantities that could be measured of the system are for example: pH, temperature, redox potential and many ions (potassium, sodium etc.). Implementation of the sensor can be given to the students.

In addition to performing experiments with the system, a pre-laboratory exercise can be given. Model the system using electric analogies (Senturia, 2001). With experiments the model parameters can be determined and the overall model verified. For students that have background in using macroworld FEM simulation packages (Lehtonen, 1999), a more careful simulation task could be assigned. When the system is planted with a suitable sensor, it can be modelled and it is also possible to design an automatic control loop for the system. At this point the system can be used as a laboratory exercise also in larger courses. This way the chemical microplant is a useful educational tool during its entire lifespan.

6. CONCLUSIONS

This paper presents a microfluidic system that is built for educational purposes. At the moment only the skeleton of the system exist i.e. there are two micropumps, micromixer and a reaction chamber that are assembled in a three dimensional rack and connected with plastic tubing in order to form a microfluidic system.

At the moment the system is merely a demonstration tool for introducing students the state of the art in commercially available microcomponents together with some microfluidic system specific phenomena. In the future the aim is to broaden the use of the system starting with the testing of the components and implementing the needed sensors and ending in the modelling and finally controlling the whole system. All this can be done by the students in well organized and supervised special projects and courses. In addition, when the control system is ready the chemical microplant can be used as a laboratory exercise in larger courses. This way the system can be used through its whole life cycle for educational purposes.

REFERENCES

Institut für Mikrotechnik Mainz Gmbh:
  http://www.imm-mainz.de/ (15.2.2003)

Lehtonen, T. (1999). *Microfluidic phenomena, components and simulation* (in Finnish), 93 p. Helsinki University of Technology, Control Engineering Laboratory.

Senturia, S., D. (2001). *Microsystem design*, 720 p. Kluwer Academic Publishers.

ThinXXS Microtechnology:
  http://www.thinxxs.de/ (15.2.2003)

SELF-ACCESS STUDYING ENVIRONMENT FOR CONTROL ENGINEERING EDUCATION

Reijo Lilja, Toni Ollikainen, Pasi Laakso

VTT Technical Research Centre of Finland, P.O. Box 1301, FIN-02044 VTT

Abstract: A new system has been developed for the self-access studying of the power plant operation. The system consists of a simulator, plant operation system, instructor application and student application. The instructor application includes tools for the management of students and courses, execution of example task by simulator, definition of the feedback criteria and preparation of the malfunction pointing. With the student application the user can execute exercises, get immediate feedback from his performance and see the total status of his progress. The system can be used without a supervising person, specific room and according to the users own schedule. With the system the control education is effectively based on the learning-by-doing method in the real circumstances. In this paper the structure and function of the system and some practical examples are presented. *Copyright © 2003 IFAC*

Keywords: computer simulation, control education, education, educational aids, simulators, training.

1. INTRODUCTION

The power plant is an example of multi-parameter relationships. It is a very complex system whose behavior is described by intricate relationships between its parameters. Unlike direct and inverse monotonic relationships, where knowledge of the change in one parameter gives the change in the other, additional information is necessary in reasoning about non-monotonic relationships. This information is typically plant-specific and it is cumulatively collected during plant construction and lifetime and mainly stored in the plant control and automation system beside the tacit knowledge of operators. Because of the advanced control and automation systems the operation of the plant is nowadays mainly observing of the monitoring systems and so the understanding of the real physical process may be very thin.

In the incidents and changing modes it is important to have deep knowledge about the behavior of the process and the function of the control system. The traditional education of control and automation systems for the operators is based on the system descriptions, textbooks and laboratory exercises, which are supervised by the full time instructor. In the cases, where the education background of the operator students is compendious, the motivation to that kind of studying is often poor and the results are unsatisfactory. The

simulator training has been the effective way to improve the skill of the operation personnel alongside with theoretical lessons. Because of the high training expenses, the use of the full-scale simulators is not very common in the education of the conventional power plant personnel. Initial costs, employment of instructors and reservation of classroom are the main reasons for the restricted use of simulators. The training arrangements are also quite laborious including collecting the trainee group, scheduling the training, measuring of individual results and maintenance of simulator.

The purpose of this work has been to develop a new system for the self-training of power plant personnel. This system is called Training Manager. With this system the student can practice the operation of the plant and study the function of sub processes, e.g. control and automation. The drawbacks of the traditional simulator training has been avoided by introducing the following features to the new system:
1. System runs on a single PC, the equipment costs are very low.
2. Instructor is not needed; system gives instant feedback automatically.
3. Classroom is unnecessary, the study may happen anywhere using the portable PC.

4. There is no need to collect a trainee group; single person can use the system according his own schedule.
5. Measure of individual learning results is impartial; system generates the results automatically using the same criteria for all students.

In the Training Manager system the learning is based on learning-by-doing method, which is an effective tool, when the high theoretical knowledge is not the goal of education. The best equipment for the learning-by-doing method is the simulator and real or almost real operation system of the plant, which are essential parts of the system.

In this paper the structure and functions of a new system for the self-access training of power plant personnel is described. The presented practical examples show how the system is used for the control education of the power plant.

2. SYSTEM DESCRIPTION

2.1 Studying environment

The system consists of a training simulator, plant operation system, instructor application and student application. Figure 1 shows the general structure and basic flow of the information. The education data and results are stored in xml-files. Via these files the two separate application are connected to the common database. Xml-files include the following data:
- information of the system users: names, roles and achievements of students
- the list of quantities, which are available in the system for the monitoring during the task
- information about the configuration of the simulator
- available malfunctions or events
- descriptions of the separate training tasks
- evaluation criteria of the tasks.

In addition to the mentioned xml-files three ascii-files are created during the exercise. The first includes the list of registered courses, the second includes the values of the quantities during the reference performance and the third includes the values of quantities during the student performance.

2.2 Instructor application

The instructor application provides tools for administration of system and to maintain, create and modify exercises. The main functions are:

Administration of the study books. The instructor can add, view and modify study books. Study books include information about the skill level of the stu-

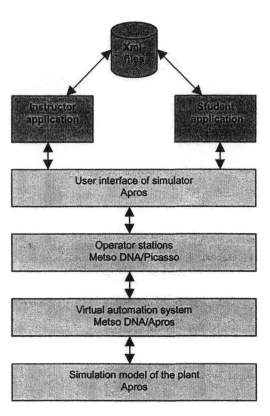

Fig. 1. Principal structure of the Training Mananager studying environment.

dents, which are registered in the system. The executed courses and gathered scores are presented.

Creating new course. By this function a new course is added to the system. The name, location and description of the course can be given. The course may contain several exercises.

Creating new exercise. A new exercise is prepared by choosing the initial condition of the plant, trend variables and malfunction. The comprehensive set of initial conditions of plant is available for the user. All exercises shall begin from these states, if necessary it is possible to add new states to the system.

The trend variables are either analog or binary quantities, whose values are inspected during the evaluation on the exercise. Binary trend variable can be e.g. state indication of automation sequence, controller, valve, pump and motor.

A malfunction may be included into the exercise to enable the creation of the more complicated tasks. The malfunctions are prepared and included in advance to the system. From the malfunction list the instructor may choose the proper incident and set the time, when the malfunction activates in the process.

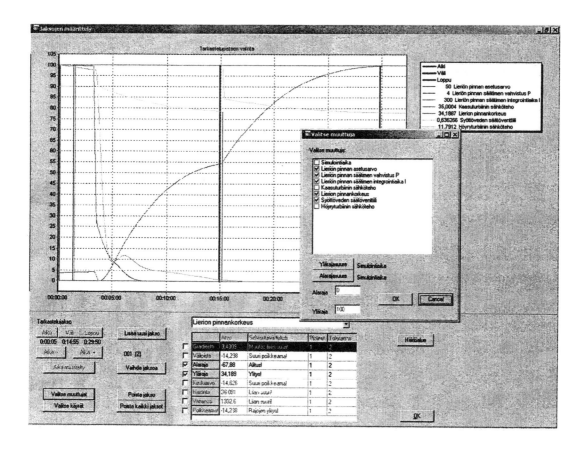

Fig. 2. Finnish version of the exercise definition sheet of the instructor application.

Execution of the reference exercise. The evaluation of the student's skill is based on the comparison between the instructor's and student's performance in the certain exercise. Before the definition of the comparison criteria the instructor must perform the reference run by the simulator. In the reference run the given operation task is executed correctly from the instructor's point of view. The simulator is operated via the simulator interface system and the plant via its own operation stations.

Definition of evaluation criteria and scores. The evaluation of the performance of the student is based on the checking of trend variables. In the Figure 2 is shown the interactive window for the definition of the evaluation criteria of the of instructor application. Choose variables –window is opened to pick criticized variables. Curves of the chosen variables are shown on the top left and criteria of the single variable on the table. Defining and navigation of checking periods are done with the buttons on the bottom left.

The system offers eight different criteria for each chosen trend variable. These criterias are the permissible
- deviation from the given value at the defined moment
- maximum value during the checking period
- minimum value during the checking period
- deviation from the reference values during the period
- gradient of the variable during the period
- average value of the variable during the period
- variance of the variable during the period
- standard deviation of the variable during the period.

The instructor chooses the proper criteria and defines the reference values, tolerances and points for those criteria.

Definition of the points in the discovery of malfunction. The discovery of the malfunction in the process is an important ability for the operator. In the instructor's application the user can include the chosen malfunction into some checking period. The instructor prepares the discovery of malfunction according to the following procedure:

1. process diagrams are selected from the drawing gallery
2. location of the malfunction is marked by selecting rectangular areas from the diagrams
3. the points for the correct timing, positioning and type of the malfunction are defined.

2.3 Student application

The student application is a self-access studying environment for the operators of the processes. The student can study the operation of the plant by executing exercises according to his own training program. The exercises are prepared by the instructor's system. The studying session is comprised of choosing, execution and feedback of exercises. Furthermore during the exercise the student can stop the simulation and point the fault in the process or get feedback about his performance.

Choosing of the exercise. After the starting the system the student can browse the list of the available courses and exercises. After the exercise has been chosen, the system offers advice to the student how to execute the task. These advices include the initial state of the plant, purposes of the exercise and possible operation guides.

Execution of the exercise. The execution of the exercise begins by the automatic starting of the operator and simulator systems. The operator displays of the real plant are available for the required tasks and monitoring. The task may be the change of the load, start-up, shutdown, discovering of the fault, changeover of process components etc…

Pointing of the discovered fault. When the student doubts that there is some fault in the process, he can stop the execution and try to point the location and type of the fault. The activated malfunction system first offers several process diagrams from which the student chooses the right one. In the next step the student points the defective component on the chosen diagram. After the pointing the system asks confirmation and if the selection was correct makes further question about the type of the fault. Finally the system presents the results of the fault discovering. The student gets points from the timing, localization of component and definition of type.

Feedback from the exercise. The feedback system can be activated both during and after the execution of the exercise. In Figure 3 is shown the feedback sheet of the Training Manager student application. The intermediate results are presented from all checking periods whose end time is smaller than the actual time of the execution. Execution of student and instructor are presented by the curves. Scores of execution are on the bottom right above navigation buttons. Literary feedback is shown on the bottom left corner.

The feedback system divides the time of performance into several sections. In each section the values of selected trend variables are compared to the reference values. The comparison is done according to the eight possible criteria, which are described in the chapter 2.2. The feedback from each active criterion includes the values of student and reference run, permissible tolerance for the correct performance, the earned points and comment. The user can browse the variables one by one to see all his results. The results are also presented in the graphical format. The reference and student curves of the selected variable are shown with different color from the start to end of the session. In addition to this the permissible limits and the borders of the section are shown on the plot area.

On the feedback form the total points of the section, sum points of all sections and the detailed points of the possible fault discovery are also presented.

2.4 Training Simulator

Training simulators have been developed for along time. Until recently full-scale training simulators has required sophisticated hardware. However during the past decade the improvement in the processor capacity and advances in software technologies have enabled creation of low-cost training systems (Porkholm, *et al.*, 1997; Ollikainen, *et al.*, 2002). This development has also made Training Manager system possible.

Typical parts of the training simulator are the simulation engine, automation system, operator displays and user interface for the simulator. Simulation engine is used to emulate the behavior of the real process. Automation system controls and "protects" the simulated process. Operator displays are used for interaction between the simulated process and operator. Ideally the functionality of the automation system and operator displays in the simulator are identical or very close to the functionality of the corresponding systems in the real plant. Some kind of simulator interface is used for controlling the training session. Typically it enables e.g. loading of initial states, starting and stopping, changing the speed and triggering malfunctions in the simulator.

3. CONTROL EDUCATION

The operation of the modern steam power station is based on the advanced automation system. The basic task of the automation system is to take care of the correct amount of the steam generation according to

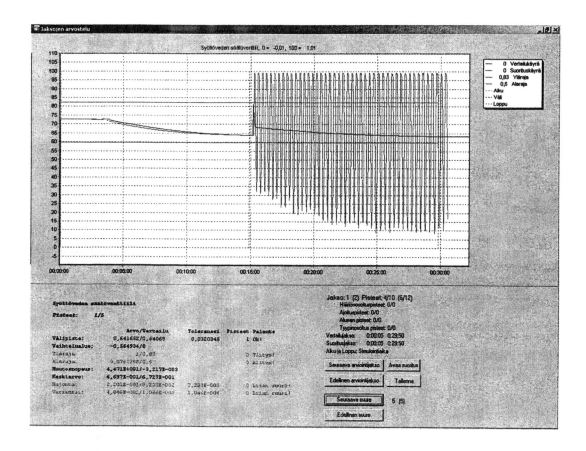

Fig. 3. Finnish version of the feedback sheet of the student application.

the load, to keep the pressure and temperature of steam inside the permissible limits during load changes and to adapt to changing load demands fast. These requirements are implemented by control circuits, which are connected via direct signals or process measurements.

The separate controllers, actuators, sensors and subprocesses compose an integrated unit, whose function is hard to perceive. In the following examples is shown how the previously described Training Manager system can be used for the control education of the combined cycle power plant personnel. The presented exercise is taken from the training course, where the tuning of the drum liquid level controller is rehearsed. Parameters of the controller only tuned according to compensating signal may not work in the set value change. In the part 1 the controller is tuned in the load reduction transient and in the part 2 the set value of the drum level is changed.

Process- and automation model is made with Apros simulation software (Porkholm, et al., 1997). Operator's displays are build with Picasso-3 user interface management system (Jokstad and Sundling, 2000).

3.1 Part 1: Load change of the plant

This exercise contains a power reduction of the plant. Diminishing heat power decreases steam consumption and less feed water is needed in the heat recovery boiler. Controller to be tuned controls feed water flow into the drum. Student has read instructions to reduce load demand of the plant and tune a controller. Tuning is done during the simulation, which is repeated needed times to find suitable parameters. Results are shown after every attempt to present the state of tuning compared to reference tuning. Feedback sheet is shown in the Figure 3, where first 15 minutes of the curves are the result of exercise 1. After succesful tuning student can continue to part 2.

3.2 Part 2: Drum level set value change

Task of the exercise is to continue from part 1 and check the parameters of the feed water controller in the step change of the set value. Tuning is not yet complete as shown in the Figure 3. Curves from 15 minutes to 30 minutes are result of the part 2. Parameters of the controller are competent for the part 1 but not for the part 2. Immediate feedback is given and student will see faults he made.

4. CONCLUSIONS

A novel system for the education of the dynamical behavior of the power plant was developed. The system can be used in the power plant personnel training and in the control engineering education. Because the system includes entire control system of the plant, it is possible to teach function of the high level control circuits, different modes of controllers, local controllers, automatic sequences and set value formation. The exercises may consists of tuning of controllers, switching control circuits on and off and discovering of the faults in the control system. Regularity, repetition and self-access use are relevant for the education by the developed system. Each student will receive a detailed feedback from system, which offers an opportunity to compensate deficiencies in basic skills.

The study of control systems via the presented system will essentially improve the understanding of the dynamical behavior of the real power plant and clarify the role of the control system as a fundamental part of it.

5. ACKNOWLEDGEMENTS

The development work of the Training Manager system was supported by the National Technology Agency TEKES, Andritz Oy, Fortum Power and Heat Ltd, Fortum Service Ltd, If Industrial Insurance Ltd, Kymenlaakso Polytechnic, Metso Paper Automation Ltd and VTT Technical Research Centre of Finland.

REFERENCES

Jokstad, H. and Sundling, C.V. (2000). Technical Overview of the Picasso User Interface Management System.

Ollikainen, T., A. Heino, K. Porkholm (2002). Generic Training Simulator for a Combined Cycle Gas Turbine Power Plant. The 43rd Conference on Simulation and Modelling (SIMS 2002). September 26-27, Oulu.

Porkholm, K., K. Honkoila, P. Nurmilaukas, H. Kontio (1997). APROS Multifunctional Simulators for Thermal and Nuclear Power Plants. 1st World Congress on Systems Simulation (WCSS 97). September 1-4, Singapore.

www.elsevier.com/locate/ifac

INTERNET TECHNOLOGY AS SUBJECT AND MEDIA IN AUTOMATION ENGINEERING EDUCATION

A. Braune, O. Sergueeva, M. Münzberg

Technische Universität Dresden
Department of Electrical Engineering
Institute of Automation, Chair for Automation Engineering
D-01062 Dresden, Germany
email: braune@ifa.et.tu-dresden.de

Abstract: The increased popularity and availability of off-the-shelf internet technologies raise a new trend for industrial automation solutions. Automation engineering education must adapt to this trend. The Institute of Automation at Technische Universität Dresden develops tele-learning experiments for teaching internet technologies via the internet. The laboratory experiments were planned based on the analysis of relevant internet technologies for automation applications. Some of these experiments exist in a first implementation version. Contents and procedures of theses experiments and first lessons learned are presented.
Copyright © 2003 IFAC

Keywords: teaching, internet, teleservice, telecontrol, interconnection networks

1. INTRODUCTION

Industrial products and systems for automation solutions are increasingly based on internet technologies. Many off-the-shelf components with new characteristics are already offered. Practical automation solutions integrate today traditional automation and current internet technologies.

Automation engineering education must adapt to this trend. Additionally to the established courses students need to be offered appropriate courses dealing with internet technologies, covering topics such as network technologies, network engineering and software engineering. One of the aims in automation education is the combination of theory and practice, i.e. it is necessary to teach the techniques and possible pitfalls of theory-based methods, when applied to technical applications (Schmid, 2001). For this reason automation engineering students have to perform also laboratory experiments with industrial internet technology to gain their own experiences.

The development of control laboratory experiments usable via Internet is the main topic of the project "LearNet - Lernen und Experimentieren an realen technischen Anlagen im Netz", which is funded by the German Federal Ministery for Education and Research (BMBF). LearNet includes seven German universities and technical colleges related to automation and it is devoted to the laboratory approach for remote experimentation with real plants via the public internet.

The Institute of Automation at Dresden University of Technology is involved in this project with a contribution to "Usage of internet-based communication services for automation solutions" i.e. teaching internet technologies via internet based tele-learning. Within the scope of the project the laboratory experiments include both instrumentation and development of industrial automation solutions with remote access via the Internet. In this approach the internet acts not only as communication media for distance learning, but its handling is a matter of laboratory experiments as well.

The current paper discusses the relevant internet-based technologies for automation applications as basis for the main topics of the planned laboratory experiments. Contents and procedures of the designed experiments and first lessons learned are presented.

2. RELEVANT INTERNET-BASED TECHNOLOGIES IN AUTOMATION APPLICATIONS

The TCP/IP (Transmission Control Protocol/ Internet Protocol) is the networking technology of the internet today. This protocol suite defines a layered protocol stack for networking (see Fig. 1) following the OSI protocol architecture defined by ISO . The IP-protocol assigns a global unique address to every packet of information. The TCP protocol suite provides a flow of data between two network nodes (hosts). The data flow is either connection oriented and reliable but expensive with TCP or connectionless and unreliable but less complex with UDP (User Datagram Protocol).

The data link layer handles the access to different types of physical media. For automation solutions the Ethernet is used commonly connected with powerful switches (Wollert 2000).

Unique services like World Wide Web (WWW), the File Transfer Protocol (FTP) or the Mail-Protocol (SMNP) define the application interface and the final data representation. At the application level automation solutions deal with several interfaces or services (see Fig. 2) in case of using internet technology (Bettenhausen et al, 2002) :

- *WWW – the standard of office automation.* In office systems the WWW is established today as the worldwide unique de facto standard and integrates other services like FTP or Mail. All Microsoft Windows™– operating systems include a standard WWW-browser. This global availability makes it very interesting for automation applications e.g. monitoring or control.
- *Open standards of automation* The requirement for interoperability between automation tools and systems of various producers enforced the definition of application layers and product specific profiles. Each automation field bus established its own standard. The increasing acceptance of TCP/IP based networks in automation resulted in the development of appropriate application layers and initiated again the discussion about a unique application layer for each communication system. One widely used and commonly accepted standard in this area is the OPC (**OLE** for **P**rocess **C**ontrol). Other drafts of OSI layer 7 (see Fig. 2) standards often include traditional field busses in addition to TCP/IP based networks. Until now however no draft could win serious recognition.
- *Specific Automation Standards.* Special application software can be overlaid directly on

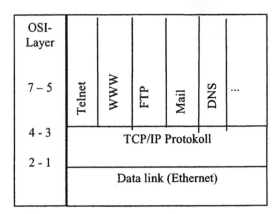

Fig. 1. Simplified layered model of the Internet

the TCP/IP protocol. Only appropriate clients can interact with their servers. Such closed client server structures are frequently implemented in proprietary company specific software systems, e.g. software development systems for programmable controllers allow with this extension full editing, loading and debugging functionality via field busses as well as via internet or intranet (Bettenhausen, 2002).

The relevant internet-based technologies determine the main topics of the laboratory experiments.

Ethernet. With its media access method Ethernet cannot guarantee real-time communication. This problem can be overcome to some extent by switched network topology and fast Ethernet with a bandwidth of 100 Mbit/s, which reduces the risk of collisions and delays. Student experiments have to deal therefore with suitable network topology, network management and communication media (electrical or optical). Tests and comparison of several network topology generally demand for changes of the experiments hardware configuration. In a tele-learning environment this would need a concurrent co-operative local operator to change the local hardware, which is far too expensive for such a purpose.

TCP/IP. Dependent on designing intranets or internets an automation engineer has to insert valid IP-addresses. He has to configure network parameters and routing functionality. Some automation applications forbid alternative routes, other ones benefit from parallel routes. Student experiments should at least integrate the network (software) configuration and some properties of TCP or UDP. For internet-based experimentation only TCP or UDP properties can be demonstrated from the practical point of view.

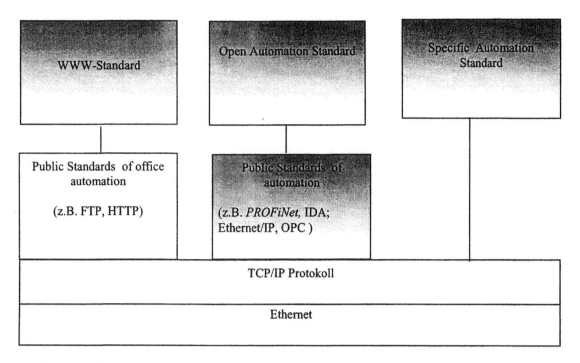

Fig. 2. Interfaces for automation solutions with internet technology

WWW. This service was designed for special requirements of office applications. The advantages and risks for automation solutions should be demonstrated by student experiments.

Open standards of automation. At present OPC is a common and open standard. Experiments should demonstrate the main functionality of the OPC Object model and its programming interfaces.

Specific Automation Standards. Experiments should show the configuration and functionality of such specific client server applications.

3. AVAILABLE STUDENT EXPERIMENTS : CONTENT AND ORGANISATION

The educational objective of the experiments at TU Dresden is to make the students familiar with the application of internet-based communication services in the field of teleservice for automated devices and processes. Engineers, who are working in this field, work in two different application profiles.

Service and field engineers. These automation engineers have to use or to maintain ready hardware or software. They need knowledge and experiences about the internet user interfaces and the relevant performance parameters of fully configured solutions. These engineers deal with the following tasks of teleservice:
- remote initiation and maintenance
- remote visualisation and operation of process data
- failure diagnosis

- remote alerting

Development engineers. In most cases the application specific software has to be implemented in existing hardware platforms. So these automation engineers should be able to design and implement complete software functions for internet-based applications on the basis of well known standards, e.g. OPC or WWW. Sound experience with object oriented software development is mandatory in this case.

Two derived educational objectives result from the application profiles:
- recognition of specific application features, of advantages and disadvantages of internet technologies for the application field of teleservice,
- development of internet applications for teleservice solutions.

According to these derived educational objectives there are two experiments as part of the "LearNet" Project of the Institute of Automation at Technische Universität Dresden.

Remote monitoring, diagnosis, alerting and control with ready (pre-configured) software using a
- WWW-Browser
- OPC-Client.

The main objectives are to make oneself familiar with the specific browser interfaces, the properties of an internet-communication (e.g. not predictable time-delays) or the object model of OPC and its configuration requirements (e.g. update time). Remote alerting experiment will demonstrate different possibilities for the usage of internet

Fig. 3: Process plant model

services like mails or telecommunication media like mobile communication to propagate event driven process alarms to remote operators.

Development of internet-based software applications:

- HTML and XML documents
- Java-Programs (e.g. Applets, Beans)
- OPC-Clients

These experiments allow the students to develop and test their own programs or applications.

All experiments use a laboratory process plant model, which is equipped with industrial components for measurement, control and process automation (see Fig. 3). The configuration allows the measurement and feedback control of temperature, fluid flow and the fluid level in the three containers. The plant is equipped with industrial automation components like programmable controllers and a process visualisation system, which allow a full local control and monitoring of the plant. This standard equipment was extended with Web-components for a direct interaction with the visualisation system as well as with the controller and with an OPC-Server. Figure 4 shows the configuration of the web-extension.

The LearNet project includes the experiments and according learning units. These learning units present online the necessary theoretical background of the experiments. Each experiment is divided into several sequenced activities :

- study of the learning units and the experimental environments;
- test of learned lessons as precondition for the experiment;
- execution of the experiment;
- final test.

4. IMPLEMENTATION AND LESSONS LEARNED

The following experiments exist in a first implementation version :

Remote monitoring, diagnosis and control with ready software

- using a *WWW-browser*. The student can download HTML-pages from a WWW-Server with embedded Java-applets for monitoring and control of the laboratory process plant model (see Fig. 5). Java-applets communicate via a server-based middleware with the local process automation system. This middleware consists

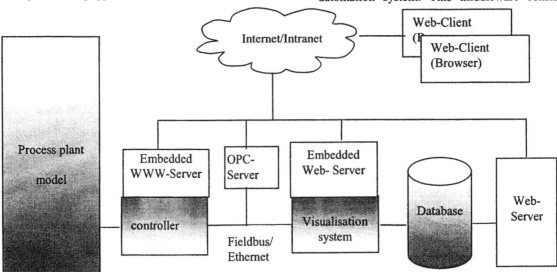

Fig. 4. Components of the experiment environment

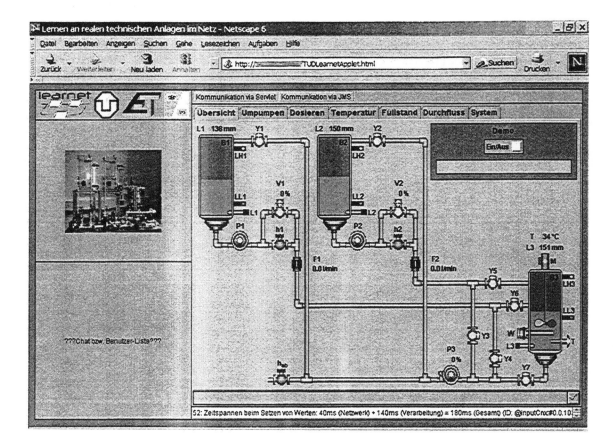

Fig. 5. Web-application for monitoring and control of the laboratory process plant model

either of a client-server based solution (Java Server Service) or of a producer-consumer concept (Java Message Service). The remote operator uses dynamic HTML pages to monitor and control the process, to measure some characteristic properties such as data rates or to test some specific design requirements, e.g. the permission of a concurrent remote access to the local automation system.

- using an *OPC-Client*. The remote operator can download an OPC-client and install it on his local computer. This OPC-client allows in an efficient way to become familiar with the OPC-interface and –its functionality by changing some
specific properties of OPC-classes such as update-rate, cyclic or event-driven read or write access.

Development of internet-based software applications:

- *HTML- and XML documents*. The student can implement HTML-pages for monitoring and control of the process plant. The necessary Java-Applets are provided as ready to use downloads. He also gets direct access to the local WWW-Server to download and test his own Web-pages.

- *OPC-client*. The programming of an OPC-client requires an appropriate software development environment. The OPC standard offers programming interfaces for Visual Basic as well as for C++ or Delphi. The idea is, to use commercial software and tools, that are popular and already available at the local computer. The current implementation is based on Microsoft Excels Visual Basic™ because of its widespread availability on PCs. The student gets some ready examples about the OPC programming interface, the "Automation Interface" and has the possibility to extend and adapt a given program to his needs.

Up to now approximately 30 students of the specialisation area of automation at TU Dresden have tested these experiments in the 2002 academic period using the university intranet. The software was already installed at the experiment PCs and a tutor was involved in execution and processing of the experiments. The students showed an extraordinary high interest in participating in the course and often prepared much more program functionality than required. They could gain first hands-on experience and many of them postulated, that but now they had understood the theory.

Using this laboratory course in a full internet far-distance training environment the following preconditions still have to be prepared:

- Implementation of a software tool for registration for the experiments (developed by a LearNet project partner)
- Installation and securing a WWW-server for remote access and downloads
- Installation of security management tools for avoidance of unauthorised or concurrent write access to the process plant model e.g. by more than one client or WWW and OPC-clients.

REFERENCES

Bettenhausen K. D., Braune A., Rieger B. (2002). Anforderungen an die Nutzung von Internettechnologien in der Automatisierung. atp 6/2002, S. 45-51.

Schmid,Ch. (2001). Web-based Remote Experimentation Preprints of 1-st IFAC Conference "Telematics Applications in Automation and Robotics, 2001, S. 443-447.

Wollert F. (2000). Ethernet in der Automatisierungstechnik Teil2. Elektronik, 21/2000, S.66-75.

ELSEVIER
IFAC
PUBLICATIONS
www.elsevier.com/locate/ifac

The use of the Control Web in Control Engineering Education

Juha Lindfors

Department of Research Methodology, University of Lapland
P.O. Box 122
96100 Rovaniemi
Finland
Tel: +358 40 576 1497
Fax: + 358 16 341 2978
juha.lindfors@urova.fi

Abstract: The teaching function in the university has been under a pressure of change in the recent years. Economical, methodological, and qualitative pressures have forced educational establishments to consider their education. One response to these pressures was to study if it is possible to build such a learning environment for control engineering that combines good characteristics of the traditional and virtual environments. Such environment could help to distribute materials and could facilitate the overall communication starting from course information until student feedback. It could also make studying more efficient enabling better follow-up of studies and enabling use of interactive functions. *Copyright © 2003 IFAC.*

Keywords: Control engineering, Education

1. INTRODUCTION

Traditional teaching at the University of Oulu is given in classrooms. Courses, consisting also of exercises, are carried out both in a classroom and laboratory. Recently, computer and network technology has developed very fast giving new possibilities for teachers and students to utilise new learning methods. The development makes it possible to utilise computer and web based learning to meet the new demands: the reduction in the resources and increasing number of students on the courses.

As the World Wide Web (later Web) service in the Internet grew on the level that it could be used to support the teaching seriously, a Web server was set up in the Control Engineering Laboratory of control engineering. Since 1997 the server has supported learning in the laboratory. From the beginning the server has been an information channel between students and teachers offering e.g., schedules for courses, results of exams and learning materials. In addition to the course administration, new activities have been implemented. Two special features have been implemented in to the laboratory Web. One is the Matlab Information Server, which is a simulation service through Web and the other is commercial learning environment called LCProfiler (Unknown B, 2000).

The basic pages in the Web server contain mainly html, although some Java code has been implemented. The main idea is to keep the structure of the pages and their construction as simple as possible. The pages can be created and maintained with simple, even free, editors. Pages were created gradually during the years, starting from simple and enhancing them when a better solution was found. The web pages have been tuned according to the student feedback.

The LCProfiler and the Matlab Web Server were used in the spring 2000 during the course Process Control II for the first time. The course was chosen because it is lectured during the last period in the spring and this causes some problems. The course starts in April and ends in May. The first problem is that the Easter holidays break the study time and the second is that the students will go to their summer work during the course. Most of them have disappeared till the end of April. In the end of the course an enquiry among students was made to collect experiences from the use of Web environments.

The learning theories applied on the course were "learning by doing" (Dewey, 1938) and "collaborative learning" (Häkkinen & al, 1998;Teasley and Roschelle, 1993)

2. REALISATION OF PROCESS CONTROL II

On completion of the course, participants should be able to:

❑ Design and carry out industrial experiments using common design methods (Hadamard matrix, Addleman Method, Central Composite Design and Taguchi).
❑ Analyse and evaluate results obtained in experiments using statistical analysis methods.

The official definition of the content of the course is as follows (Opinto-opas 2000):

❑ Basics of experimental modelling;
❑ Statistical analysis of experiment results;
❑ Experiment design;
❑ Variance and regression analysis; and
❑ Dynamic models.

These foci reflect the subjects that form the core curriculum of the course. The core curriculum represents those standards of learning that are essential for all students. They are the ideas, concepts, and skills that provide a foundation on which subsequent learning may be built. The following list presents the domain subjects that belong to the core curriculum.

Basics of experimental modelling. The student can state decisions, define the risks, establish an objective decision criterion, and compute the sample size.

Statistical analysis experiment results. The student knows the population mean, variance, and correlation coefficient, and can apply the t-test, Normal, t-, and f-distributions. Also a basic knowledge of the graphical methods is given. The graphical methods covered include the use of histograms and trends.

Experiment design. The two-level multivariable experiment design methods; matrix methods, especially the Hadamard matrix method and multi-level experiment design methods such as Addleman and Central composite designs are presented in this section. Exercises consist of experiment design projects. The students carry out the project in teams of at most five students. Four seems to be optimal number. The teams study a process, which is usually a rotary dryer. The teams go into details of the model and a dryer simulator. Based on this study, they define the variables they want to use design their experiments using a Hadamard matrix, and then carry these out using the simulator. Results are presented to the other students on the last day of exercises and also in the WWW environment.

Regression analysis. In exercises, students study how to make a model using the least-squares method. They should be able to apply linear, multivariable, exponential, and polynomial models after the course. The students also learn the kinds of problems to which the models can be applied.

Dynamic models. Two cases of dynamic model definition methods are introduced: definition of transfer function parameters from step response and impulse response. In the first case, the theory of Ziegler-Nichols, Miller, Broida, and Sten methods is studied. Students apply their knowledge in a laboratory exercise where they study and define the parameters of a thermal process.

For additional knowledge, the students study the basics of the Taguchi Method, Greco-Latin Squares, and random-strategy experiments and familiarise themselves with the DOE PC, a computer program for the design and analysis of experiments.

The course consists several sections. The main sections in exercises are as follows: simulation, additional examples, and independent study.

Simulation exercises are done partly independently and partly in a laboratory. In the independent part of the simulation, the students are expected to do their preliminary tests on the WWW and, according to the results, choose the variables to study and design the experiments.

Additional examples are problems that are given to the students to solve at home. The examples prepare the students for the simulation. Two examples concern regression analysis. To solve the examples, students have to make a regression model for a given problem. The students can get help from the assistant if they have difficulties. They are given just enough help to overcome the problem, after which they have to continue on their own again.

Independent study is done in the distance part. The students can use the WWW materials to help. Independent study means that the students deepen their knowledge on the subject. This part also contains the preparation of a written report on the experiment design project and familiarisation with the network learning environment. There was no time for collective practice with the environment. The assistant helped students when they needed it, which proved to be quite a good strategy. In the beginning, the assistant got only few emails concerning the use of the environment; in other words, the use of the environment seemed to be straightforward for the students. The use of the environment was studied once in the exercises, which took about one hour of time.

The course is given in the last period of the academic year (in the spring.) Three main problems have affected the realisation. Firstly, the students begin to leave for their summer jobs during the course. Thus they are spread all over the country, with some of them even abroad. The network part binds them to the course. Secondly, holidays during the period disrupt studies and, finally, May 1 activities at the end of the period further distract the students.

The exercises are completed as projects. Students form teams that decide on their own timetables according to their other commitments. There are, however, common lectures where the team or at least one member of it must participate.

The aim of the exercises is to acquaint students with experiment design and to have them study a process, design experiments for a process, carry out these experiments using a simulator, analyse the results, and, finally, present their work to other teams. The results of the experiments are summarised and compared with each other.

To motivate students, extra points are given on the course examination according to the work that they have done. Usually, there are five additional exercises, of which the experiment design accounts for three.

Usually, the first additional exercise deals with analysis of data using analysing software, most often Microsoft Excel. The data to be analysed is provided for download on a Web page on the laboratory Web. If a team does not have the possibility to download the data, it is supplied on a floppy disk. The fifth exercise deals with the sizing of control valves.

The number of exercise days varies according to the calendar. There are usually 8-10 days for exercises. The first is the motivation day. General aspects of the exercises are taken up, such as how to carry out the exercises, form the teams, additional exercises, what programs are needed, where to find them, and how to find things on the Web. The participation of all students is recommended. The same applies to the second day, when the experiment design part begins.

After the first two days, the teams go on with their work according to their own schedule. However, there are certain days when the assistant is available to the students if they want to do something under guidance or ask advice. Participation in the exercises is not compulsory, but is highly valued. It has been observed that students who are absent miss out important information. Email has been found to be a good tool to communicate with the teams during their work with the project.

3. THE CONTROL WEB

Building Control Web started as early as 1995. The hierarchical structure chosen is chosen that described in Figure 1. Development of the structure has been based on student feedback, which has given guidelines for the structure and functions included in the environment. Some guidelines have been adopted from other environments as the results of other studies of similar Web services have become available (Manninen 1999, Lifländer 1999). Also closed learning environments (TELSIPro and Proto) were studied when they were introduced on the market.

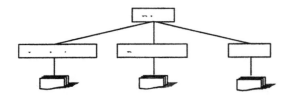

Fig. 1. Structure of the educational part of Control Web (Lindfors, 2002).

Such functions as study guidelines are given on the pages of University and the department. An expert directory is not implemented, but a list of interesting WWW pages that refer to the course in question can be found on the course page. Guidelines for studying are given in Adobe pdf documents that can be downloaded from the pages or read directly using the browser with the reader plug-in. A plug-in is a program that enables a feature to be used directly in a browser. Public discussion forums are implemented on a common C++ program base and modified in the laboratory. A library is implemented as a set of course materials and list of links where more information about the course can be found.

Course results were also requested by the students, this was the first service implemented in our approach.

Control Web consists of WWW pages that have links which connect all the materials together. The structure is a hierarchical, treelike system (Lindfors *et al.*, 2000). There is a main page for Control Web, which includes the link to the educational pages. Figure 1 illustrates the overall structure of the educational part of Control Web.

The structure of the Web courses can be based on functions or on courses. The "functions approach" means that separate pages correspond to functions of the courses such as course descriptions, schedules, and the results of examinations. A division by functions means that, for example, the administration links for all the courses are on one page.

The other method is to keep the courses intact and include all the functions and information concerning a specific course in one place. According to the feedback, the division by courses is clear to students, and it is used in this application. The division-by-courses method was chosen initially because it was believed to be as good as division by functions. Division by functions was not implemented in this study.

According to the responses on the questionnaires, the structure was clear and functional for the users. Moreover, it was easy for the administrator, because the folder structure in the Internet information server is similar. In the physical folder structure each course has its own folder where the HTML files (Web pages) of the course are located.

One disadvantage of a division by courses is that it may lead to the repetition of pages. For example, in our solution each course has its own page for grades, whereas in the division-by-functions solution grades would be on a single page. In the division-by-course solution, pages are connected by hyperlinks to each other and the solution uses shared pages whenever possible.

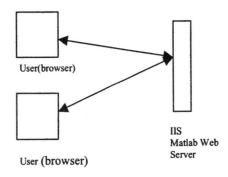

Fig. 2. Simplified presentation of communication with Matlab over the WWW (Lindfors, 2002).

4. THE MATLAB WEB SERVER

To run the Matlab simulations, the Matlab Web Server was configured to run together with the Microsoft IIS server. The Matlab Web Server enables one to make Matlab applications that use the capabilities of the WWW to send data to Matlab for computation and to display the results in a Web browser (The Mathworks 1999). The basic configuration of the server is a Web browser running on a client workstation with Matlab, the Matlab Web Server, and the IIS running on another PC. (See Figure 2). Simultaneous use may be limited depending on the performance of the IIS and PC. The server is designed to run continuously in the background as a WinNT service.

The Matlab Web Server consists of a set of programs that enable Matlab programmers to create Matlab applications and access them on the

WWW (The Mathworks 1999). Matlabserver manages the communication between a WWW application and Matlab. Matlabserver is a multithreaded TCP/IP server. It runs a Matlab program (m-file) specified in a hidden field (mlm-file) in the initial HTML document. Matlabserver invokes Matweb, which in turn runs the m-file. Matlabserver can be configured to listen on any legal TCP/IP port by editing the matlabserver.conf file. The number of simultaneous sessions is specified there.

Matweb.exe is a TCP/IP client of Matlabserver. This program uses the CGI to extract data from HTML documents and transfer it to Matlabserver. Matweb calls the m-file that a user wants the WWW application to run. Matweb.conf is a configuration file that Matweb needs to connecting to Matlabserver. Applications that can be run must be listed in the matweb.conf file. Otherwise, the user cannot run an application on the Web site.

The Matlab Web Server file consists of three parts: input, actual program, and output. In the input part, the variables are received from the browser and old images are cleared, among other actions. The output part creates the images and places them on the output Web page.

5. THE LCPROFILER

LCProfiler is a commercial learning environment software developed in the University of Oulu (Unknown B, 2000). It is a closed learning environment and run in a different server than the laboratory Web. Closed in this context means that the student needs a password to login the LCProfiler. The learning space in the LCProfiler is divided to five subareas. The subareas are as follows: Project Office, Workshop, Communication, Library, and Administration. The Administration subarea is not available for students. The idea behind the LCProfiler is that students will have a study project, during which they make documents and the other students will comment these documents and discuss about them.

6. RESULTS

Two questionnaires were made in the end of the course. One was the traditional enquiry that will concentrate on the contents and teaching methods and materials, including also a question of learning

environments. The second enquiry was made in the spring 2000. This enquiry contained only questions of the Control Web environment.

According to the enquiries, most of the students liked the use of the environments, although they had some technical problems. Their opinion was that the use of Web was important, motivating and exciting method and they would like to use it more in the future.

Besides the feedback, grades during 1993 to 2000 were also studied. There was a possibility to take the exam in the May, but because most of the students were out doing their summer work, only few students took the exam. There were two possibilities in the autumn of 2000. The analysis of the exam results shows that the students performed as well as the average year. There were fewer fails, but also fewer the highest degree. The study shows also that there is a shift to better grades during the years the course was under development.

7. DISCUSSION

Simple Web pages can be enough to support learning. There is no need to construct complicated learning environments. The simple Web pages are easy to maintain and students can download them quickly, depending on the resources they have. When data transfer rates increase, more complicated objects and services can be put on the pages.

One thing that influenced to the speed of taking in the use of the new methods was the rapid development of networks. In the end of 1999 a network connection was installed into every apartment in student dormitories at the campus area.

The educational subarea of the laboratory Web pages is divided by courses. Division could be done e.g. according to the subject, but this division has been found clear. Division by the subject was not tested.

The LCProfiler was used as a closed environment where the teams can discuss of problems and look at the work of the other teams. "Closed" means that one needs a password to login the environment. To teach how to use the environment took a couple of hours from the exercise time. In the tight schedule that may be too much and in the future time for studying the environment has to be reserved.

Only two teams from ten did not publish their work in the LCProfiler. The one said they have no resources to use the environment and the other said they had problems to insert files in their documents. However, the second team used cut and paste technique to make a document in the environment.

One problem was that the teams did not want to publish their exercises. The reason was that the students did not want the other teams to copy their work or ideas. In the end of the course a day when all teams published their work was set. That was too late to start a discussion. However, the teacher could read all the documents and comment them.

In the future the exercises will be changed so that the teams will do their own exercises, not the same exercise as was before. To open the discussion, the teacher has to make some questions beforehand. For example, there could be one question for each week.

Although the students did not want to publish their work, it was good that the teacher saw their work and could comment and tutor the students if they were going to wrong direction.

The time the teacher used for the course more than doubled. The time applied to the course can be divided to three parts: first part is the normal administration of the course, the second part is the time to read and comment the documents in LCProfiler, the third part was just time to be in reserve in the chat if a student wanted to discuss with the teacher. In the LCProfiler, the first and third parts could be done at the same time.

REFERENCES

Dewey, J. (1938). *Experience and Education*, New York: Collier MacMillan.

Häkkinen, P. & Eteläpelto, A. & Rasku-Puttonen H. (1998). Project-based science learning in networked environment: analysing cognitive and social processes in constructing shared knowledge spaces. In S. Järvelä & E. Kunelius (Eds.) *Learning & Technology – Dimensions to Learning Processes in Different Learning Environments*. Oulu: University of Oulu, 35-49 p.

Lifländer, V-P. (1999) Verkko-oppiminen - Yhteistoiminnallinen projektioppiminen verkossa, Edita 1999, Helsinki.

Lindfors, J. (2002) A modern learning environment for Control Engineering. D.Sc. (Tech.) Thesis, Acta Unversitatis Ouluensis C 178, Oulu, Finland, p. 214.

Lindfors, J. & Yliniemi, L. & Jaako, J. (2000). Home page of Control Engineering Laboratory. http://ntsat.oulu.fi/ 16. Aug.2002.

Manninen T (1999) Computer supported learning and training centre for engineering education. Acta Brahea, Raahe. p. 93.

Opinto opas (2000) http://www.oulu.fi 20.08.2000

Teasley, S.D. & Roschelle, J. (1993). Constructing a joint problem space: The computer as a tool for sharing knowledge. In S.P. Lajoie & S.J. Derry (Eds.) *Computers as cognitive tools*. Hillsdale, NJ: Lawrence Erlbaum Associates.

The MathWorks Inc. (1999). Matlab Web Server Manual.

Unknown, A. (2000). http://www.ttk.oulu.fi/ opinto-opas/ 16. Aug.2002. (In Finnish).

Unknown, B. (2000). http://www.lcprof.com. 15. Feb. 2003.

Copyright © IFAC Advances in Control Education
Oulu, Finland, 2003

IMPROVING CONTROL DIDACTICS USING MULTIMEDIA

Reimar Schumann and Olaf Kriewald

University of Applied Sciences and Arts, Fachgebiet Automatisierungstechnik,
Ricklinger Stadtweg 120, D-30459 Hannover, Germany, FAX +49-(0)511-9296-1111
email:reimar.schumann@mbau.fh-hannover.de

Abstract. The technical evolution of university teaching has been triggered by multimedia because this new technology extends the presentation alternatives enormously. However, the didactical challenge to improve university teaching cannot be met by introducing multimedia alone but must be complemented by a thorough analysis of and by a subject specific adaptation to the didactical situation in which it is to be applied. Only if the specific didactical requirements are detected a concept for the efficient and combined use of multimedia based and traditional didactical techniques can be developed which yields the chance to make the learning process more efficient and to improve the learning results. – In this contribution several typical didactical scenarios in teaching control at the university are described and analysed with respect to didactical deficits, and possible improvements are outlined which can be obtained by modifying the traditional didactical approach and introducing multimedia based techniques. *Copyright © 2003 IFAC*

Keywords: Control Education, Computer Aided Instruction, Multimedia

1. INTRODUCTION

Multimedia has more or less become a synonym for the evolution of university education although it describes merely the new technical media which may (or may not) improve the presentation of teaching material. Using notebook and beamer it is possible to show a literally unlimited number of pictures, video clips and animations, to present simulations and to access teaching material on the internet including virtual labs or life video from production facilities, see e.g. Schmid (1998a, b).

At the 1999 IFAC congress the technical potential of multimedia was discussed in Schumann et al. (1999) for improving the presentation techniques of teaching material in the traditional university course set up. In this paper didactical aspects for using multimedia in university teaching are dis-cussed along typical scenarios for a university course on feedback control. The focus hereby is on the detection of potential presentation and learning problems which cannot be overcome with traditional methods and call for the application of multimedia techniques.

The paper is organised as follows: First personal and technical preconditions are discussed which are assumed as base for the following considerations. After a short description of the general set up of a conventional university course on feedback control, typical scenarios for lecture, exercise and laboratory are described pointing at potential didactical shortcomings. For every scenario possible solutions are outlined utilising multimedia presentation techniques for improving the didactical set up and technical and didactical chances and problems are discussed. General remarks on using multimedia for teaching control conclude the paper.

2. PERSONAL AND TECHNICAL PRECONDITIONS

One major problem for the introduction of multimedia into university teaching is the required qualification of lecturers and the provision of a standard technical multimedia environment in lecture rooms and laboratories. This is why in Schumann et al. (1999) a strategy was proposed which introduces lecturers into the use of multimedia gradually such that they become able to produce and present multimedia teaching material in principal without being required to become multimedia experts. Following this proposal the following minimum preconditions are set for the realisation of the subsequent scenarios:

2.1 Personal Preconditions

The basic abilities of the lecturer may include

- *Multimedia text processing:* The lecturer is able to produce and use standard text processing tools like MS-Word™ and to use its multimedia features (hyperlinks, video and photo integration, conversion to HTML etc.).

- *Multimedia format handling:* The lecturer basically is able to use HTML production tools like MS Frontpage™ and to convert multimedia material into PDF format using e.g. ADOBE Acrobat™.

- *Video and photo handling:* The lecturer knows basically how to produce, process, organise and present digital video clips and photo catalogues using tools like ULEAD Videostudio™ or ADOBE Photoshop™.

2.2 Technical Preconditions

The technical multimedia infrastructure may include (Fig. 1)

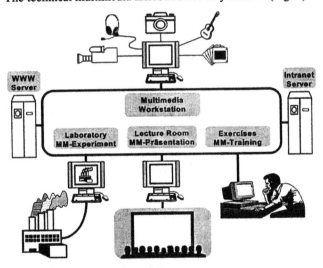

Fig. 1. Technical multimedia infrastructure

- *Multimedia workstations:* These are available in computer rooms or at the lecturer's office to produce multimedia teaching material.

- *Multimedia presentation facilities:* In the lecture room video projection facilities are available for multimedia

computers.

- *Multimedia laboratory computers:* In the laboratories multimedia computers are available for online assistance during experiments.

- *Internet infrastructure:* Internet and Intranet servers are available to provide storage and access to multimedia teaching material as lecture script or training material.

The following considerations are based on the assumption that the above preconditions are fulfilled.

3. TRADITIONAL SET UP OF A UNIVERSITY COURSE ON FEEDBACK CONTROL

The traditional set up of a control course comprises lecture, theoretical and laboratory exercises. Although there are great variations from university to university and also between the different countries a rough sketch is outlined here as a foundation for the following considerations.

Lecture: Typically at the beginning students are introduced into the concept of control as part of automation and into typical application fields (industry, private home, automotive systems etc.). Then control theory is presented including topics like the general concept of modelling, time and frequency domain models, system analysis methods, control system design methods, and complemented by a review of control realisation technologies. Depending on the time available typical extensions may include modelling techniques (theoretical, experimental), nonlinear control, multivariable control or digital control systems – just to name a few possibilities. Teaching is traditionally done by oral presentation using chalk and blackboard and/or slides, teaching material may include lecture script (prepared by the lecturer), handouts and/or textbooks.

Theoretical exercises: In the exercises typical control design tasks have to be solved theoretically in modelling, system analysis and control system design. This is done in taught exercises by the lecturer using chalk and blackboard and also in stand alone exercises carried out by the students themselves either at the university supervised by an assistant or at home. Exercise specific teaching material may include exercise scripts listing and describing the tasks to be solved.

Laboratory experiments: In the laboratory typical control design tasks have to be solved practically starting with the physical set up of control systems including laboratory process, instrumentation and control devices. Practical control system design comprises process analysis and modelling, control system design including selection and tuning of the controller and test of the realised control system. The laboratory experiments may be presented by the lecturer or carried out by the students stand alone in small groups supervised by a laboratory assistant. Teaching material provided for the students may consist of laboratory scripts describing the control design tasks to be solved and including technical descriptions of the laboratory equipment used.

4. LECTURE SCENARIO 1: ILLUSTRATION OF IN-DUSTRIAL CONTROL APPLICATIONS

The introduction of newcomers into the field of feedback control includes in general the introduction of the feedback control concept and the illustration of typical applications in various fields. Whereas the feedback control principle can be explained using well known applications like air conditioning or heating systems the illustration of industrial or automotive applications (to name just some popular application areas) can become rather problematic: control courses are taught on a course level where most students are not yet familiar even with the application areas of their specific engineering discipline for which they are educated. This means that the technical background is not sufficiently known for which feedback control applications are to be described. This lack cannot really be overcome by verbal descriptions or explanations.

This is the *first lecture scenario* for which a didactical improvement is proposed using multimedia techniques. A typical solution for introducing an unknown application field for control using multimedia may employ the following three description levels easily available with a multimedia computer, see Fig. 2:

1) *Pictures and video clip:* The technical background may be introduced using various pictures or a video clip illustrating the general purpose of the technical environment. Within the picture or video clip the control instrumentation should become detectable in a sense that the lecturer can point at the devices of interest in the technical context.

2) *Schematic drawing:* From the picture a first abstraction level can be realised by turning the picture into a schematic drawing or optionally, more illustrative but also more demanding, by producing an animation from the video clip. In both cases the interaction of instrumentation and control devices should be highlighted within the technical environment.

3) *Block diagram:* From the schematic drawing (or animation) a block diagram can be derived as a second abstraction level where all devices of the control instrumentation and the I/O behaviour of the process are represented by individual blocks. Optionally the interaction of process and control can be illustrated here by using a simulation program.

The production of the video clip or the pictures should be done with digital cameras and stored directly in a multimedia computer using standard recipes, see Schumann et al. (1999). The schematic drawing can be produced from the picture (or the video clip) by using standard photo processing tools for blurring, framing etc. – and highlighting the interaction lines graphically. The block diagram may be developed interactively using a block diagram editor or programs like MATLAB/SIMULINK™ which besides the drawing supports also the simulation of the block diagram.

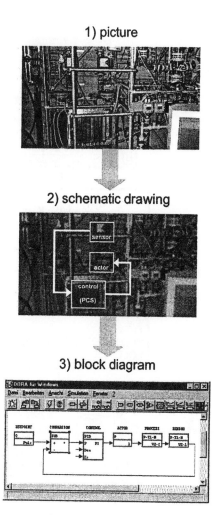

1) picture

2) schematic drawing

3) block diagram

Fig. 2 Instrumentation within industrial application

Having solved the technical problems the didactical advantages become obvious: The students can be introduced step by step, first into the technical environment presented in real pictures or video clips, then into the role of instrumentation and control by detecting the respective devices within the technical environment and last but not least into the functional interaction of process and control using the block diagram.

5. LECTURE SCENARIO 2: MODEL CONCEPT

The *second lecture scenario* is built up for the introduction of the concept of a mathematical model. Starting from the physical set up of a process, in simple cases it is rather straightforward to derive a mathematical model equation for a specific dynamic input-output relation. Even if the mathematical model is a simple linear differential equation the task to explain how this equation really "functions" turns out to be a didactical challenge. The mathematical answer is to solve the differential equation for specific initial conditions and for a special input signal. However, this mathematical answer is not sufficient for the intuitive understanding of the model which is able to process arbitrary input signals - there is a lack between mathematical abstraction and intuitive understanding.

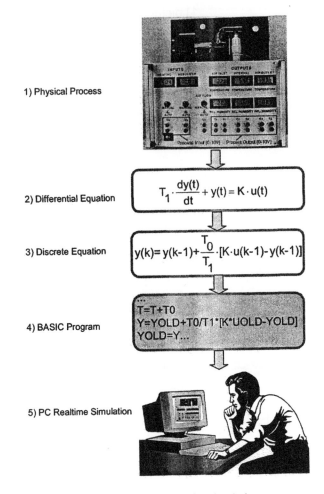

1) Physical Process

2) Differential Equation
$$T_1 \cdot \frac{dy(t)}{dt} + y(t) = K \cdot u(t)$$

3) Discrete Equation
$$y(k) = y(k-1) + \frac{T_0}{T_1} \cdot [K \cdot u(k-1) - y(k-1)]$$

4) BASIC Program
```
...
T=T+T0
Y=YOLD+T0/T1*[K*UOLD-YOLD]
YOLD=Y...
```

5) PC Realtime Simulation

Fig. 3. Mathematical model used for simulation

To fill this lack the multimedia computer in the lecture room can be utilised for the following demonstration procedure, see Fig. 3:

1) *Model derivation:* For a simple process a mathematical model in form of a first order linear differential equation may be derived.

2) *Discrete model:* By replacing differentials by differences the differential equation is transformed into a simple discrete equation which is able to predict the next output signal value. This simple prediction algorithm helps to build an intuitive understanding of "what the model does".

3) *Model simulation:* Within a simple BASIC (QBASIC, GWBASIC,..) program the prediction algorithm can be implemented as a single program line which is calculated recursively. By calculating the algorithm with the computer it can be illustrated how the mathematical model produces step by step output signal values based on arbitrary input signal values provided.

4) *Realtime simulation*: Optionally also the concept of realtime can be explained. By running the prediction algorithm at fixed time intervals (equal to the discretisation time interval) using the timer function of BASIC it can be shown that the computer produces the output signal

values at the same time as they could be measured at the real process – this means in real time.

Using the computer for these simple programming and simulation tasks adds just a new facet to the multimedia presentation features made available in the lecture room. The didactical advantage of this approach is that students can see and understand the simple means by which a mathematical model in form of a differential equation can be programmed on a computer and how it can make the computer act like the process in the sense of a true model.

6. EXERCISE SCENARIO

Exercises are used to train students in solving control design tasks based on the theoretical knowledge hopefully acquired in the control lecture. For this purpose typical exercise tasks are handed to the students on exercise sheets (→ exercise script) and solved either in taught exercises where the lecturer develops the solution or in stand alone exercises where the students have to solve the tasks individually either at the university supervised by an assistant or alone at home. The typical set up for exercises comprises the exercise script where the exercise tasks are listed and described and the lecture script providing the underlying theory. Typically the students produce individually a formula collection (collection of concentrated notes including formula and general information on solution schemes etc.) as the traditional working tool for exercises and examinations.

The *exercise scenario* regards the stand alone exercises. At the university the students are assisted in solving the exercise tasks by assistants. Depending on the numbers of students the supervision for the students may become a considerable teaching load. Furthermore it may be desirable to provide some additional assistance also for the home exercises. So, to help the students in their stand alone exercises it may become desirable to provide more multimedia based teaching support using natural extensions of the exercise set up. This may be done by converting the exercise script into hypertext format and by introducing multimedia assistance in the following steps, see Fig. 4a and b:

1) *Help extension 1 - hyperlinks to lecture script:* By reading the exercise script now the student can access and repeat the required theory online if necessary by using hyperlinks to the related sections of the lecture script providing the underlying theory for the solution of the exercise tasks. For this purpose appropriate hyperlinks must be added to the exercise script and the lecture script must be made available as hypertext on a web server or otherwise accessible e.g. on a special CD together with the exercise script.

2) *Help extension 2 - solution schemes:* For standard control design tasks specific solution schemes may be produced and stored as hypertext on a web server or on CD as well. Within the exercise script hyperlinks may be added to point to the solution schemes along which the student may become able to solve the specific task.

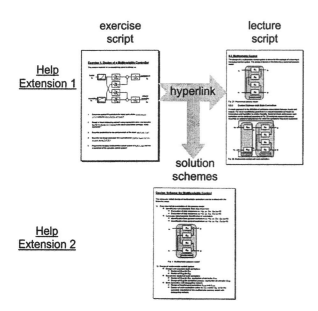

Fig. 4a. Hyperlinks as help extensions for exercises

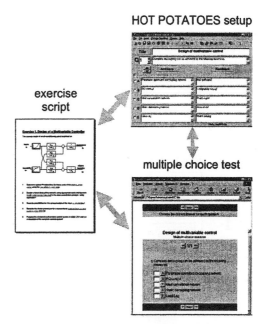

Fig. 4b. Teachware extension 1 – result checking

3) *Teachware extension 1 – result checking:* The hypertext exercise script may be extended with interactive elements to become a teachware system. A first simple step is the addition of a result check which evaluates solutions to the exercise tasks provided by the student. For this purpose the exercise script itself has to be modified such that simple checks are made possible – e.g. multiple choice questions which can be accomplished e.g. by applying simple tools like HOT POTATOES™.

4) *Optional teachware extension 2 - interaction:* The next step could provide a general teachware set up with interactive elements reacting to individual learning situations. This requires an integrated rework of lecture script and exercise script to become an interacting set of teachware course material where the learning progress can be measured e.g. by the correct completion of exercises and the sequence and repetition of lecture sections is proposed by the system. This, however, requires by far more expertise than assumed in the personal preconditions set above and sophisticated authoring tools must be used to produce such a teachware system

The technical effort required for the rework of exercise script and lecture script increases from step to step in the above proposal such that from a technical point of view the technical advantages (less supervision effort) are balanced against the workload for the rework. The didactical question, however, concentrates on the question by which measures the teaching results can be improved. By providing additional assistance the students may be motivated to train themselves more frequently (activating effect), however, by providing too much assistance the students may not become able to solve the task on their own but may always rely on external assistance provided automatically (passivating effect). So the didactical question is to find a good balance between sufficient and too much assistance for the exercises.

7. LABORATORY SCENARIO

The set up of laboratory experiments may be organised as follows: The laboratory experiments are demonstrated to the students by the lecturer or carried out stand alone by small groups of students supervised by a laboratory assistant. A laboratory script is handed out to the students in advance for preparation providing specific information about the laboratory equipment used and the specific tasks to be carried out. After the laboratory experiment the experimental results may be summarised and discussed by the students at home in a protocol document.

The *laboratory scenario* regards the situation in the stand alone experiment case. Here the supervision task for the laboratory assistant includes a check of the students' preparations before the experiment, the assistance of the students during the experiment by providing handling hints and additional technical information, and the evaluation and rating of the protocol documents prepared after the experiment. To reduce the supervision load of the assistant and to optimise the learning results for the students it makes sense to introduce multimedia assistance for the laboratory experiments in the following way, see Fig. 5:

1) *Assistance level 1 – hyperlinks to lecture script:* The preparation of the students for the laboratory experiments can be made more efficient by adding hyperlinks to the laboratory script for direct access to related sections of the lecture script which allows them to collect and repeat the underlying theory easily.

2) *Assistance level 2 - computer assisted experimenting:* During the laboratory experiment the laboratory script is made available on a multimedia computer guiding the students through the experiment. To accomplish this the laboratory script should provide in addition hyperlink

access to additional handling instructions for the laboratory devices and experimental tasks. These handling instructions may include video clips or photo sequences showing e.g. the electrical wiring of lab devices or the manipulation of mechanical components. This provides fast online experimenting support for the students and relieves the supervision load from the laboratory assistant.

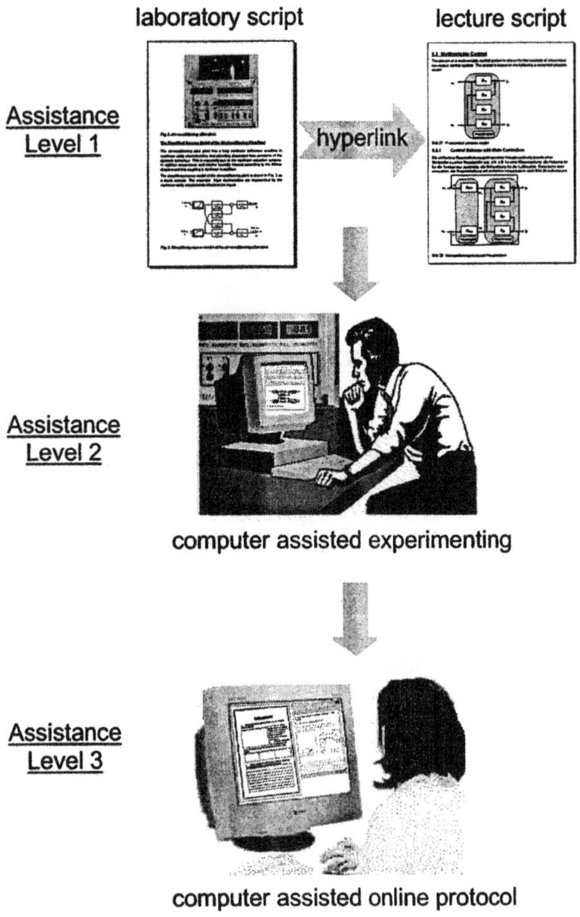

Fig. 5. Computer assisted laboratory exercise

3) *Assistance level 3 - computer assisted online protocol:* The compilation and discussion of experimental results should be done online during the experiment. For this purpose a protocol frame can be provided on the multimedia computer for the laboratory experiment - in the simplest case just a preformatted WORD document. The students have to compile the experimental results directly into the document and add their comments online. This means that all experimental results must be collected with the computer online (the computer records and processes all relevant signals, or results recorded with external devices are scanned into the computer). This procedure forces the students to be well prepared – otherwise they are not able to produce sensible comments online - but relieves them from the time consuming rework of the experimental results at home.

The technical realisation of the multimedia assistance for laboratory experiments requires additional technical effort which has to be balanced against the reduction of workload on the laboratory assistant. The didactical advantage of the described approach lies especially in the online protocol which forces students to start the laboratory well prepared such that they really know what they do during the experiment rather than start thinking about the experiment at home during the protocol writing.

8. CONCLUSIONS

The use of multimedia to improve the set up of teaching control at the university requires some technical knowledge. In Schumann (1999) a concept was proposed for the introduction of nonexpert lecturers and assistants into the production and use of multimedia teaching elements. However, for an efficient use of multimedia elements in teaching control even in the traditional set up, the didactical situation must be analysed carefully with respect to the learning success and the teaching effort required. Only if this analyses points to considerable improvements the technical effort for the rework of the traditional set up using multimedia becomes justifiable.

In case of the scenarios described in this paper the analysis of the technical and didactical situation points to a considerable improvement potential using the extended presentation possibilities a multimedia computer can provide in the lecture room, in the laboratory or at the students' home. Nevertheless in every case it remains essential not to follow just the amazing technical possibilities multimedia has to offer but to stay always aware of the didactical situation and the expected effect on the learning process – a perfect presentation is not always a guarantee for the activation of students which must be one primary goal for the teacher.

REFERENCES.

Schmid, Chr. (1998a). Using the World Wide Web for Control Engineering Education. *Journal of Electrical Engineering, 49* (1998), No. 7-8, pp. 205-214.

Schmid, Chr. (1998b). The Virtual Control Lab VCLab for Education on the Web. Proc. 1998. American Control Conference, Philadelphia, USA, Paper WP-16.

Schumann, R., B. Syska, M. Simcic and B. Zupancic (1999). Making Multimedia Work For Control Education. IFAC-Congress, Peking, 1999.

Online-References

Adobe Acrobat and Adobe Photoshop.
  http://www.adobe.com/products
Hot Potatoes. *http://web.uvic.ca/hrd/halfbaked/*
Microsoft Office (Frontpage, Powerpoint, Word).
  http://www.microsoft.com/office/
Ulead Videostudio. *http://www.ulead.com/vs/runme.htm*

ELSEVIER

IFAC
PUBLICATIONS
www.elsevier.com/locate/ifac

RAPID CONTROLLER PROTOTYPING WITH MATLAB/SIMULINK AND LINUX

Roberto Bucher * Silvano Balemi *

* University of Applied Sciences of Southern Switzerland
(SUPSI), Department of computer science and electronics,
CH-6928 Lugano-Manno, bucher@die.supsi.ch

Abstract: This paper presents a Rapid Controller Prototyping System based
on Matlab, Simulink and the Real-Time Workshop toolbox. Executable code is
automatically generated for Linux RTAI, a hard real-time extension of the Linux
Operating System. The produced code runs as a kernel module on a standard PC
with the modification of the Linux Operating System.
This environment can be used in a teaching laboratory to quickly implement
real-time controllers. Students have the possibility to follow all the phases of
the controller design within a unique environment (analysis, design, simulation,
implementation) and can concentrate on the design aspects without having to
deal with programming issues.
Some applications are presented to demonstrate the capabilities and the performance of this environment. *Copyright © 2003 IFAC*

Keywords: Control applications, Education, Rapid Controller Prototyping

1. INTRODUCTION

Rapid prototyping can be successfully used in
a control laboratory but usually at the cost of
expensive software and hardware packages. These
costs can be reduced by using free Linux software
on a single standard PC and the commercial
Matlab suite. Then, students can work within
the same environment from the analysis to the
controller implementation phase (see figure 1).

The considered "hardware in the loop" is a standard PC in which a common Linux OS is modified
with the RTAI extension from the Dipartimento
di Ingegneria Aerospaziale of the Politecnico of
Milan (DIAPM). This extension adds hard real-
time capabilities to a Linux OS, allowing sample
frequencies up to several thousands of cycles per
second with a jitter of just a few microseconds.

The RTAI extension was created as an environment for implementing low cost data acquisition

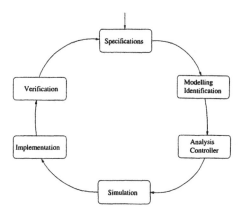

Fig. 1. Design phases

and digital controller systems ([D. Beal and Papacharalambous (April 2000)]). It has already reached a good level of maturity and has been widely
used in different applications ([Bianchi and Dozio (2000)], [E. Bianchi and Ghiringhelli (1999)],
[E. Bianchi and Mantegazza (1999)]). RTAI allows

to implement processes both in the kernel and in the user area.

Matlab/Simulink is a design and simulation tool used at most universities. Simulink allows the user to create models for dynamic systems simply by connecting blocks from given libraries. Among others, some blocks implement linear systems given as transfer functions or state space realizations both in continuous and discrete time. The Real-Time Workshop toolbox (RTW) generates C-code from a Simulink model without the need of any programming knowledge.

Starting from the C-code generated by the Real-Time Workshop toolbox, the subsequent generation of executable code for processes in the user area of the extended Linux OS has been demonstrated ([Quaranta and Mantegazza (2001)]). The present work shows how to automatically generate code for the kernel area instead. This guarantees that processes run with highest priority, without being influenced by various other tasks running on the PC.

Such an integrated system is very valuable for control system laboratories. Students can design and implement controllers without having to explicitly program hardware. Moreover, the control hardware is given by standard PCs, thus limiting the investment needs and allowing later inexpensive upgrades of the control system hardware with newer, faster computers when higher performances are needed.

2. THE ENVIRONMENT

2.1 Hardware and software requirements

The system can be installed on a PC with following features:

- Linux with kernel 2.4.x modified with the RTAI extensions,
- Matlab, Simulink and Real-Time Workshop,
- C source code for the generation of the kernel modules, and the respective Matlab TLC (target language compiler) files and Template Makefiles,
- Acquisition and interface boards (analog, digital, encoders, ...),
- Simulink library with the drivers of the I/O boards.

2.2 RTAI extensions

The RTAI extensions can be downloaded from the web ([www.aero.polimi.it]). The Linux OS can be modified to a hard real-time OS following these steps:

- The kernel has to be modified with the RTAI extension files, re-compiled and installed.
- The RTAI specific modules have to be compiled and installed.
- The RTAI devices (FIFOs, shared memories) have to be created.

These activities require less than one hour effort.

2.3 Matlab, Simulink and Real-Time Workshop

The Matlab/Simulink tool is very useful when teaching students how to analyze and design control systems. In spite of its cost, RTW is one of the best solutions to automatically generate C-code, and can be quickly adjusted to work with different targets. Mathworks Inc. allows Universities to acquire this software at special conditions.

2.4 The "rtwrtai" toolbox

This software represents the main part of our work. The toolbox can be freely downloaded at www.die.supsi.ch/~bucher. A set of modules allows the integration of the code generated by Matlab, Simulink and RTW with the RTAI Linux environment. The most important modules are:

- The main module `rt_proc.c` which integrates the kernel specific functions and the Interrupt Service Routine.
- An intermediate module which combines the kernel specific task of the main module with the non-kernel code generated by RTW.
- The source code for specific Simulink modules (I/O drivers, scope) implemented as C-mex S-functions.
- The source code to access the I/O boards and the scope from the kernel module.
- The source code to implement specific functions not present at the kernel level (malloc, strcmp, memcpy, etc.).

2.5 Matlab TLC and Template Makefile

RTW requires two special files to allow the code generation for a specific target:

- The Target Language Compiler file to control the code generation.
- The Template Makefile to create the project specific Makefile which contains all directives needed to compile and link the C-code generated by RTW.

As a starting point for the creation of the TLC files we could use the default Target Language Compiler file with little modifications. The main problem was to obtain a Template Makefile which was able to link kernel specific code (RTAI) with user specific code (RTW). This was only possible with the use of compilation flags handling the compilation of the different types of modules. The resulting Template Makefile allows also the use of several Matlab toolboxes (Signal Processing Toolbox, Fuzzy Control Toolbox, etc.).

The resulting files support both continuous and discrete-time blocks, which remarkably increases the potentiality of the tool.

2.6 I/O Boards

The use of acquisition and interface boards is essential when controlling real plants. For each function provided by a board (AD-conversion, DA-conversion, ...), two modules have to be programmed:

- A C-mex S-function to integrate the I/O in the Simulink environment.
- A C-module, linked to the RTAI kernel module, to access the board.

The C-mex S-function only handles the module parameters (number of inputs, number of outputs, board address, sampling time, etc.). The C-module handles all I/O functions (register programming), for instance:

- writing values to a D/A converter,
- controlling a A/D conversion and reading the resulting values.
- controlling a digital I/O.
- reading the counter of an incremental encoder.

2.7 RTAI library for Simulink

Figure 2 shows the Simulink library currently available.

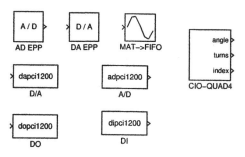

Fig. 2. RTAI specific Simulink Library

At present, the project supports the following boards from Computerboards:

- PCI-DAS1200 with 16 330KHz 12-Bit A/D converters with 3μs Burst Mode & Prog Gain and 2 D/A converters.

- CIO-QUAD04, 4 Channel quadrature encoder board.

The block "MAT->FIFO" allows to exchange data from the kernel area to the user area.

3. SOME EXAMPLES

3.1 Example 1

Figure 3 shows a simple Simulink model with two sinusoidal input signals.

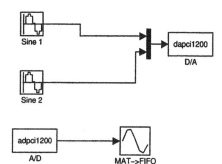

Fig. 3. Example 1 - Simulink Block Diagram with sinusoidal input signals

The signals are sent to the DAQ board and converted to analog voltages. The two voltages are read by the A/D converters of the same board and sent to the scope. The scope saves the values into a FIFO, which allows to retrieve the values in the user area. The signals are eventually plotted using "gnuplot" (see Figure 4).

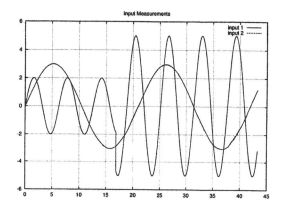

Fig. 4. Example 1 - Data plot

Parameters can be modified online during the execution of the real-time code. This feature has been programmed using the "Parameters Tuning" capability implemented in RTW. A real "External Mode" has not been implemented yet.

3.2 Example 2

This example, similar to the previous one, shows the possibility to implement blocks with multiple sampling rates. The first signal is sampled at 1KHz, the second at 5Hz. Figure 5 shows the data plotted after the acquisition.

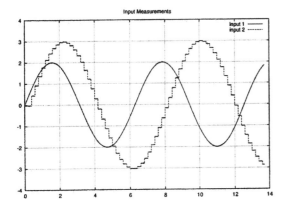

Fig. 5. Example 2 - Data plot

3.3 Example 3

This example shows how students can control an inverted pendulum. A state-feedback controller is used, the unmeasured states of the plant are estimated using a reduced order observer. Figure 6 shows the Simulink model used by the students to simulate the controlled plant.

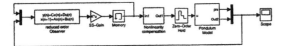

Fig. 6. Example 3 - Simulation model for the inverted pendulum

Figure 7 shows the Simulink model used to generate the real-time control program: there are very few modifications with respect to the simulation model shown in Figure 6.

The model of the plant is simply substituted by the I/O modules and the controller is now ready for implementation. Writing by hand the code for this controller can be very time-consuming. Automatic code generation in this case, can remarkably reduce the implementation time. Students can concentrate on variants of the controller without spending a lot of time in programming.

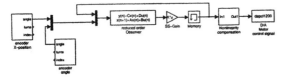

Fig. 7. Example 3 - Controller implementation

4. STUDENT FEEDBACK

Student feedback is very positive. Above all, students appreciate the different possibilities to design, implement and test a controller whithin the

same environment. Furthermore they can start implementing controllers very early, even without knowledge of discrete-time systems: continuous and discrete-time blocks can be mixed in the same Simulink model.

4.1 Continuous-time controllers

The system allows the implementation of continuous-time transfer functions. An integration algorithm is automatically integrated into the generated code if continuous blocks are defined. As soon as the students have the first knowledge in analysis and design of classical controllers, they can implement the first controllers and test how they work with real plants.

4.2 Identification

Possibility is given to collect I/O data for identification. Different input signals (step, random, PBRS,...) can be applied to the plant in order to collect input and output signal sets. Afterwards these data can be analyzed offline in order to identify the plant. Figure 8 shows a simple block diagramm used to collect data for identification.

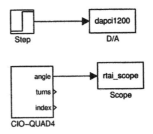

Fig. 8. A simple schema to collect data

4.3 Controller algorithms

One of the drawbacks of the use of rapid controller prototyping systems in teaching activities is that students are not faced to the issues raised by writing and implementing controller algorithms by hand. However, this system allows to simply extend generated code with code parts written by hand. The code generated by RTW can be modified in order to call external procedures.

For example a discrete-time PID controller written in C by hand in a file can be simply merged into the RTW code thus avoiding tedious and inefficient work (e.g. writing or integrating code for card drivers). The code corresponding to the block diagram of figure 9 given by the following lines

```
rtb_temp9 = (rtB.Pulse_Generator -
    rtB.S_Function_o1) * rtP.controller_Gain;
```

can be modified in order to call the external PID controller code as follows.

```
err = (rtB.Pulse_Generator - rtB.S_Function_o1);
rtb_temp9 = reg_pid(err);
```

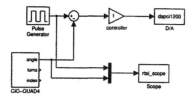

Fig. 9. Schema to implement self written code

4.4 Student involvement in the project

This project begun with a student project focused on a feasibility study. Now students participate in the further development of the environment by implementing new drivers during laboratory activities and student projects.

5. CONCLUSIONS

A simple and inexpensive tool for use in control system laboratories has been presented. This tool allows to reach sampling frequencies up to approximately 10KHz, which is more than enough for controlling most plants in student laboratories.

Using the "rtwrtai" toolbox, the Matlab/Simulink environment can be used for acquiring data to identify the plant, for analysis and for simulation. Just with a few steps, the Simulink model used for the simulation can be transformed into a real-time controller running on a standard PC.

The present work provides a valid tool for rapid controller prototyping systems while minimizing the cost of the control hardware.

REFERENCES

E. Bianchi and L. Dozio. Some experience in fast hard real-time control in user space with rtai-lxrt. In *Real Time Linux Workshop*, Orlando, 2000.

L. Dozio S. Hughes P. Mantegazza D. Beal, E. Bianchi and S. Papacharalambous. Rtai: real time applications interface. *Linux Journal*, April 2000.

D. Martini E. Bianchi, L. Dozio and P. Mantegazza. Applications of a hard real-time support in digital control of complex aerospace systems. In *AIDAA Congress*, Torino, Italy, 1999.

P. Mantegazza E. Bianchi, L. Dozio and G. L. Ghiringhelli. Complex control system, application of diapm-rtai at diapm. In *Real Time Linux Workshop*, Vienna, 1999.

G. Quaranta and P. Mantegazza. Using matlab-simulink rtw to build real time control application in user space with rtai-lxrt. In *Real Time Linux Workshop*, Milano, 2001. www.aero.polimi.it.

www.elsevier.com/locate/ifac

LabVIEW AS A TEACHING AID FOR CONTROL ENGINEERING

Dr. Petra ARADI, Dr. György LIPOVSZKI

Systems and Control Engineering Group
Department of Information Engineering
Faculty of Mechanical Engineering
Budapest University of Technology and Economics
Goldmann Gy. tér 3., Budapest, H-1111, HUNGARY
{petra,lipovszki}@rit.bme.hu

Abstract: LabVIEW-based simulation programs help staff and students in various systems and control engineering courses at BUTE. This paper reports on the development and application of two LabVIEW-based simulation systems TUBSIM and .:PSim:. The importance of interactive, modifiable simulation programs lies both in coursework and in individual study. The courses taught, theses written and defended during the last years have proven grounds to this kind of applied simulation. *Copyright © 2003 IFAC*

Keywords: educational aids, control engineering, systems engineering, simulation, control system analysis, design and synthesis

1. INTRODUCTION

Computer-based materials are increasingly used in education, nowadays. The style of these materials range from simple hand-written lecture notes, through computerised documents, to interactive, hypermedia-type online systems. Each of these styles has its uniqueness, advantage and disadvantage. Hand-written notes are usually scanned and made available in a standard document format (PS, PDF, RTF, DOC, etc.). While such a document has a more personal touch from its author, it clearly has lots of drawbacks: it is rather difficult to correct the incidental mistakes, it has no interactivity at all, but has limits in extensions. Lecture and lab notes generated with word processors have the advantage of clear reproducibility, the ease of correction and further development. The above-mentioned two categories can be classified as static documents: once they are written, they are usually made available on the Internet, and can be downloaded and printed. Their aim is similar to those of textbooks.

The next evolutionary step is the extension of static documents with dynamic content, such as hypertext capabilities. Such documents can still be of certain common formats (e.g. PDF, DOC), but with internal references implemented. A more common format for hypertext documents is HTML, namely web pages. HTML documents are usually publicly available, and are mostly intended for online reading. It is however possible, to get the contents of such a web-site for offline browsing.

Further enhancements to education materials are embedded animations, video, and interactive sample programs, usually simulations. This approach is suitable for the expansion of documents from both the second and third types mentioned above. Animated illustrations and video clips help the students in understanding real world processes, that are otherwise very difficult or plain impossible to demonstrate. The application programs can serve multiple purposes from simply illustrating systems and processes, to providing the users with interactive access to the process itself. For disciplines involving experimentation, computing and the Internet further widens the horizon. There are examples of virtual laboratories on the net, that simulate a real world experiment, and there are laboratories with experiments set up for interactive usage over the Internet.

A number of interactive, computer-based course materials in various disciplines and subjects, from all over the world are available through the Internet. These courseware materials have diverse target audiences both in level and type of students, e.g. from primary school to university, from regular weekly lectures to distant-learning.

This paper focuses on systems and control engineering course materials developed and used in various courses and study levels at the Faculty of Mechanical Engineering of BUTE.

2. SIMULATION SYSTEMS USED IN SYSTEMS AND CONTROL ENGINEERING COURSEWARE

Systems Engineering courses are taught prior to Control Engineering courses at BUTE, to introduce modelling and simulation theory, methods and applications, not only for control, but for other disciplines as well. Modelling and simulation theories, methods and techniques – parallel with and serving the needs of systems and control engineering – have undergone a significant development during the last century. Measurement technology achieved important milestones, too. Up-to-date instruments and principles; accurate and fast measurements characterise today's measurement technology. Modern computing resources increase data management speed and capacity. Scientific results of the above mentioned fields are joined together in interdisciplinary applications. The most recent information should be included in education, at the same time the basics have to be taught, too.

Keeping the "old" knowledge and including the new improvements poses a problem for educators. They have to find the balance between the amount and quality of information relayed to students. In systems and control engineering frequency-domain techniques are still taught, however their importance have been reduced by the improvement in computing and time-domain simulation. They are however necessary, because of the insight and handy techniques they give to students. At the same time modern modelling and control knowledge, such as soft computing, adaptive and optimal control have to be included into the curriculum. To satisfy both tasks a number of programs have been developed to help visualise the conventional information (text, diagrams, drawings, and equations) in textbooks and lecture notes. These programs could easily be considered as computer games. It is widely proven that learning-by-doing is a leader among the approaches of learning. Playing with these simulations radically improve the understanding and future application in real world situations.

2.1. Simulation Programs

During the years various tools have been used in systems and control engineering courses, from the first text-based simulation systems running in DOS to up-to-date programs with user-friendly graphical interfaces. Most of these programs are written for simulation only, however there are add-ons to use with general-purpose programming environments. The mostly used programming languages at the Systems and Control Engineering Group today are Pascal/Delphi and LabVIEW. From 1994 National Instruments' LabVIEW is extensively used both in courses and for research. The next part describes the properties of LabVIEW, and emphasises its ease of use and versatility. In 1994 when it was time to buy a new software for data acquisition, simulation, teaching and research purposes, LabVIEW was the only one to provide an ideal environment for all of these tasks.

2.2. LabVIEW

LabVIEW stands for Laboratory Virtual Instrumentation Engineering Workbench. It is a graphical programming environment, originally developed for measurement, analysis and process instrumentation. Recent upgrades allow the integration of databases, data-acquisition and control over the Internet. The programming language itself is called G. LabVIEW has numerous built-in functions that make creating user interfaces, data processing, file I/O and communication with various instruments easy. LabVIEW is available for multiple platforms, such as Microsoft Windows, Macintosh and UNIX-variants. In LabVIEW terminology programs are called virtual instruments (VIs). Except for VIs with OS specific functions VIs can be transferred between platforms without any problem. Each LabVIEW VI consists of two parts: the Front Panel is the graphical user interface, the Block Diagram is the actual program (Fig. 1). Unlike conventional text-oriented programming languages, G uses data-flow programming. The program elements are structures (e.g. loops, case structures), and LabVIEW functions (basic mathematical, logical and matrix operations, file and instrument I/O functions, user defined subroutines, etc.). The VIs are "wired together" and these wires carry the data-flow. LabVIEW has several basic data types (numeric, string, Boolean, etc.) and special types (for example Graphs, Intensity Charts). These data types may be bundled into user-defined clusters.

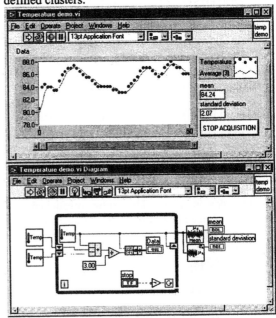

Fig. 1. Front Panel and Block Diagram of a simple LabVIEW program

Two simulation packages – TUBSIM and .:PSim:. – have been developed for LabVIEW at the (former) Department of Systems and Control Engineering by the authors of this paper. Both packages were initially developed according to the CSSL (Continuous System Simulation Language) recommendations.

2.3. TUBSIM for LabVIEW

TUBSIM. TUBSIM was originally written as a simulation extension to Pascal for DOS and Windows. Later it was implemented, as a set of LabVIEW VIs. TUBSIM is the interpretation of an analogue computer in the LabVIEW graphical programming environment. Analogue computers were widely used for simulation, but they had many disadvantages (e.g. the size of the computer grows with the size of the model). When digital computers became universally available, analogue computers and analogue simulation were soon replaced with digital simulation. One group of the digital simulation systems uses the same principles as analogue simulation does. TUBSIM belongs to this group. The TUBSIM VI Library (Fig. 2) contains the basic blocks typical to analogue computers (summers, integrators, potentiometers, signal generators). In addition TUBSIM has different Boolean blocks, typical systems engineering elements (for example first order element, continuous time controllers – like PI, PID –, time delay block) and discrete time blocks.

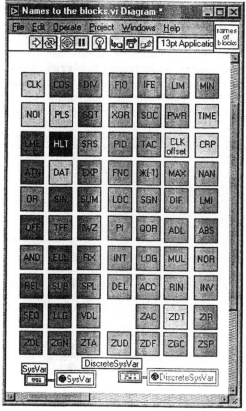

Fig. 2. TUBSIM VIs in LabVIEW

TUBSIM has successfully been used in various applications from teaching aids to large-scale industrial processes (e.g. the secondary side water chemistry model of the Nuclear Power Plant in Paks, Hungary).

2.4. .:PSim:. for LabVIEW

.:PSim:., like TUBSIM, was first developed as a set of LabVIEW VIs for simulating continuous time processes, according to the CSSL (Continuous System Simulation Language) recommendations. Through the design and implementation of .:PSim:. the experience from the development and application of TUBSIM helped a lot and served as a basis. .:PSim:. however includes not only time-domain methods, but frequency domain methods as well, thus making it useful for demonstration purposes. Further additions to .:PSim:. include libraries for soft computing methods, specifically fuzzy systems and neural networks to handle complex non-linear models or models based on measured data. Bondgraph models, discrete-time systems, identification and stability analysis are also recent .:PSim:. enhancements. The most recent addition is .:PSim:.Compartment, a set of VIs for compartment modelling (used e.g. in biomedical, physiological and pharmacokinetical simulations). A compartment model is basically a state-space model that is why it can be used not just in the above-mentioned areas, but for systems and control engineering, too. Compartment models require the ability to simulate non-linear, time-dependent dynamic state-space models with time delay, so .:PSim:.Compartment implements such models.

.:PSim:. blocks are functionally organised into LabVIEW libraries. These libraries are made directly available for easy programming, as Diagram Panel menus (Fig.3).

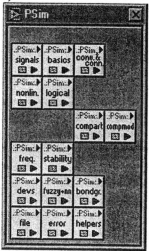

Fig. 3. .:PSim:. VI libraries as function menus in LabVIEW

While TUBSIM implements model definition similar to analogue computer programs (block diagrams with simulation elements) and transfer functions, .:PSim:. offers more. Model definitions and conversion among models have much similarity with MATLAB conventions. Time-domain models may be defined as transfer functions, in pole-zero-gain format, as state-space models, and can be build from basic elements like integrators, proportional and first-order blocks etc. To facilitate model definition, there are Front Panel Controls for these formats available, as well as predefined diagrams for Bode, Nyquist and Nichols plots (Fig. 4).

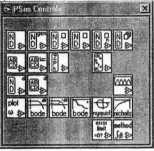

Fig. 4. .:PSim:. Front Panel Controls

3. SIMULATION APPLICATIONS

A continuously increasing number of TUBSIM and .:PSim:. simulation applications are used in both introductory and advanced systems and control engineering courses at BUTE. There are other advanced studies (e.g. Computer Controlled Systems, MSc and PhD theses) too, which utilise similar applications.

Some examples are shown in the following figures. One of the first models introduced in a systems engineering course is the problem of a bouncing ball (Fig. 5).

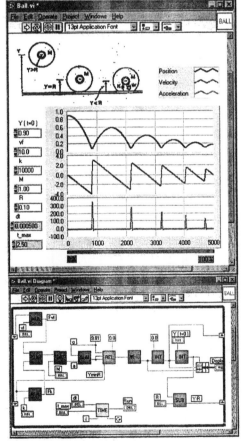

Fig. 5. Bouncing Ball – a TUBSIM application

The first task is to set up the model of the ball-ground system according to the governing physical laws. The second task is to solve the mathematical model with different parameter and input settings. This task is best done with a simulation tool. Introducing computer simulation in systems engineering courses help students understand the necessity and advantages of such an approach, compared to conventional numerical solution techniques. Graphical model definition is a great plus, too, because of its ease of use. Basic LabVIEW programming skills are relatively easy to acquire, so using TUBSIM or .:PSim:. for simulation does not pose a big problem for students. When there is no need to show the subtleties of modelling and simulation with TUBSIM or .:PSim:., there are simulation-shells available, that students can use to solve their particular problems with.

Further examples from the area of systems engineering are frequency-domain representations and state-space models. It is imperative to emphasise the connection between time- and frequency domain models of systems. Naturally, it can be worked out with classical methods (paper-pencil, or blackboard-chalk), however interactive, simulation-based approaches are far more effective (Fig. 6). This method is supported by the improvement of facilities in auditoriums. An increasing number of auditoriums are equipped with computers and multimedia projectors. Partly these equipment force the continuous development of educational materials into animated, interactive "documents".

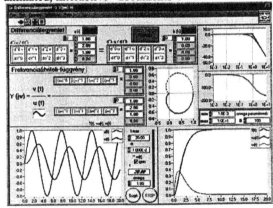

Fig. 6. .:PSim:. application to illustrate time- and frequency domain properties of systems

State-space models are particularly hard to solve with conventional methods, especially when non-linearity and time-delays are involved. An example of such a complex non-linear system from the area of biomedical engineering is the compartment model of entherohepatic circulation (Fig 7). Biomedical simulations are especially useful when "control systems" have to be designed to compensate the effects of illnesses. One such example is the development of medication regimes for patients with diabetes.

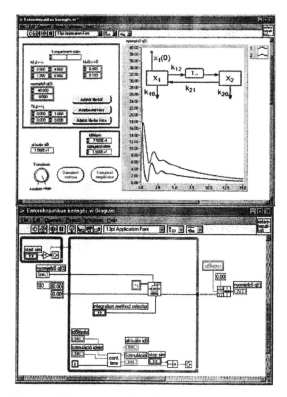

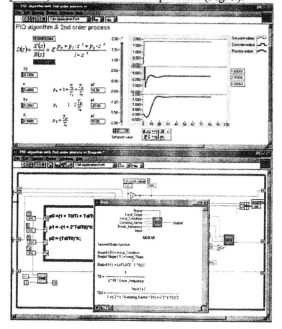

Fig. 7. .:PSim:. model of entherohepatic circulation

Simulation is a very appropriate way for presenting and comparing different types of controllers. The application of knowledge studied is basic control theory courses is best tested with simulation programs. One application to be presented deals with PID controllers, specifically analogue and digital implementations (Fig. 8). Another widely used application is the computer model of a real world process, a water-level control problem (Fig. 9).

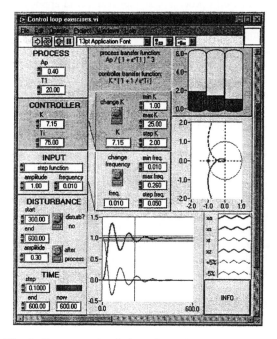

Fig. 8. illustrates LabVIEW's VI help feature: each VI can have a small help window, that shows the VIs connection points and a short description. Both TUBSIM and .:PSim:. VIs are equipped with such help windows.

Fig. 9. Computer simulation of a real world process

The three-tank level-control system is a real world process, so students have a chance to see and operate the process in life, and to use the simulator (Fig. 9). This particular application stresses the importance of simulation, namely in cost- and risk-reduction. Fig. 9 is also an example of simultaneously presenting time-domain and frequency-domain parameters and properties of control loops.

Soft computing applications are also presented, such as the fuzzy logic control of an automatic transmission system, which was developed as an MSc project, using TUBSIM and .:PSim:. simulation and fuzzy VIs (Fig. 10).

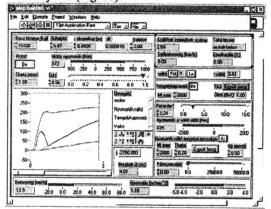

Fig. 10. Fuzzy control of automatic transmission

There are LabVIEW applications directly connected to real world process models, too. One such application is a National Instruments Process Demo Box with a fan, a light bulb, a thermometer and a

Fig. 8. TUBSIM application for introducing continuous and discrete time PID controller implementations

microphone. The utilisation of this box is dual. First of all it serves as an example of data acquisition and process control with the help of a National Instruments DAQ board. Besides, this system allows the combination of simulation and data acquisition in a real world process. Another real world application is a LEGO Dacta system, which is connected to the computer through an RS232 port. The versatility of LEGO blocks, sensors and motors allow the construction of various models. These LEGO models are controlled from LabVIEW, too. LabVIEW simulations serve as a substitute for real world processes in some PLC-controlled applications used by the students in laboratory exercises.

Some of the simulation applications presented in this paper are used as illustrations in lectures and labs, others are publicly available on the Internet for studying and practising. In each semester newer additions are developed both by staff and students, so the list of simulation examples grows almost exponentially. More and more students, not just from the Faculty of Mechanical Engineering, but from other Faculties as well, become interested not just in control engineering, but simulation as well, and use their simulation knowledge in individual and team projects and theses.

Currently the speed of LabVIEW applications, especially the ones with a large need of run-time calculation is considered. When for example a fuzzy controller has to be optimised with a genetic algorithm (as was the case in a recent MSc thesis), DLLs (dynamic link libraries) are used to speed up the calculation. Although each algorithm could be implemented in LabVIEW, it is worth to take into account the speed of a native LabVIEW application and a compiled function library. The advantage of developing native LabVIEW programs lies in the ease and speed of development, however the speed of calculation is best boosted with the use of external function calls from a DLL. DLLs open the world of object-oriented programming to LabVIEW, as well. Objects can be dynamically created and destroyed inside the simulation (especially in discrete-event simulation).

4. CONCLUSION

This paper reviews the use of simulation as teaching aid for systems and control engineering education. LabVIEW is the mostly used programming environment for these applications, with two simulation packages – TUBSIM and .:PSim:. – developed at the (former) Department of Systems and Control Engineering, by the authors of these paper. The courses taught, theses written and defended during the last years have proven grounds to this kind of applied simulation.

Educators use prepared simulations to illustrate their lectures. Students can download simulations from the internet to help them understand and practice the theory presented in lectures and labs. The use of the presented simulations improved the standards of information transfer both in classroom and individual studies.

Current research and development work includes the development and integration of special purpose DLLs, for fuzzy logic, neural network and genetic algorithm applications. The introduction of the recently developed discrete-event simulation system into the education is planned for the next semester. According to the internalisation needs of educational materials, everything developed at the Systems and Control Engineering Group has to be at least bi-lingual, Hungarian and English.

REFERENCES

Åström, K.J., Wittenmark B. (1997) *Computer-Controlled Systems*, Prentice-Hall

Bárdossy, A., Duckstein, L. (1995*) Fuzzy Rule-Based Modelling with Applications to Geophysical, Biological and Engineering Systems*, CRC Press

Bequette, B.W. (1998) *Process Dynamics*, Prentice Hall

Bronzino, J.D. (Editor-in-Chief) (1995) *The Biomedical Engineering Handbook*, CRC Press

Dorf, R.C., Bishop, R.H. (1998) *Modern Control Systems*, Addison Wesley Longman

Gordon, G. (1969) *System Simulation*, Prentice Hall

Hartley, T.T., Beale, G.O., Chicatelli, S.P. (1994) *Digital Simulation of Dynamic Systems – A Control Theory Approach*, Prentice Hall

Johnson, GW (1994) *LabVIEW Graphical Programming. Practical Applications in Instrumentation and Control*, McGraw-Hill

Kheir, N.A. (editor) (1995) *Systems Modeling and Computer Simulation*, Marcel Dekker, Inc.

Law, A.M., Kelton W.D. (1991) *Simulation Modeling & Analysis*, McGraw-Hill

Man, KF, Tang, KS and Kwong, S. (1999) *Genetic Algorithms*, Springer

Monsef, Y. (1997) *Modelling and Simulation of Complex Systems. Concepts, Methods and Tools*, Society for Computer Simulation International

National Instruments (2002) *LabVIEW User Manual* at http://www.ni.com/pdf/manuals/320999d.pdf

Wells, L.K.; Travis J. (1995) *LabVIEW for EveryOne – Graphical Programming Made Even Easier*, Prentice Hall

Zeigler, B.P., Praehofer, H., Kim T.G. (2000) *Theory of Modelling and Simulation*, Academic Press

www.elsevier.com/locate/ifac

DATABASE TECHNOLOGY AND INDUSTRIAL TRANSACTION TECHNIQUES IN CONTROL EDUCATION

Jana Flochová, Miroslav Galbavý

*Department of Automatic Control Systems, Department of Computer Science,
Faculty of Electrical Engineering and Information Technology
Slovak University of Technology, Ilkovičova 3, 812 19 Bratislava, Slovak Republic
flochova@kasr.elf.stuba.sk, galbavy@dcs.elf.stuba.sk*

Abstract: New developments in computer networks and communications provide new possibilities for control purposes. Control systems for highly complex plants are themselves very complex and heterogeneous. A new software infrastructure is needed that exploits these new emerging software technologies. An essential activity to be pursued is that of education and of including new technology in control education. The aim of this paper is to describe some of our activities and experiences in teaching of information techniques and technologies included in open control platforms of complex control systems. *Copyright © 2003 IFAC.*

Keywords : Information technology, Database management systems, Relational databases, Control education, Human-machine interface

1. INTRODUCTION

Today's word is changing rapidly; the global market generates a need of global technical support. Information technology (IT) and telecommunication technology enable remote access to equipments and provide a large number of opportunities for enhancing speed and quality of the support process. Rapid developments that are taking place in the areas of computer science and communications influence the field of computer control. The complexity of systems to be controlled (and control systems too) increases rapidly. Systems are subjected to many constraints concerning energy consumption, safety and reliability conditions, environment protection, next to ever-increasing demands on economical production and trading-results. A key issue in control engineering is the application to highly complex systems: the coupling of complicated and heterogeneous systems.

The primary goals of the distributed control technology are to accommodate rapidly changing applications requirement, easily incorporate new technologies interoperate and maintain viability in heterogeneous changing environments (Vebruggen et al., 2002). Notions of "control" are expanding from the traditional loop-control concept to include such others functionalities as supervision, coordination and planning, situation awareness, diagnostics, and optimization. Complex dynamic distributed systems are demanding new capabilities that traditional control technology is not offering. Among these capabilities the following software issues will be of major importance in this area:

Openness, adaptability/dynamic reconfigurability, real-time operating systems functionality for control and supervisory control, networking, plug-and-play extensibility, remote diagnosis, **embedded databases and database systems for look up tables, process archives and production analysis,** specific solutions for the plants, multi-process control task and adaptive learning. Internet techniques play and will play an important role in this area, such as internet-based communication, web-based remote sensing, monitoring and management.

The issue of open control architecture is extremely important for future research. One needs to work closer with various groups of applications engineers, and to take into account the real needs of applications in developing of appropriate solutions. To take full advantage of telecommunication and IT innovations, man has to be able to insert new technology into the control system architecture without redesigning the components already in the systems. An essential activity to be pursued is that of education and of including new technologies in control education. The aim of this paper is to describe some of our activities and experiences in teaching industrial information system design and IT included in open control platforms of distributed control systems.

The paper is organized as follows: in the second chapter we briefly introduce real-time database systems (DBS) and discuss their roles in software of distributed control systems, the third chapter depicts some features and facilities of an industrial transaction system, in the fourth chapter we describe the PLC and IT laboratories of our department, and the concept of our courses with several examples of students projects.

2. THE DATABASE TECHNOLOGIES OF CONTROL SYSTEMS

A database system (DBS) consists of a database (collection of data) and a database management system (programs that create, access and organize data). The **relational database** was invented by E. F. Codd at IBM in 1970, and represents a collection of data items organized as a set of formally-described tables (which are sometimes called **relations**) from which data can be accessed or reassembled in many different ways without having to reorganize the database tables themselves. The definition of a relational database (King, Haritsa, Ramamrithan 2000) results in a table of metadata or formal descriptions of the **tables, columns, domains,** and **constraints.** A database management system (DBMS) is a program that lets one or more computer users create and access data in a database. Typically, a database manager provides users the capabilities of controlling read/write access, specifying report generation, and analysing usage with help of a non-procedural data access language based on algebra or logic. Programs could be written in terms of the "abstract model" of the data, rather than the actual database design; thus, programs were insensitive to changes in the database design. They manage user requests (and requests from other programs) so that users are free from having to understand where the data is physically located on storage media. In handling user requests, the DBMS ensures the integrity and the security of the data. A

transaction usually means a sequence of information exchange and related work (such as database updating) that is treated as a unit for the purposes of satisfying a request and for ensuring database integrity. Databases and database managers are prevalent in large mainframe systems, but are also present in smaller distributed workstation or mid-range systems and on personal computers. A **distributed database** is one that can be dispersed or replicated among different points in a network. **Real-time relational DBS** are designed to interface with large amounts of data, to enforce data integrity constraints and to satisfy application-timing constraints simultaneously. An **object-oriented database management system** (OODBMS), sometimes shortened to ODBMS for object database management system), is a database management system (DBMS) that supports the modelling and creation of data as objects. Several big RDBMS companies (Oracle, Informix Software, IBM, Microsoft and Sybase) began to look for ways to incorporate complex data types so that they could also be manipulated, parsed for content, validated, indexed, and searched. The answer provided by the major RDBMS companies is now called a universal database (or universal server). This is a hybrid of a relational system that one-way or another incorporates complex data as objects. These systems are more correctly labelled object-relational DBMS (ORDBMS).

Among software industries, the database industry is second only to operating system software and it is growing at 35% per year. The research community (both industry and university) embraced the relational data model and extended it. Most significantly, researchers showed that a high-level relational database query language could give performance comparable to the best record-oriented database systems. The SQL (structured query language) relational database language was standardized between 1982 and 1992. By 1990, virtually all database systems provided an SQL interface (including network, hierarchical and object-oriented database systems). Meanwhile the database research agenda moved on to geographically distributed databases and to parallel data access. Today, all the major database systems offer the ability to distribute and replicate data among nodes of a computer network.

Manufacturers of modern control systems guarantee absolute openness and modular structure of their products. They often use the latest IT (OCX, ActiveX, OLE, OPC-OLE for process control, OLE DB, COM+, ADO, ADC etc), integrate real-time SQL databases and provides access to leading database programs including Microsoft SQL Server, Oracle, Sybase, dBase and others that support the Open Data Base Connectivity (ODBC) standard. In the cases where the conventional relational database technology is not suited to high-speed acquisition and storage of plant data, the manufacturers offer their own high-quality features and technologies providing fast real-time database access and high-volume data collection and retrieval delivery. The exploitation of RDBS can be found not only in their traditional applications – components of control systems like history collections, storage option packages, alarm, event summaries, reporting packages, statistical analysis packages, but in real-time process data acquisitions and real-time data managers and security managers too.

List of several products, manufacturers and their integrated RDBS follows. (The list is by no means exhaustive; it is used only to show various possibilities.)

Table 1 Relational database systems integrated in the software of control systems

Product(s)	Manufacturer	Integrated DBS
WINCC, WINAC	Siemens	Sybase SQL AnyWhere
RSView32, RSViewStudio	Rockwell sw	SQL AnyWhere, Dbase
RSBizware Historian	RSBizWare	MS SQL Server
IndustrialSQL Server (FactorySuite 2000)	Wonderware	MS SQL Server
Security Manager	Honeywell	MS SQL Server
I/A Series software (Unix)	Foxboro	Informix-SQL (ISQL) Informix ESQL/C
I/A Series I-SCADA NT	Invensys Foxboro	MS SQL Server
IFIX, iHistorian	Intellution	MS SQL Server
Cimview	Aspentech	IBM DB2

3. RSSQL – AN INDUSTRIAL TRANSACTION PLATFORM.

An industrial transaction links control systems to database systems, so that they act as one. A true, end-to-end link is established that provides the level of reliability required to support enterprise-wide integration. RSSQL^{Tm} (RSSQL) is a WindowsNT/2000 based industrial transactions processing system for sharing manufacturing data between enterprise systems and shop-floor control systems. As part of the RSBizWare framework, RSSQL provides a bi-directional link between control systems and enterprise database systems. Its architecture consists of four primary components: a graphical interface (GUI) and three NT/2000 services (Transaction Manager, Control Connection, and Enterprise connection). The Transaction Manager executes transactions, controlling the collection, manipulation, and storage of data. The Control Connections are the interfaces to the process control

systems; the Enterprise Connections provide links to the relational database management systems.

On the control side, RS SQL can connect to RSLinx, RSView32 or RSView Studio or any AdvanceDDE or OPC server. On the enterprise side, RSSQL can connect to any ODBC compliant-databases including Microsoft SQL Server, Access, Sybase, Informix and others or to Oracle via their direct callable interface (OCI) – a native connectivity to Oracle on any of the supported OS including UNIX and AS-400. The services connect to each other using TCP/IP sockets. This provides the ability to operate as a single system even when the components are distributed over multiple computers on a network. RSSQL supports bi-directional transactions in one of two ways and provides three primary ways to trigger transactions; internal scheduler trigger, control execution trigger and external trigger. It can be configured to execute a transaction at time-base events or at regular intervals. The control system can also control transaction execution, RSSQL provides the ability to trigger a transaction when a control point changes, when it goes high (or low) only, or it can be configured to run at defined intervals while a control points is high. Finally an external function (used from Visual Basic, 'C', PowerBuilder or any other application package) call is provided to allow any application to trigger transaction. As final check on a transaction prior to going to a database, RSSQL provides the ability to analyze several criteria prior to completing the transaction, commits the data, and performs several optional commit procedures. The software can manipulate data prior to passing it to a table or a stored procedure (aggregations of multiple points – Avg(), Min(),...Diff()), provides full complement of arithmetic, bitwise, logical operators and filters, and includes several Wizards to easy system setup and configuration. The tool has many safe guards in place to ensure system and database integrity, its services have the ability to send e-mail notification of failures including lost connections and failed transactions.

4. THE PLC AND IT LABORATORIES AT THE DEPARTMENT OF ACS FEI STU

The five and half years' study at our faculty is divided into three periods. The basic stage gives students theoretical knowledge (Mathematics, Physics and Control Theory). The second period that ends with the Bachelor's degree provides a basis for a specific discipline. In the last study period, which ends with obtaining the title of Engineer (Master's degree), students submit their diploma theses and take the final (state) examination in three main subjects.

Various laboratories two of them being the PLC (Allen Bradley) and the Industrial information systems laboratory support the courses. Allen Bradley Laboratory was established at our Department seven years ago. It was possible due to the Global Development Program of Rockwell automation, inc.

and due to the help and support of the Allen Bradley product authorized distributor in the Slovak Republic. At the beginning, the AB Laboratory was equipped with 3 racks of SLC5/01 systems and the software package APS. Later, the systems were extended with racks containing the SLC5/02, 5/04, 5/05 processors, two PLC 5 processors, I/O and Flex I/O modules and four Micrologix systems. Students coming to our Allen Bradley Laboratory (in 6th, 7th, 9th semester) are already familiar with control systems hardware, plant equipment, operating systems, real time programming and basic principles of the real-time control. The laboratory is used for teaching of following subjects: Control System Software (90-100 students per year), Control System Design I+II, Data Processing in Control Systems, and Control of Discrete Event Dynamical Systems (40-60 Students in each course per year and per subject). Thus, the total number of students taught there is about 210-280 each year. The present laboratory equipment covers PCs Pentium, Windows NT/2000 OS, six SLC500 systems, four Micrologix systems, two PLC 5 processors, PanelView Operator Terminals, corresponding communication drivers, control, emulation and SCADA/HMI software, **RSSQL** and models of real plants. There are also workplaces for elaborating of student projects and diploma theses. The DH485, DH+, ControlNet, DeviceNet and Ethernet networks have been built in the laboratory to establish the communications of all PLC and PC nodes.

An Ethernet link connects the AB laboratory and the laboratory for Industrial information systems design and information technologies. The equipment of IT laboratory consists of OS Windows 2000, MSOffice2000, Visual Studio, Microsoft SQLServer, RSSQL, Sybase PowerDesigner, CAD tools, Siemens SCADA/HMI application WinCC and Yokogawa Centum CS3000. Both laboratories have been used for teaching PLC programming techniques, IEC61131-3, SCADA/HMI design, industrial information and communication technologies; the concepts of RDBS and real-time DBS; IT laboratory for teaching CAD, their use in control systems design and logistic principles.

The course "Data processing in control systems" has been taught since 1992 and covers the following topics:
Principles and mathematical background of RDBS, DBS architectures, data processing in RDBS and real-time distributed control systems (DCS) , industrial transactions, linking of control systems and RDBS, data integrity, DBS in SCADA/HMI and their design, data archives, statistical analysis and WEB monitoring of the colleted data. Concepts taught through lectures are complemented by laboratory exercises and projects. The exercises are built on the present laboratory equipments and on the hardware and software (simulations of several simplified real plants)

models. The course provides students with the skills needed to function productively with relational databases, tables, queries, SQL basic, relations, replication techniques, industrial transaction configuration, monitoring and management. The no obligatory course gained students interests and has been subscribed by 65-70% of control engineering students in the last tree years.

The first four weeks of exercises are spent with database skills and techniques, during the second period students work on software projects. Any group of two students elaborate two projects in the areas of classical RDBS, SCADA real-time databases, RSSQL, RDBS process archives and WEB-real time monitoring of a plant model. Several models of simplified plants (among others conveyor and flexible manufacturing system model, level control, simplified batch reactor and ethylene storage tank technology) have been designed and built for the AB laboratory, the courses taught there, and student diploma theses (Hrúz et al. 1997, Flochová, Hüber, 2000). Interactive experiments on real plants, real-time acquisition, monitoring and retrieval improve the student motivation and develop an engineering approach in solving problems.

The following figures 1.-4. represent a collection of student projects. Any group of students elaborate two projects in the areas of classical RDBS, SCADA real-time databases, RSSQL, RDBS process archives and WEB real-time monitoring of a plant model.

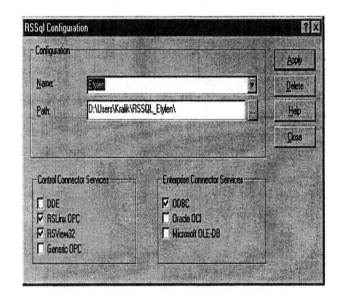

Fig. 1 A RSSQL configuration.

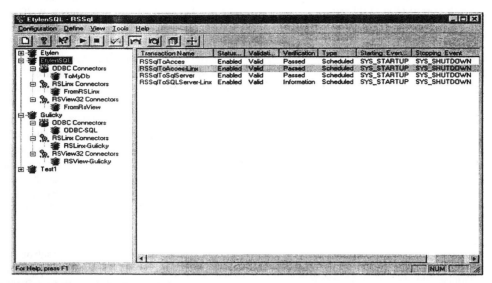

Fig.2 Industrial transactions monitoring.

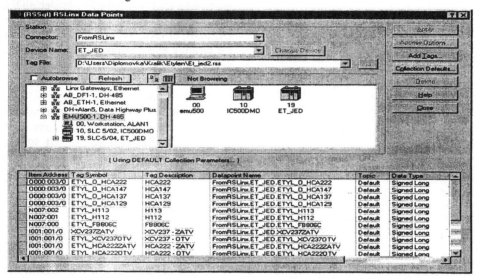

Fig.3 RSLinx Data points processed in RSSQL transactions.

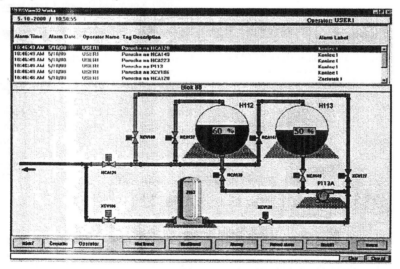

Fig. 4   RSView32 SCADA/HMI, the plant model has been used as a source for real-time process data for a RSSQL configuration and in two WEB-based monitoring applications

An RSSQL Configuration Report follows:
Report generated on Mon Dec 03 12:30:53.310 2001
Using RSSQL Version 2.10.02, Build 2.326
File Messages
Type(s) of message conditions logged:
    Fatal errors
    Severe errors
    Warnings
    Informational messages
The files will be stored in 'D:\Users\ Kralik\RSEtylenSQL\' and will contain no more than 10000 messages and be smaller than 1000000 bytes
The following display options have been configured:
Float and Double data types are displayed in the data base:
The following services have been configured:
Transaction Manager will run on the host named 'ALAN1', on port 400 using username 'ALAN1\kralik', password '*****'
RSView32 Connector 'FromRsView' (instance 0) will run on the host named 'ALAN1' using username 'ALAN1\kralik', password '*****'

ODBC Connector 'ToMyDb' (instance 1) will run on the host named 'ALAN1' using username 'ALAN1\kralik', password '*****'
Data Base Tables Options:
    use table owner when accessing data base tables
Performance Options:
    maximum number of real time threads is set to 1
    SQL buffer size is set to 4 kilobytes
Cached Transaction Files Options:
    cached transaction files will be stored in 'D:\Users\Kralik\RSEtylenSQL\' with a base name of 'ToMyDb'
    cached transaction files will be processed when 10 transactions have been logged or 30 second(s) have passed since the last log file has been processed
RSLinx OPC Connector 'FromRSLinx' (instance 2) will run on the host named 'ALAN1' using username 'ALAN1\kralik', password '*****'

The following transactions have been configured:
Transaction 'RSSQLToAcces' (id 2) is enabled, valid, and no verification messages exist defined as:
Solicited and scheduled to run every 1 minute(s)
    from starting event 'SYS_STARTUP'
    until stopping event 'SYS_SHUTDOWN'
    Times out after 1 minute(s) and it always stores its values
    Uses the real time thread option to store the transaction
    Data will be stored using Connector 'ToMyDb' via Data Object 'ToAccess':
    using system DSN 'RSsql_Acces', user '', password '*****'

    new records will be created in table 'Table3'
The following bindings have been configured:
    Column 'Txt' type is String, size is 50 NULLs are allowed IS BOUND TO Expression "RSView_Etyl"
    Column 'FB806C' type is Signed Short, size is 2 NULLs are allowed IS BOUND TO. DataPoint FromRsView.work1.bl85_LRA806' from Connector 'FromRsView' via server 'alan1', device '', access path 'work1' tagfile '' scheduled via hot link
    retrieval timeout of 2 second(s)
    valid timeout of 0 second(s)
    substitute specific value of 'NULL'

....

similar DataPoints definitions follow

5. CONCLUSIONS

Developments in computer networks and communications provide new possibilities for control purposes. New software infrastructure for control systems is needed that exploits these new emerging software technologies. An open control platform (OCP) for complex systems, and the issues of new information and communication technologies will be of major importance in the areas of control engineering and of control education. The students need a better background of newest IT among others the background in the field of database systems, industrial transaction management of real-time databases and Internet based monitoring. It will help them to identify potential problem areas, analyze the failures in control systems, minimize errors, improve control efficiency and other key performance indicators, get the data to optimize and improve the manufacturing effectiveness. Courses "Data processing in Control systems" and "Industrial system design I, II" were included in the Bachelor's education of FEI STU, and cover the topics of new IT, real-time database systems, transaction managers, SCADA/HMI, and WEB based monitoring.

RSSQL – an industrial transaction system has been described in the paper. The software of transaction manager interfaces with PLC from many manufacturers using industry standards OPC or DDE data servers, and with many databases using ODBC, OLE-DB, or OCI. The software has powerful features for creating and managing transactions, including the ability to support unidirectional and bi-directional transactions. Its transaction-based approach focuses on the movement of data between DS and control systems with a high degree of control, reliability, and flexibility. That makes it suitable for building wide range of manufacturing applications connecting DBS and control systems of various manufacturers and for real-time transaction techniques teaching.

This work was supported by Slovak grants VG 1/7630/00 and VG 1/0161/03.

REFERENCES

Flochová, J., Hüber, M.: Utilization of RSLogixEmulate500 in process simulation ant in SCADA/HMI design teaching. In: *Proceeding of international conference Řip 2000*, Kouty nad Desnou, Czech republic, pp. 231-236.

Haritsa, J. R., Ramamritham, K.: Real-Time Data-bases in the New Millenium, In: *Real-Time Systems*, **19** (3), pp.205-208, 2000.

Hrúz, B., Ondráš, J., Flochová,J.: Discrete event systems-an approach to education, In: *Proceeding of the 4th symposium on Advanced in control education*, Turkey, Istanbul, 1997, pp.283-288.

King, N.H.: *Object DBMSs: Now or Never*. http://www.dbmsmag.com/9707d13.html, 1997.

Vebruggen, H.B. et al.: IFAC 2002 Milestone report on computer control, In: *Preprints of 15th Triennial World Congress*, Barcelona, Spain, July 2002, pp.233-241.

http://www.software.rockwell.com
http://www.ab.com/manuals/
http://www.ad.siemens.de/
http://www.sybase.com
http://content.honeywell.com/sensing/control/products/software.stm
http://content.honeywell.com/ebi/Security_Manager.htm
I/A Series ® Software Real-time Database manager PSS 21S-4K1 B3; Foxboro, 1995
http://194.184.64.99/SCADA_function.htm
http://194.184.64.99/i_a_series_scada_nt1.htm; 2001.

ELSEVIER
IFAC
PUBLICATIONS
www.elsevier.com/locate/ifac

INTERACTIVE CONTROL SYSTEM DESIGN

J.P. Keller

*Institute of Automation, University of Applied Science, Solothurn,
Bittertenstr. 15, 4702 Oensingen, Switzerland
juerg.keller@fhso.ch*

Abstract: This contribution describes a LabVIEW based interactive computer aided control system design tool. It offers a set of ready to use solutions of typical control system design problems. Instead of a syntax a menu driven user interface with convenient system editors and analysis tools enables the user to interactively attain the design goals. The tool allows the student to efficiently do the control system design from plant identification to controller implementation on the same platform. *Copyright © 2003 IFAC*

Keywords: Computer-aided control system design; Interactive programs; User interfaces; Educational aids; Identification; LQG control; Implementation;

1. INTRODUCTION

In control engineering there is still a wide gap between what a student has to learn during his studies and what he actually uses in the subsequent industrial work. There are two main reasons for this discrepancy. First, adequate equipment is often absent in an industrial environment. Even basic approaches, like PID-tuning with step response are not widely used because tools to measure a step response are not at hand. So there is no reason to expect that more sophisticated methods, like for example loop shaping controller design, will be used. Only if the control problem is not solvable with a PID controller by trial and error, a controller design project is started causing considerable development costs. Most likely, the control engineer is faced with a heterogeneous environment. Signal generation and data acquisition equipment has to be installed to get the necessary data for plant modelling. The following steps, i.e. parameter estimation and controller design, are carried out on a PC or a Workstation. Controller implementation and final testing are done on the industrial plant control system. Usually, no workflow is available and experiment design, for example, relies

solely on the expertise of the engineer. With increasing capabilities of control systems, including the industrial communication networks, it is possible to integrate controller design tools into the capabilities of industrial control systems.

The second reason for the methodological gap between education and daily industrial routine is that control education should be focused more on engineering abilities than on research skills. Considerable time is lost on teaching mathematical formalisms and syntax of CACSD-systems. Consequently there is not enough time left to acquire expertise and routine in the complete design process including all necessary steps from problem analysis to controller implementation. Methods for successful plant start up, plant operation as well as handling emergency situations should also be mandatory topics in control engineering education. The student must primarily acquire expertise in selecting and applying the appropriate methods. This can be achieved by providing the student with opportunities to solve several design problems. Evidently, available course time is always too short but the program can be focused on the topics mentioned above by means of suitable tools.

The tool presented in this paper supports the student in the complete controller design process. It consists of the following modules:

- system modelling including parameter estimation,
- loop shaping controller design
- state-feedback with observer either with LQ or pole-placement
- support for loop transfer recovery
- controller implementation either for PID-type (Aström et al.,1997) or state-space controllers.

The tool hides mathematical formalism as much as possible. It is interactive with respect to parameter variations and design path. If controller or plant parameters are changed, the results are immediately visible on the selected analysis displays. Preferably, system parameters are modified in their native representation. This means, that plant parameter variation is done by changing values of physical parameters in an algebraic plant model, whereas a controller transfer function is modified by changing pole or zero locations. Whenever reasonable, parameter values can be changed in graphical editors.

A second aspect of interactivity is interactive wizard support. After completion of a design step, there might be different path to follow and the user has to choose the most suitable. Sensible design paths are modelled in a state sequential function chart. Its animated graph is displayed to the user. So, at every decision point, she is aware of the consequences for the following design steps. In order to simplify interaction with the tool, the user interface is object oriented. Consequently, each similar system type, whether it is a plant, a controller or a closed loop, has the same appearance, menu bars and analysis methods.

The tool offers a set of ready to use solutions of typical controller design problems. This increases motivation to seek a more sophisticated controller design because there are no tedious control system calculations and there is no CACSD-syntax to learn. Obviously, only problems within the provided set of solutions can be easily solved.

The tool is based on LabVIEW, National Instruments, which is particularly well suited for this purpose. It can run on different platforms, for example on a Notebook or an industrial PC with real-time operating system, and is one of the most powerful tools to create user interfaces. A large library for signal processing and mathematical functions can be combined with a complete set of industrial data acquisition and signal generation hardware.

Ideas for modern interactive loop shaping were proposed by Johansson (1998). A useful MATLAB based tool for Control Design education is available for the book of Astrom et al. (1997). Typical design problems can be interactively explored. In MATLAB's Controller design toolbox an interactive tool for SISO controller design is available. (MATLAB, 2002) . In Kottmann (2000), the work on an object-oriented CACSD-tool at the ETH-Zürich is summarised. The proposed tool provide some interactivity in basic editors. A very effective feature is the action tree, providing the capabilities of workflow based scenarios for controller design. An application of the tool to a mechanical system is described in Qiu et al. (1999). An ambitious project for web-based control education is the Dynamit project (Löhl, 1999). A virtual control lab is provided on an elaborate web interface.

In this paper, the main ideas of the interactive, computer aided control system design tool (i-CACSD-tool) are presented. In the first chapter general properties are explained. The next chapters describe the main modules, i.e. plant modelling and identification, controller design and finally controller implementation. Practical experience and student feedback are summarised in the last chapter.

2. THE INTERACTIVE CACSD-TOOL

2.1 General Properties

The modular structure of the i-CACSD-Tool represents the 3 major steps in control system design: plant modelling and parameter identification, controller design and controller implementation. In order to make the tool easy to use, the GUI-entities are standardised. In Figure 1, the most frequent objects are shown.

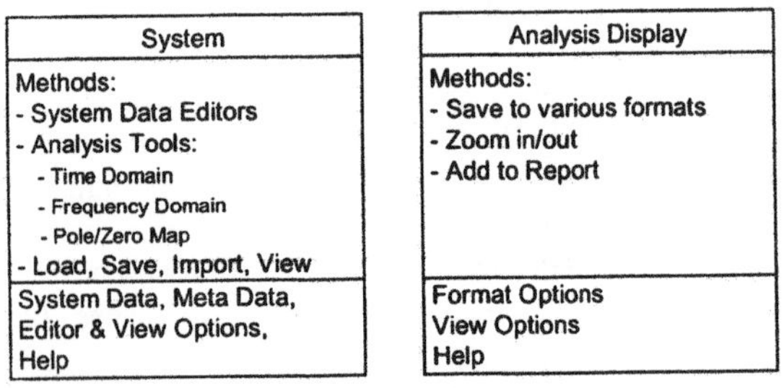

System
Methods:
- System Data Editors
- Analysis Tools:
- Time Domain
- Frequency Domain
- Pole/Zero Map
- Load, Save, Import, View
System Data, Meta Data, Editor & View Options, Help

Analysis Display
Methods:
- Save to various formats
- Zoom in/out
- Add to Report
Format Options View Options Help

Fig. 1. System and analysis display class

Most elements in a block diagram are of the class 'system'. They have their appropriate editors, the same tools for analysis and standardised methods to load and save the data. System data consists of the model data, for example transfer function coefficients, and names of inputs, outputs and states. This allows an easy selection of signals in the analysis tools. The methods for the class 'Analysis Display' are shown in Figure 1 on the right. The results can be saved to various formats or added to a report. Each design can be easily documented with a report tool into a html-document. Additional plots and comments can be manually appended to a standard report.

Supplementary information, henceforth called meta data, can also be managed. It contains data like user name, date and time, name of source data and some text. This is very convenient for identifying data in the sequel, even years later. Meta data can be previewed in the tool's file dialog without importing the data.

2.2 Plant modelling and identification

The availability of plant models is one of the main reasons that model based controller design methods are not broadly used. Since sophisticated physical models are not easy to derive and require a considerable amount of time and money, simpler methods have to be available. Methods for frequency response analysis are well known (Ljung, 1987) and can easily be used on industrial plants if the dominant time constants are reasonably short, preferably shorter than 1 minute. Since controller design based on a non parametric frequency response model is limited with respect to time domain analysis and controller design methods, i.e. to loop shaping, it is reasonable to approximate the measured frequency response with a plant model. This might be a black box transfer function or a physical model. There results a two step approach: first the frequency response is measured and in a second step approximated by a plant model. Obviously, this approach is limited to SISO-Systems. The system is preferably stable, although it can also be applied to systems with integrators.

Frequency response identification

For frequency response identification several experimental settings must determined. A wizard supports the user in specifying appropriate signal levels, experimental frequencies and optimal sampling time. Wizards usually offer a sequential navigation with back and next buttons. The user can not see what the next or previous step will do. A suitable representation of a wizard sequence is the sequential function chart (SFC). The representation is similar to the decision tree representation proposed in Qiu et al (1999). The SFC of an identification wizard is shown in Figure 2. With a glance at the wizard diagram, the user is aware of what she is currently doing – the highlighted wizard step – and what she will be doing next. From Figure 2 it can be depicted that identification starts with experiments to get the appropriate signal levels. Next, the dominant time constant is determined to attain knowledge about dominant poles. This allows the wizard to propose a frequency pattern for first experiments. The frequency response is measured using either single frequency scans or a periodic, multi-sine signal. Excitation with sine signals are chosen to guarantee good signal to noise ratio at the investigated frequencies. For multi-sine excitation frequency pattern and phase shift are optimised to get an excitation signal with minimal peak value and maximal amplitude for each sine component within the admissible signal range. As can be

seen in the wizard SFC (Figure 2), estimation can be aborted to specify new signal levels, left to model fitting or improving the frequency response measurements by refining the frequency pattern. Based on the first results, a new frequency pattern is

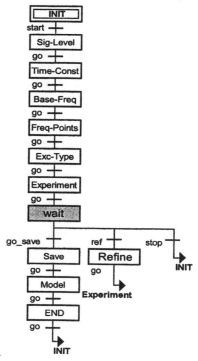

Fig. 2. Wizard sequential function chart

determined so that phase changes between two measured frequencies are small. This is ideal for several reasons. Technically, phase unwrap can be done unambiguously. From the plant identification point of view it follows the suggestive hint (Ljung 1987) that inputs should be chosen in order to sensitise the output with respect to parameter changes. Dominant plant poles and zeros always lead to obvious phase changes. If plant phase is well measured, i.e. if there are no large phase changes between two measured frequency points, plant magnitude is also well defined at the pole and zero locations. Figure 3 shows how a new frequency pattern is proposed for the identification of a PT2-type system. The vertical dark lines are at the new frequencies. All frequencies are integer multiples of the base frequency shown on the left to ensure a perfectly periodic signal. This is necessary to avoid spectral leakage.

Plant modelling

An algebraic plant model can be formulated either as a transfer function or a state space model. Each element is an algebraic term consisting of known and unknown physical parameters. The unknown parameters can be obtained by manually fitting the frequency response of the plant to the measured frequency response. Although manual approximation may not be very scientific, it has several appealing advantages over numerical identification methods. First of all, it uses the human skills to weigh data and

335

interpret outliers. When optimising physical parameters, the engineer gets a good feeling of the frequency response sensitivity with respect to parameter changes. This is particularly useful if the plant has to be modified in order to achieve some requirements. In addition, an inadequate model structure becomes immediately evident and it is the responsibility of the engineer that parameter values remain within a physically sensible range. Furthermore, it does not require any knowledge about identification methods. The resulting model can be used for state space controller design methods and for loop shaping.

The i-CACSD tool also provides the possibility to fit a black box transfer function. An initial transfer function has to be specified by the number of integrators, the relative degree and a guess of the order. Poles and zeros are dragged to their optimal locations. On a scenario basis, poles and zeros can be provisionally added. Both the system with and without the singularity are displayed, enabling the user to accept or discard the changes. Furthermore, DO and UNDO are available. or if the physical model is wrong or too simple to fit the frequency response. The model has to saved for the next step, the controller design.

The wizard SFC can be modified using a convenient editor. This editor allows the user either to specify sequential function charts for plant control or to define a workflow for some design task., see Keller (2001).

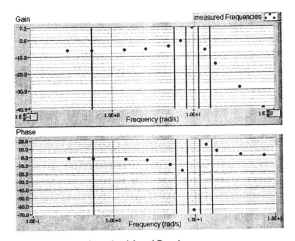

Fig. 3. Improving the identification

2.3 Controller Design

At present the controller design tool consists of two design modules. The first is simple loop shaping design and the second is state feedback with observer, with either LQR/LQG or pole placement. For the loop-shaping tool the only major differences to the MATLAB tool are outlined in the next paragraph. With more details, the state-feedback tool is described in the following paragraph.

Simple Loop Shaping Controller Design

Loop shaping controller design can be done for the system shown in Figure 4. Many design problem can be formalised into this simple structure. The loop shaping design tool is similar to MATLAB's 'SISO controller design tool'. In both tools, the idea is to vary controller parameters to get the desired open loop frequency response. This can be done either by editing the controller parameters directly within the open loop frequency response plot or by editing the frequency response of the controller transfer function while the open loop frequency response is immediately adapted to the controller changes. In the MATLAB Toolbox the first approach was chosen. In the proposed controller design tool, the second approach is favoured. Experience shows, that it does not make sense to freely shape the open loop without monitoring the properties of the resulting controller. This easily results in controllers with non optimal and unrealistic lead-elements leading to non acceptable stress on the actuators due to high controller gain at high frequencies.

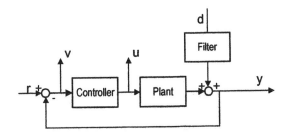

Fig. 4. SISO control system

State Feedback With Observer

As pointed out in Johansson (1998), there is also a need for an interactive controller design tool for state space methods. Controller design consists of the following steps (Geering 2001): Determine a state feedback gain so that closed-loop requirements are satisfied , then design an observer and modify the observer until loop transfer recovery is satisfactory. State feedback and observer design can be done either with linear quadratic (LQ) methods or with pole placement. In many situations integral action is also required for tracking control. The resulting control system structure is as proposed in Pierre (1994) and is shown in Figure 5.

The complexity of the controller design requires the user interface to be well structured. The control system in Figure 5 is divided into subsystems. Each system is realised with a similar object consisting of methods and data. This is shown in Figure 6. This allows to user to first analyse the plant. This can be done with the available system analysis tools. In the following a state feedback controller can be designed. Properties of the system under state feedback can be interactively explored. When using pole placement, the poles can be graphically moved to the desired location on the complex plane. In order to reposition complex pole pairs, the position of the pole with positive imaginary part has to be changed. Real poles remain on the real axis until two real poles are moved to the same location. A change of colour indicates that the poles can be moved onto the complex plain. The reverse action is also possible. For pole-placement, a very robust method proposed by Roppenecker (1990) is used. For systems with more than one plant input, the additional degrees of free-

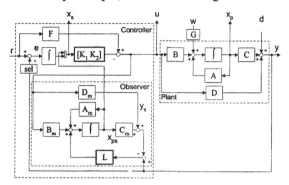

Fig. 5. State feedback with integral action and observer

dom appear in parameter vectors (Roppenecker 1990). At present, no parameter optimisation is realised but they can be manually modified.

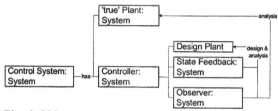

Fig. 6. Object structure

For LQR design weighting matrices for states, inputs and optionally cross-terms have to be specified. To set the values of the state weighting, the following options are available: diagonal, C^TVC with C being the measurement matrix for y or no special structure. Furthermore, an additional matrix C_{opt} can be defined resulting in a state weighting $C_{opt}^TVC_{opt}$. With the matrix C_{opt} one can think of an additional outputs, $y_{opt} = C_{opt} x$ which is subject to optimisation. This could also be represented with a 2-port model of the plant with outputs y and y_{opt}. Experience shows that y_{opt} often has to be modified during state feedback design. It is therefore reasonable to specify C_{opt} in the LQR-design. The observer can be designed in a similar way. Dual to C_{opt} a matrix G (see Figure 5) can be defined for LQE design. Again, parameter changes lead immediately to a recalculation of the observer properties resulting in a true interactive design.

State Feedback and observer design are based on a design plant. The resulting design can be analysed with both the design plant and the 'true' plant. With

the 'true' plant parameter uncertainties or changes can be imitated. This enables a user to examine design robustness.

Since loop transfer recovery is a property of the control system, analysis tools are available in the control system object. At present, the open-loop transfer function opened at the control value u can be analysed. Observer parameters can be changed according to Doyle (1981) to recover the open loop frequency response. Frequency responses of the open loop with and without observer can be immediately compared on a plot. An example is shown in Figure 7. Without additional programming effort, the controller frequency response and the control signal's response to measurement noise can be monitored during loop transfer recovery. This may uncover the consequences of a state feedback design with unrealistically large bandwidth because in most practical applications, it is not realistic to let a LQ-controller increase plant phase by more then 90 degrees unless disturbances are minimal. Properties of the resulting controller can be analysed in the controller object.

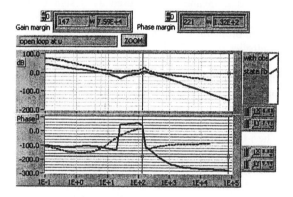

Fig. 7. Loop transfer recovery plot

2.4 Controller Implementation
After successful controller design, the controller has to be discretised and implemented. Simple lead-lag controllers are discretised and implemented as PID-controllers as proposed in Aström 1997. The resulting controller can be tested on the same system. State feedback controllers are discretised and implemented in modal form. Integrators are equipped with anti-windup strategies and actuator saturation is taken into consideration (Aström 1997). When testing the controller with the true plant, predicted observer states and outputs are drawn in a chart in order to verify the observer design.

3. HARDWARE ENVIRONMENT

The i-CACSD-Tool can run on different industrial plattforms, i.e. on industrial PC, Fieldpoint modules or PXI-Systems. National Instruments offers a large range of products for industrial automation. The

Fieldpoint modules with a real-time operating system is a PLC-like system, which is suitable for distributed process control. The PXI-Systems allow the implementation real-time feedback control at high sampling rates. All systems have the capabilities to run the real-time part of the i-CACSD-Tool in addition to the plant control tasks. There are no barriers as in common PLC-systems, that prevent an engineer from integrating more sophisticated controllers into plant control software.

4. EXPERIENCE AND STUDENT FEEDBACK

The CACSD tool has been used in the advanced control system course since 3 years. It is used as demonstration tool in control lectures, as design tool in exercises and laboratory courses. As a demonstration tool, it is very suitable for demonstrating system properties and design rules. The effect of parameter changes on the plant's frequency response, the consequences of a small gain margin or the effect of changing LQ-weightings can be demonstrated in an impressive manner. When used as design tool in exercises, the student's work is focussed on controller design problems - he can easily explore the world of controller design. With the report generation tool, a complete documentation of the controller design can easily be produced. In the laboratory course, a complete controller design from plant modelling to controller implementation and test is performed.

The i-CACSD-tool had a large impact on the course concept. The reason for this can be imagined from a colleague's comment: 'with this tool every idiot can design a controller'. The exercises had to be redesigned in such away that not only a documented controller design had to be done, but additional problems had to be solved. A challenging problem is to let the student find general design principles. A typical example is controller design for plants with a resonance peak close to crossing over frequency. Faced with this problem, it requires special skills to derive general design principles. With the i-CACSD tool, the student can easily verify his ideas. Also, a comparison of loop shaping with state feedback controller design can be achieved in a reasonable amount of time.

The students' acceptance was reflected in the tool selection. For loop shaping controller design about half of the class used MATLAB. The other half used the LabVIEW tool and added a valuable contribution by testing the tool. For state feedback controller design, most of the students used the LabVIEW tool. In the laboratory the students ran the tool on their notebooks and used the lab-monitor as second display. This indicates a shortcoming of all the design tools: either you have several plots as small as a stamp or your screen is too small.

5. CONCLUSIONS

This contribution presented an interactive controller design tool suitable for control engineering education. It offers the opportunity to focus education on learning important aspects of control system design and minimises the effort required to master short-time valued syntax of CACSD-systems. The underlying object-oriented interface structure combined with wizard support simplifies user interaction. Graphical editors, immediate update of analysis panels to parameter changes, automatic report generation and no tedious control system calculations motivate students to an increased commitment to control education. The tool is based on LabVIEW and makes use of the wide variety of available industrial process interfaces provided by National Instruments.

6. REFERENCES

Aström K.J. and Wittenmark B. (1997). Computer Controlled Systems Theory and Design. 3rd ed, Prentice Hall, Upper Saddle River, New Jersey.

Doyle J. and Stein. G. (1981). Multivariable Feedback Design: Concepts for a Classical/Modern Synthesis. IEEE Trans. on Automat. Contr., AC-26, 4-16.

Geering H.P. (2001). Robuste Regelung. IMRT-Press, ETH-Zürich.

Johansson M., Gäfvert, M. and Aström, K.J. (1998). Interactive Tools for Education in Automatic Control. IEEE Control Systems, 18, 3, 33-40.

Keller J.P. (2001). Programming Sequential Function Charts in LabVIEW. Proc. NIWeek 2001, Austin.

Kottmann, M., Qiu X. and Schaufelberger W. (2000). Simulation and Computer Aided Control Systems Design using Objec-Orientation. vdf Hochschulverlag, Zürich.

Ljung L. (1987). System Identification, Theory for the user. Prentice Hall. ISBN 0-13-881640-9

Löhl T., Pegel S., Klatt K.-U., Engell S., Schmid C. and Ali A. (1999). DYNAMIT – Internet based education using CACSD. Proc. of the 1999 IEEE International Symposium on CACSD, 273-278.

Pierre, D.A. (1994). Discrete-Time Enhanced Linear Quadratic Control Systems In: Control and Dynamic Systems, 66, Academic Press.

Qiu X., Schaufelberger W., Wang J., Keller J.P. and Sun Y. (1999). Object-Oriented Analysis and Design of 03CACSD using OMT. Proceedings of IFAC World Congress, Bejing.

Roppenecker G. (1990). Zeitbereichsentwurf linearer Regelungen: Grundlegende Strukturen und eine Allgemeine Methodik ihrer Parametrierung. Oldenbourg, München, ISBN 3-486-21640-6.

The Mathworks. (2002). Inc. MATLAB – Control system toolbox. Reference Manual.

AUTHOR INDEX

Printed and bound by CPI Group (UK) Ltd, Croydon, CR0 4YY

08/05/2025

01864925-0001